NEW DIRECTIONS IN ATTRIBUTION RESEARCH

Volume 2

NEW DIRECTIONS IN ATTRIBUTION RESEARCH

Volume 2

Edited by

JOHN H. HARVEY
Vanderbilt University

WILLIAM ICKES
University of Wisconsin

ROBERT F. KIDD
Boston University

 LAWRENCE ERLBAUM ASSOCIATES, PUBLISHERS
1978   Hillsdale, New Jersey

DISTRIBUTED BY THE HALSTED PRESS DIVISION OF

JOHN WILEY & SONS
New York   Toronto   London   Sydney

Lawrence Erlbaum Associates, Inc., Publishers
62 Maria Drive
Hillsdale, New Jersey 07642

Distributed solely by Halsted Press Division
John Wiley & Sons, Inc., New York

Library of Congress Cataloging in Publication Data (Revised)

Main entry under title:

New directions in attribution research.

Includes bibliographies and indexes.
1. Attribution (Social psychology) — Addresses,
essays, lectures.   I. Harvey, John H., 1943-
II. Ickes, William John.   III. Kidd, Robert F.
HM132.N475   153.7'34   76-26028
ISBN 0-470-98910-6 (v. 1)
     0-470-26372-5 (v. 2)

Printed in the United States of America

This book is dedicated to **Edward E. Jones** and **Harold H. Kelley** for their influential contributions to the development of attribution theory and research.

Contents

Preface

At the time of the publication of the first volume of *New Directions in Attribution Research* (1976), a concern was expressed about the imbalance that existed between theory and research in the area of attribution. At that time, theory apparently had outstripped empirical work, a condition rarely encountered in social psychology. To some degree, Volume 1 offset this imbalance by consolidating much of the work that had been conducted along relatively traditional lines and by pointing to some interesting new directions. Volume 2 is a continuation of the first effort in that it provides a perspective on past theory and research and extends attributional concepts into other areas within social psychology. The basis for these volumes is the great amount of high-quality research on attributional processes that continues to be done by investigators in the area.

The image of man as a "naive scientist" still pervades much of the work on attribution. According to this conception, individuals process information from a variety of sources by applying inferential rules that enable them to structure and "explain" the events they experience in terms of relatively invariant features and relationships. Early attribution theorists such as Heider, Jones, and Kelley attempted to specify the fundamental processes by which attributions are derived. Most recent research simply has assumed the presence of these fundamental cognitive processes and, rather than attempting to elaborate or refine them, has focused on how such processes mediate attributional effects. As Jones and Kelley suggest in their discussion in this volume, these latter approaches are better described as attributional theories inasmuch as they tell us little about how inferences are made but a great deal about the cognitive and behavioral implications of attributional processes in various social-psychological domains.

Traditional attributional analyses have not gone unchallenged, however, as some of the papers in this volume reveal. The important process statements of the 1960s involved the assumption that people generally are rational, logical, and reflective. Although these assumptions are still implicit in much of the current work on attribution and often are employed as guides in orienting future research, they are beginning to be challenged by a number of investigators working in the field. Increasingly, questions such as the following are being raised: Are people always as thoughtful as attribution theories suppose? Do self-serving and motivational biases sometimes affect the logical, cognitive processes implied by the various theories? What are the information-processing and attentional mechanisms that limit the person's ability to make causal inferences? How do relevant social contexts and concerns shape the course and content of attributional analysis by the naive attributor? These questions, among others, are addressed in the chapters contained in this volume.

To provide some continuity between the first and the second volumes, we have adopted an organizational scheme similar to the one employed in Volume 1. The chapters concerned with current issues and research have been divided into two main sections dealing with attribution at the personal and interpersonal levels. The third section contains papers devoted to theoretical integration and analysis. The volume concludes with an interview of two of the most influential attribution theorists, Edward E. Jones and Harold H. Kelley. In this interview, Jones and Kelley discuss the present status of attributional concepts and speculate about the course of future work in the area.

Several people provided invaluable assistance in the preparation of this volume. We are most grateful to Larry Erlbaum for his continuing support and enthusiastic encouragement. We also would like to express our appreciation to the Departments of Psychology at Vanderbilt and Wisconsin for their help and to our colleagues at these institutions for critical feedback during the collection and organization of the manuscripts. Completion of this volume was assisted by generous and continued assistance from the Vanderbilt Research Council to John Harvey and by grants from the Graduate School of the University of Wisconsin and from the National Institute of Mental Health (NIMH) to William Ickes.

JOHN H. HARVEY
WILLIAM ICKES
ROBERT F. KIDD

NEW DIRECTIONS IN ATTRIBUTION RESEARCH

Volume 2

ATTRIBUTION AT
THE PERSONAL LEVEL

At the personal level, attributional processes are central to the individual's construction of a personal reality. To a large extent, the structure and meaning of the events experienced by an individual derive from attributional analyses that are often subtle and complex. Phenomenologically, such analyses may at times appear to be fully represented in the person's consciousness. These occasions may typify only one end of a continuum, however; since on other occasions the analyses seem to occur partially, if not entirely, out of conscious awareness. The meanings and consequent behaviors following from such analyses often indicate that a logical, rational integration of information has occurred. Frequently, however, the attributional process may appear to be less rational and at times may even be characterized as "biased."

Whether apparent biases in attribution reflect real distortions in information processing or merely the individual's attempt to rationally integrate information about the external environment with information supplied by various internal states is a question of much current interest. It is also a question that, for the most part, has yet to be answered. The available evidence increasingly suggests that in order to achieve an adequate understanding of attribution at the personal level, both internal and external informational inputs must be taken into account.

As the chapters in the present and subsequent sections of this volume indicate, attention to both sources of information will not only help to elucidate the more general processes by which attributions are made, but may also help to elucidate the more unique and idiosyncratic aspects of personal experience that have typically been characterized as phenomenological.

The phenomenological emphasis is perhaps most evident in the first chapter of this section, "Is It Real?," in which Philip Brickman poses an attributional question of the most elemental sort: "How do people decide whether something is real or not?" Brickman contends that the answer to this question, far from being a strictly metaphysical concern, is actually of central interest and importance to the field of social psychology. After reviewing the kinds of errors people commonly make in their attribution of reality to various events, Brickman suggests that such attributions are based on the implicit application of two basic criteria — internal correspondence and external correspondence — to one's experience of the event in question. His elaboration of these theoretical criteria is quite provocative, indicating not only how they may provide the basis for such experiential states as alienation and fantasy but also how their relative importance in defining reality may vary as a function of such factors as age, sex, and the difference between the actor and the observer perspectives. Brickman's chapter concludes with a discussion of the importance of establishing a sense of reality of "phenomenological validity" in social psychological research — a discussion in which he emphasizes the potential value of games for the study of human behavior.

Brickman has noted a bit wryly that research in the area of attribution reveals that "people make attributions about almost everything" — the category "almost everything" now including the subjective experience of "reality." But are these manifold inferences always made at the level of conscious awareness? Ellen Langer, in her chapter "Rethinking the Role of Thought in Social Interaction," suggests that they are not. In sharp contrast to the implicit and widely held assumption that attributional processes must be represented in a consciously elaborated sequence in order to influence behavior, Langer's assumption is that much of human behavior is "mindless" and largely unconscious — even behavior that would seem to follow only from a complex sequence of information processing. She contends, in line with the *script* approach proposed by Robert Abelson, that habitually recurring sequences of behavior are gradually "chunked together" to form a smaller set of cognitive units that the individual can overlearn. Following extensive experience, the person's awareness of certain key features that typify and cognitively symbolize the essence of common behavioral sequences eventually becomes sufficient, in and of itself, to trigger automatically a complex series of responses that may appear quite "mindful" to an outside observer but are actually quite mindless from the standpoint of the actor's awareness of them. Langer shows us the other side of this coin, however, by describing a series of studies which indicate that individuals may behave inappropriately and thus reveal their "thoughtlessness," when they rely only on the scripted aspects of situations to guide them on occasions in which these scripted cues are misleading or inappropriate.

The notion that cognitive cues may elicit fairly stereotyped responses is also apparent in the subsequent chapter by Bernard Weiner, Dan Russell, and David Lerman. In "Affective Consequences of Causal Ascriptions," these authors propose that stereotyped affective reactions to achievement-related outcomes may be directly elicited by the ascription of the outcomes to specific causal factors. The results of their preliminary research suggest, for example, that the typical affective responses to a success outcome may vary greatly. A person may report a feeling of competence when the outcome is attributed to his own ability, self-enhancement when it is attributed to his personality, gratitude when it is attributed to another person, and surprise when it is attributed to luck. Similarly, the person may report a wide range of affective reactions to a failure outcome as well. He may feel incompetent if he attributes it to his lack of ability, resigned if he attributes it to his personality, aggressive if he attributes it to another person, and surprised if he attributes it to luck. In addition to discussing the possible relevance of their findings to the contemporary study of emotion, Weiner, Russell, and Lerman examine their implications for clinical psychology and the study of achievement motivation and suggest some intriguing directions for future research.

If people's emotions and motivations are affected by their attributions, the reverse may also be true — attributions may be influenced by motivational factors. In their chapter, "Attributional Egotism," Melvin Snyder, Walter Stephan, and David Rosenfield examine motivation to take credit for success and deny blame for failure. The motive to make an egocentric attribution depends on the presence of two necessary factors: (1) the attribution of a valanced (+ or −) outcome to the person; and (2) the perceived relevance of the attribution to the person's self-esteem. They argue that whether egotism actually occurs depends not only on the presence of the motive but also on the perceived probability of successfully protecting or enhancing self-esteem and on the strength of conflicting motives. It is also important to consider where the attributor's sympathies lie. The chapter concludes with a section in which the two-factor model of egotism is examined in terms of theoretical perspectives provided by control motivation, dissonance and balance theories, and the theory of objective self-awareness.

In the final chapter of this section, "Attributional Styles," William Ickes and Mary Anne Layden offer additional evidence implicating personal dispositions in the attribution process. Their research, when considered in the context of the many other studies they review, indicates that individuals exhibit characteristic styles of attributional preference that appear to be integrally related to such core aspects of the self-concept as sex and level of self-esteem. In general, masculinity and high self-esteem are associated with the internal ascription of positive or "success" outcomes and the external ascription of negative or "failure" outcomes.

By contrast, femininity and low self-esteem are associated with a relatively greater externalization of success and internalization of failure. The results of the authors' studies suggest not only that attributional style has often been confounded with sex and self-esteem in previous investigations of attribution and performance; they also suggest that the unconfounding of these factors in an experimentally induced failure situation reveals strong and additive performance deficits that are associated with sex and attributional style but are unrelated to level of self-esteem per se. Moreover, attributional style and self-esteem appear to be dynamically related such that directional changes in one variable are associated with corresponding changes in the other. The authors discuss the possible relevance of these findings to current theory and research in the area of human depression and learned helplessness.

1 Is It Real?

Philip Brickman
Northwestern University

> *Think it over what you just said.*
> *Think it over in your pretty little head.*
> *Are you sure that I'm not the one?*
> *Is your love real or just for fun?*
> — From the Buddy Holly song,
> "Think It Over."

How do people decide whether something is real or not? The question is a venerable one in psychology, the subject of one of William James's most famous papers (1869). It is also an important one. Our sense that a person can correctly decide what is real and what is not may be the most important element in our judgment of them. Attributing reality to things that are not real, or denying reality to things that are real, is taken as the most serious sign of mental illness. Undermining a person's sense of both physical and social reality is a standard element of modern torture. Defying other people's sense of reality, in reporting UFOs or contradicting their apparent perception of the lengths of lines, requires a definite degree of courage.

A decade of research in social psychology has focused on how people make attributions about almost everything (Harvey, Ickes, & Kidd, 1976) — but not how they judge whether or not something is real. Attribution research probably assumes that people are making judgments only about things they think are real and passes over the question of the initial attribution of reality. Perhaps social psychologists feel that questions of reality refer first to physical reality and as such are more appropriate for physicists and philosophers. Perhaps they feel that questions of social reality are too vague for precise research and are thus best left for sociologists whose analyses may be brilliant but whose taste in research is different (Berger & Luckmann, 1966; Goffman, 1974).

5

It is the purpose of this paper to show, on the contrary, that questions about what is real are already central to social psychology; second, that it is possible to develop a series of researchable propositions on the attribution of reality; and finally, that by focusing explicit attention on this topic we can improve our understanding of validity in social research and our ability to do research that is responsive to human concerns. First we turn to a consideration of how reasonable mistakes that people make in attributing reality are already central to method and theory in social psychology.

ERRORS IN ATTRIBUTING REALITY

We are always fascinated when people are made to treat as real things that they know are not, as with hypnosis, sensory deprivation, optical illusions, and magic. These involve, at their most dramatic, alterations in the appearance of physical reality. The equivalent for social reality are cases in which people come to treat as real a role or a relationship that they know is not real. The center of the psychoanalytic relationship is the patient's endowing the analyst with feelings and powers that the analyst does not in fact possess but that belonged to another figure in the patient's life, a process called transference. Perhaps the most famous example in social psychology of something initially unreal becoming real is the Zimbardo, Haney, and Banks (1973) prison experiment, in which the guards, the prisoners, and even the experimenters found themselves caught up in the situation far more than they had expected. Note that saying the Zimbardo experiment became real in this sense is not saying that the experiment was a true or precise simulation of a prison, a point on which the experimenters themselves are not always clear. The experiment was a prison for those imprisoned in it, but of a sort that may or may not have corresponded to those that exist outside.

Theater provides perhaps the most common instance of our capacity to treat as real events that we know are not. Indeed, we require that actors in a movie or play appear to be caught up in their roles. In addition, however, I think we have an extra degree of fascination with movies in which the actors' roles become blurred with or extensions of their own lives. Actors may give their best performances under these circumstances. Examples include Frank Sinatra in *From Here to Eternity*, Richard Burton and Elizabeth Taylor in *Who's Afraid of Virginia Woolf*, Warren Beatty in *Shampoo*, Sylvester Stallone in *Rocky*, and any number of Bogart, Hepburn, and Garbo films. Incidentally, it may be helpful to note what keeps actors from being overwhelmed by their roles in the way the much less talented players in the Zimbardo production were. Actors are continually interrupted between scenes and performances by reminders that what they are doing is not real, whereas the Zimbardo experiment ran continuously 24 hours a day. Dramas that come closest to continuity are the soap operas, in which actors play the same role day after day for years. These productions are notoriously difficult for both viewers and actors to keep separate from reality.

Through dissonance theory (Wicklund & Brehm, 1976) and self-perception theory (Bem, 1972), social psychology also provides examples of our capacity to treat as real events we initially felt were not. The central drama with which these theories are concerned is a situation in which a person is asked to say some-think he or she does not believe. Under specifiable circumstances, people making these statements are likely to give some credence to what they initially thought was false. In particular, if people do not have good external reasons for their playing a role (such as being paid to do so), they are more likely to see their behavior in that role as expressing their internal needs and beliefs. Moreover, as people change in response to different roles and different situations, they deny that they are in any way insincere in so doing or even that they are in fact changing (Gergen, 1972; Goethals & Reckman, 1973).

Sometimes people begin by thinking that a situation is unreal and are gradually convinced by cues in the environment that what they thought was only an experiment or a rehearsal is in fact the real thing. A Northwestern University dormitory resident taking part in a fire drill noticed that there were fire engines and firemen busy on the scene. It must be a real fire, she thought, and how clever of them to tell us it was only a drill to keep people from panicking. In fact it was only a drill, for the firemen as well as the dorm residents. A similar layering of reality is being employed by Gamson and his colleagues at the Universtiy of Michigan in a dramatic study in which subjects believe they have been recruited to participate in an experiment but are gradually allowed to discover that they are being manipulated for political purposes by a major oil company (Rytina, Cohn, Fireman, Gamson, Morgan, & Taylor, no date). Their reactions to this discovery and the stages by which a group comes to protest against what they feel is an illegitimate authority are the major dependent variables of the research. In fact, of course, it is only an elaborately staged experiment (set in a hotel rather than on campus), with the apparent ulterior purposes of the oil company simply a second layer of deception in the experiment.

We have now arrived at perhaps the central methodological problem of experiments in social psychology, namely, what it means to people to know that they are subjects in an experiment. How often do they remind themselves that, "This is only an experiment, not for real"; and what can be done to keep them from continually editing their responses through the knowledge that they are in an experiment? The most elegant solutions are to convince subjects that they are not in an experiment after all, as in the examples of the previous paragraph, or not to let them know that they are in an experiment in the first place. Both of these are often impractical, especially if the subject is recruited through a subject pool or brought to a psychological laboratory, and both also threaten the ethical requirement that subjects give informed consent to their being studied. Ideal or not, a third course is most often followed by experimenters. Subjects are told that they are in an experiment, but an effort is made to have the experiment establish its own reality and to involve subjects in it. This is what Aronson and Carlsmith (1968) call experimental realism.

Curiously, on this level experiments with deception are no different from the role play exercises they criticize (for a recent statement of the debate, see Forward, Canter, & Kirsch, 1976; and Cooper, 1976). Both require subjects to act as if their involvement were real — as if certain variables were real and as if the situation were not just a role play or just an experiment. Arguments that minimize the distance between role play and deception experiments are usually taken as arguments in favor of role playing and against deception. As Harré and Secord (1972) write: "We used to think that through the use of deception, people in experiments were prevented from playing a role. But now it seems that possibly the result of our deceptions has been to prevent the experimenter from knowing which role the participant is playing [p. 317]." The point, however, is not that a powerful experiment in which people are led to believe that they are administering dangerous shocks to another subject (Milgram, 1965) can be duplicated by asking people to imagine the situation. The point is that the Milgram experiment was itself a role play situation of a most sophisticated and successful sort (cf. Hamilton, 1976; Mixon, 1976), in which subjects were induced to take seriously their role in a particular experimental situation.

Merely asking subjects to imagine playing such a role creates a much weaker reality. When asked, people not surprisingly deny any willingness to hurt the victim, although they do not deny abdicating responsibility for the victim's welfare to the experimenter (Mixon, 1972), behavior still quite conducive to atrocities. To take this weak exercise of imagination as our paradigm of role play makes role play seem much less real than its alternatives. To take the Zimbardo experiment as our paradigm of role play, on the other hand, makes it clear that subjects can be involved as deeply in a role play experiment as in any other kind of experiment. It may be easier to deceive people into treating a situation as real than to involve them in treating it as real through the cumulative effects of their role playing; but deception can fail to create a reality, and role play can succeed. It is the attribution of reality to the situation, not deception or role play, that is the key (cf. Cooper, 1976).

The critical point in this commonality between deception and role play situations is the strange mixture of suspicion and gullibility that subjects bring to experiments. On the one hand, subjects know (or quickly learn) that nothing an experimenter says is to be trusted or taken at face value. On the other hand, if the experiment is done properly, subjects will ordinarily have no better explanation for what is happening than the explanation the experimenter gives them. Why not then act as if what the experimenter says were true or as if it were not an experiment? From this perspective, the state of mind of a subject in an experiment, acting as if things were true, is not so different from the state of mind of a subject in a role play, also acting as if things were true (cf. Mixon, 1977). Nor is this mixture of suspicion and gullibility unusual. It is the same state of mind with which people approach carnival side shows or political elections, and promoters like Barnum have long known that people can as easily be engrossed in a situation by playing on their suspicions as by playing on their credulity. Indeed, magicians claim that a suspicious audience is more easily fooled than a naive one, because their attention is more easily misdirected toward things the magician wants them to watch. This mixture of suspicion and gullibility, and the compromise solution of acting as if what people said were true until proven otherwise, is probably also the state of mind with which we approach ordinary social interaction (Goffman, 1959). If suspicion were removed entirely from a psychological experiment, the experiment might lose both its mundane and experimental realism!

What would happen to a person who tried to resist treating an ongoing interaction as real or taking it seriously? This is, of course, what we would do if we became prisoners in a replication of the Zimbardo experiment, knowing what we do now about the original. It is also what we would do if we were given modest sentences to a "real" prison — keep trying to remind ourselves that our involvement was temporary and should be allowed to touch us only superficially. Can we doubt, however, that this was what prisoners in the original Zimbardo experiment did, or what prisoners now going to jails in the outside world do — or that we would be any more successful than they in keeping our distance? In other words, though the Zimbardo prison may have been quite different in many ways from an outside prison, the process by which participants came to treat that prison experience as real may be quite similar to the process by which participants in other prisons come to treat them as real. The process involves people playing roles, taking them seriously, and sanctioning severely others who do not take them seriously.

If nothing else, being punished for not treating a situation as real can make that situation real. I have a friend who in pre-Vietnam days trained as a member of the Army Special Forces, a training he eventually abandoned when he discovered he had some taste and aptitude for it. In a field exercise, he was assigned the task of allowing himself to be captured by an enemy team, discovering what he could about their position, and escaping to bring back this information. He was smiling and about to say, "Hey, guys, let's take it easy," when his first captor smashed him in the face with a rifle butt for trying to talk out of turn. The rage and the desire to kill stimulated by this act no doubt helped him to carry out the rest of his assignment, just as these emotions (properly managed, to be sure) would help him to carry out a real assignment. Punishment can certainly provide the jolt of arousal ("Thanks — I needed that!") that helps make things real. Indeed, much of the pain in the world may simply reflect the fact that through pain we can make other people feel something for us and treat situations we are involved in as real when we cannot touch them with love, joy, or other forms of communication.

The point is that all roles, in a game, in an experiment, or in the outside world, are unreal at first and become progressively, ineluctably more real through our own behavior and other people's responses. They do not start out as real. Only over time do people come to feel that they are really professors or really married. It might be interesting to trace out the time course of identification with a new role or what determines how quickly people come to feel that a new name or a new title is really them. It may be that mere exposure has something to do not only with how much we like things (Zajonc, 1968) but also with how real we think they are. However, although one can learn roles by observation (cf. Bandura, 1977), it is likely that only enacting them can make them real. All roles are also strange when first created or introduced into a culture. They may become progressively more real and serious to subsequent generations for whom the reality they create is familiar rather than novel. Here is Morton Hunt's (1959) account of a game that created a new and enduring reality in the most profound of human relationships:

> Toward the end of the eleventh century A.D., a handful of poets and noble-men in southern France concocted a set of love sentiments most of which had no precedent in Western civilization, and out of them constructed a new and quite original relationship between man and women known as l'amour courtios or courtly love. [It] began as a game and a literary con-ceit, but unexpectedly grew into a social philosophy that shaped the manners and morals of the West. It started as a playful exercise in flattery, but became a spiritual force guiding the flatterers; it was first a private sport of the feudal aristocracy, but became finally the ideal of the middle classes . . . and men and women throughout the Western world still live by and take for granted a number of its principal concepts [p. 131].

Nowhere is social psychology further apart from public consciousness than in its understanding of how things become real for people. The evidence is the Patty Hearst case. People could understand if she had simply been a dedicated revolutionary and only pretended to be a carefree heiress or if she had helped stage the kidnapping in the first place. They could understand if she had simply been a carefree heiress and only pretended to cooperate with her SLA captors. What could not be understood was that she could be really an heiress, really a revolutionary, and then perhaps really an heiress again. Some of social psy-chology's most eminent analysts, like Martin Orne and Robert Jay Lifton, testified to this effect; but the jury chose to disregard all social science testimony, perhaps sensing that it threatened current concepts of legal responsibility. Social psychology still has a lot to learn about what makes things real for people, certainly before it can communicate effectively on this topic to the public. In the next section we outline a framework within which this learning might take place.

INTERNAL AND EXTERNAL CORRESPONDENCE

Questions that begin as a methodological nuisance often evolve into matters of major substantive interest (McGuire, 1969). It is time for the question of when people treat things as real to make this transformation from frog to prince. Showing that this question is not simply a methodological nuisance requires us to develop a model and a series of researchable propositions for predicting when people will attribute reality. This is the purpose of the current and following section.

The central feature of our model is the idea that behavior may have both internal correspondence (correspondence with feelings) and external correspondence (correspondence with consequences) and that both are necessary for attribution of reality. It may be helpful to begin with an illustration. In praising auto racing, David Carradine says: "It's real, you can't fake it, people actually die" (Windeler, 1977). The first element of this assertion ("You can't fake it") specifies what we mean by internal correspondence, that the actor's behavior reflects and corresponds to an internal state. The second element ("People actually die") specifies what we mean by external correspondence, that the actor's behavior determines and corresponds to important consequences. More formally, experiences will be considered real to the extent that each of the following holds:

1. *Internal correspondence.* Behavior corresponds with feelings. This means that a person's behavior expresses feelings that are both substantial and appropriate to the behavior. If a situation is only a game and an actor shoots someone, the actor only pretends to be angry or does not act angry at all. If a situation is real and an actor shoots someone, the actor is expressing angry or hostile feelings toward the person shot. If a situation is only a rehearsal and actor makes a mistake, he is not upset. If a situation is real, making a mistake or anticipating a mistake will mean the actor feels upset and embarrassed.

When we say a situation is real, we mean in part that there is a good deal of emotion in the situation. With no feelings attached to behavior, internal correspondence is necessarily low, and we speak of a person as acting like a robot or in a trance. The song "What Now My Love" contains the lines: "Once I could see, Once I could feel, Now I am numb, Now I'm unreal." It should be noted, however, that internal correspondence can be low even when feelings are substantial if these feelings are inappropriate, unexpressed, or bear no relation to the behavior. A person experiencing arousal, anger, or euphoria in someone else's company may not express these feelings but will in consequence find the interaction seeming unreal.

2. *External correspondence.* Behavior corresponds with consequences. This means that a person's behavior elicits responses that are both substantial and appropriate to the behavior. If a situation is only a game and an actor shoots someone, the person apparently shot only pretends to be hurt or does not act hurt at all. If a situation is real and an actor shoots someone, the person shot is hurt. If the situation is only a rehearsal and an actor makes a mistake, there is no audience to be misled, amused, or annoyed; and the actor can simply repeat the line until it is done correctly. If a situation is real and an actor makes a mistake, an audience is affected and the consequences cannot be undone.

When we say a situation is real, we mean in part that there is a good deal at stake in the situation. With nothing at stake or no consequences attached to behavior, external correspondence is necessarily low. However, it should be noted that external correspondence can be low even if consequences are substantial if these consequences are inappropriate or bear no relation to the behavior. People may get mad at a person who is unaware of having done anything to cause it or may fail to get mad at a person who has supplied sufficient provocation. Adolescents may mature either faster or more slowly than their peers and come to feel that others — peers and adults both — no longer respond appropriately to their behavior. The behavior elicits consequences, but inappropriate ones. Hence external correspondence is low, and the situation will seem unreal. Detaching behavior from consequences is also a critical treatment for making people feel helpless (Seligman, 1975), suggesting that a sense of unreality may be one correlate of learned helplessness, and vice versa.

Testing whether a situation is real or not may involve procedures that in the ordinary case occur automatically, outside of awareness (Langer, chapter 2, this volume). People may or may not be able to describe correctly why a particular situation seemed real or unreal to them any more than they can judge correctly how they arrived at other attributions (Nisbett & Wilson, 1977).

Three of the most highly valued words of the English language all refer to internal correspondence: being free, being honest, and being high. Internal correspondence can be absent either for reasons external to the actor or internal to the actor. If people's behavior fails to correspond to their feelings for reasons beyond their control, we say that they are unfree. If people's behavior fails to correspond to their feelings for reasons of their own choosing, we say that they are dishonest. Finally, the state of being high refers to an unusually powerful degree of correspondence between a person's feelings and their behavior. This is certainly how people describe the euphoric experience of "flow" in dancing, rock climbing, and other physical activities (Csikszentmihalyi, 1975). As a common element in our attributions of freedom, sincerity, and joy, the sense of reality derived from internal correspondence may be a good starting point from which to derive much of our entire Western value system.

The distinction between internal and external correspondence parallels a distinction that frequently arises in defining social constructs. For example, when Bandura and Walters (1963) are trying to decide what behavior is "really" aggression, they find it necessary to refer both to the feelings associated with the behavior and the consequences associated with the behavior. Similar issues arise in defining altruism, conformity, and so forth.

The distinction between internal and external correspondence may call to mind the distinction between intrinsic and extrinsic motivation (see Deci, 1975; Ross, 1976). A person is intrinsically motivated if he is performing an activity for its own sake, for its consummatory value, because it is intrinsically enjoyable. A person is extrinsically motivated if he is performing an activity for the sake of the reward attached to it, for its instrumental value, not because it is intrinsically enjoyable. There is a relationship, but not a simple one, between intrinsic and extrinsic motivation and the pursuit of internal and external correspondence. A person must be intrinsically motivated to care about internal correspondence.

There are more rewards for acting in a pleasing or socially desirable way than for trying to match one's behavior with one's feelings, especially if a person is concerned with being liked by others (Jones & Wortman, 1973). On the other hand, a person can pursue external correspondence for either intrinsic or extrinsic reasons. An intrinsically motivated person is here one concerned with competence for its own sake (cf. Atkinson, 1957). An extrinsically motivated person is one concerned with competence only as a means to social rewards.

It may be especially helpful in understanding our ideas of internal and external correspondance to consider cases that are high in internal correspondence but low in external correspondence, and cases that are high in external correspondence but low in internal correspondence. A summary of the four possible combinations of internal and external correspondence is presented in Table 1.1, along with an example involving positive affect (loving) and an example involving negative affect (fighting). In our analysis, a situation is felt to be real only if both internal and external correspondence are present. Ordinary social interaction, including loving and fighting, is real. Role play in which participants play at loving or fighting (without either the associated emotions or consequences) is unreal.

External correspondence without internal correspondence characterizes alienation. A person who is paid to love or paid to kill may enact these behaviors without feelings or without the feelings that are ordinarily associated with them. Nonetheless, their actions will still cause other people to fall in love, feel pain, or die, just as would the same actions performed with feeling. In some instances, people seek or choose to cut themselves off from their feelings or to

TABLE 1.1
Elements of Internal and External Correspondence

External Correspondence: Actions elicit correspondent consequences	Internal Correspondence: Actions express correspondent feelings	
	Yes	*No*
Yes	Real	Unreal
	Ordinary interaction	Alienation
	Examples:	*Examples:*
	Loving	Deceiving a lover about one's love
	Fighting	Killing as a profession
No	Unreal	Unreal
	Fantasy	Role play
	Examples:	*Examples:*
	Having an imaginary lover	Playing at being in love
	Making hostile comments in private	Acting out a fight in a movie

affect others without involving themselves in the process. Professional fighters and prostitutes do this in part to maintain control of the situations they find themselves in. In other cases, society may alienate people by forcing them to perform acts that are meaningless or distasteful and preventing them from performing acts or seeking relationships that would be meaningful (Etzioni, 1968). In either case, such people usually think of themselves or are thought of as "just doing their job."

An activity that begins as unreal may become real through first generating a sense of external correspondence or consequences that subsequently involve the person's feelings. Most of us have had the experience of arguing a position just for fun, finding that others responded with feeling, and winding up becoming far more involved in that position than we had intended or imagined. Even if people are told that an argument a person reports may not reflect that person's views, they still tend to respond to the person as if those were his views (Manis, Cornell, & Moore, 1974; Snyder & Jones, 1974), setting in motion all the machinery of self-fulfilling prophecy (Merton, 1948; Rosenthal, 1966). Stars of soap operas are often treated by the public as if they were the characters they portray, and Leonard Nimoy of "Star Trek" felt obliged to entitle his book, *I am Not Spock* (the name of his character in the space series). Police spies have repeatedly been found not only to observe illegal behavior but to instigate or solicit it, for which they are rightly condemned as agents provocateurs. Surprisingly little attention has been given to the possibility that they may have been entrapped in their roles in a somewhat different sense from the way they were endeavoring to entrap others.

Internal correspondence without external correspondence characterizes fantasy. This includes what people might like to do or say but avoid because they fear the consequences as well as things they might like to do that simply seem impossible. A game or a role may seem especially real when the pretense of fantasy enables actors to behave in ways that correspond more closely to their feelings than they can outside the game or the role. Understanding this led Freud to his interest in the analysis of jokes and dreams and Goffman to his interest in the apparent accidents of everyday life. Ordinarily, the high internal correspondence of a game in which people act out their fantasies would be more than offset by the loss of external correspondence through the knowledge that none of these actions had serious consequences. In some instances, however, a game may come to seem more real than life outside the game, as in the remarkable short stories by Harry Mark Petrakis (1968) and Milan Kundera (1974). In the Petrakis story, a young prostitute enacts her fantasy of returning to her hometown. A man gets caught up in helping her, only to discover her identity and angrily reject her. He soon regrets his anger and returns to confess his own deceptions. But he has humiliated what was most real to her, and she will now offer him only what was real to him — the bleak identity of prostitute. Likewise, people playing the transactional games Berne (1964) describes, like alcoholic or schlemiel, have feelings that correspond more strongly to their roles in the game than to anything outside it. They will accept almost any criticism or abuse of themselves in the role of alcoholic or schlemiel, but will lash out violently at a therapist or friend who dares to suggest that they are just playing a game for its hidden payoffs, or that the game is not real.

An activity may also become real through first generating a sense of internal correspondence or emotions in the actor that in turn lead the actor to demand that others treat him as if the role were true. This may describe the sequence by which the guards became caught up in the Zimbardo prison experiment (the prisoners probably followed a different sequence), and also the process by which leaders come to make more and more unreasonable demands for glorification by their followers. Conversely, the sense of internal correspondence may be the last to go after social change has deprived behavior of the consequences it once had. This is the essence of our feeling for empty rituals and our sense of nostalgia for songs and gestures that had meanings in other years that they no longer carry. Such rituals are perhaps most visible among remnants of upper classes who continue to behave as if their liaisons or clubs had the same importance for society as they once did.

Ordinarily we want things to be real rather than unreal. We may exaggerate the correspondence between our behavior and both our feelings and subsequent events in the service of this need to believe that things are real. Occasionally, experiencing a situation as unreal may be a psychologically adaptive defense mechanism (cf. Laing, 1967), a form of existential dissonance reduction. More generally, however, a sense that our behavior does not reflect our feelings or is detached from consequences is symptomatic of psychological disorder. Indeed, much psychotherapy may have only the essential function of re-establishing a sense of internal and external correspondence, as could be accomplished through jogging.

On the other hand, there are risks involved as either internal correspondence or external correspondence becomes very high. If internal correspondence is perfect, actions are unambiguous indicators of feelings and can never be excused. If external correspondence is perfect, actions have irredeemable consequences and can never afford a mistake. Furthermore, people are free to pursue internal correspondence to any limit only if they disregard the question of what consequences (if any) their actions have. Likewise, they can pursue external correspondence without limit only if they are willing to sacrifice or disregard the question of whether their actions correspond with their feelings. This means that people must often choose between acting in a way that maximizes the extent to which their behavior expresses their feelings and acting in a way that maximizes the extent to which their behavior will affect others. This tension may be picked up in the overjustification effect (e.g., Ross, 1976), in which people are found to see less correspondence between their feelings and the behavior of engaging in a pleasurable activity when external correspondence is increased, or when they are paid for that activity. Indeed, in our analysis of external correspondence and internal correspondence, we may think of ourselves as bringing together perhaps the two major concerns of current research in social psychology: concern for effectiveness or control (external correspondence) and concern for intrinsic motivation or internal attribution (internal correspondence). Not the least interesting feature of our analysis is the researchable proposition that people must occasionally choose between the demands of external correspondence (effectance) and internal correspondence (intrinsic motivation).

We have gathered preliminary evidence that both internal and external correspondence are important in the attribution of reality. Thirteen groups of students ranging in size from 5 to 10 played a simple game in which different players made proposals (on any topic) that the group then discussed and voted on. All members of the group started out with an equal number of chips, which they used to represent the extent to which they were for or against each proposal by placing them in an appropriate spot on the table. People could add to the number of chips they had placed on the board at any time during the discussion, but could not remove any chips. At a point determined by chance, a final count was taken to determine whether the group had accepted or rejected the proposal being discussed. Players voting on the winning side of an issue gained a few chips, thus increasing somewhat their ability to influence subsequent issues. Play was stopped after an hour.

Afterwards, players filled out a short questionnaire in which they described the emerging distribution of power in their group, whether they felt they had discovered anything about their feelings toward power during the game, and how much they liked the game. They also answered yes or no to the following questions: Was this game real for you? Did you act in this game in a way that expressed your feelings? Did other people in this game respond to your actions with feeling? A parallel set of three questions asked whether power (or lack of power) in the game was real for them. These results are reported in Table 1.2.

As can be seen, an overwhelming majority of players reported that the game was real and that power was real for them when they felt both that they had expressed their feelings and that others had responded. An intermediate proportion felt the situation was real when only one of these conditions held, and only a small minority felt things were real when neither internal nor external correspondence held. The pattern of results suggests an additive model in which both self-expressiveness and others' responsiveness make independent contributions to the attribution of reality. The most reliable difference, however, is between the cell in which both conditions are true and cells in which only one or none of the conditions is true. The difference between the proportion of people saying that the game was real for them in the cell in which both conditions held and the cells in which only one condition held was significant with a $X^2(1)$ = 8.99, $p < .01$. The same difference for the proportions saying that power was real for them was also significant with a $X^2(1)$ = 12.50, $p < .01$. The difference between the proportion of people saying that the game was real for them when neither internal nor external correspondence held and the proportion

TABLE 1.2
Proportion of Respondents in Power Game
Making an Attribution of Reality

External Correspondence: Person's actions elicited responses from others	Internal Correspondence: Person's actions expressed feelings	
	Yes	No
Yes	Game real: 79.0% (62)[a]	Game real: 28.6% (7)
	Power real: 90.9% (44)	Power real: 55.9% (9)
No	Game real: 50.0% (8)	Game real: 11.1% (9)
	Power real: 50.0% (12)	Power real: 21.7% (23)

[a]Numbers in parentheses represent cell N's. The slight discrepancy in the total N for the game and power ratings is due to a few participants who omitted one of the questions.

saying that the game was real for them when one of these conditions held was not significant. The same difference for the proportions saying that power was real for them was significant with a $X^2(1) = 4.45$, $p < .05$. It should be noted that we have found a similar pattern of results for two other games we have tested. These data, of course, only show that internal and external correspondence are associated with attributions of reality, not that they are causes of these attributions. For that we need an experiment in which we manipulate internal correspondence (by instructions or incentives given to subjects) and external correspondence (by programming of confederates) and measure their effects on subjects' attributions.

INDIVIDUAL AND SITUATIONAL DIFFERENCES IN THE ATTRIBUTION OF REALITY

The distinction between internal correspondence and external correspondence will not always be sharp. Most situations will be high in both or low in both. Nevertheless, it should be possible to specify cases or perspectives in which one component will be more important than the other. We offer the following six hypotheses:

1. Observers will emphasize external correspondence more and actors will emphasize internal correspondence more. This should follow from the fact that actors have access to internal cues that observers do not. However, it does not imply that the actors see their feelings or dispositions as the main cause of behavior in the situation (cf. Jones & Nisbett, 1971; Monson & Snyder, 1977),

only that they see them (and their relationship to their behavior) as a main determinant of whether the situation is real or not.

This could account for why games, encounter groups, and revival meetings seem more real to participants than to observers or outsiders, with whom these experiences may be difficult or impossible to share. It could also account for the fierce attacks on people in these situations who choose an observer role. Refusal to participate is often a specific device to deny or minimize the reality of a situation, something other participants are likely to resent. If others attack the observer, choosing the observer role has great consequences, and the rather paradoxical effect of making the situation more real and involving for all concerned than would otherwise have been the case (Hampden-Turner, 1976; Watzlawick, Beavin, & Jackson, 1967). Attacking people is the explicit device used by EST training (Brewer, 1975) to transform people from observers to actors both within the training sessions and, allegedly, in their own lives. In many situations, however, a familiar but inactive participant elicits little response, so that their behavior is low in external correspondence. Furthermore, observing is socially passive behavior whose correspondent emotions are either weak (if the person is comfortable with the role) or inappropriate (if the person is uncomfortable with the role). Thus we would expect that shy people or people who are characteristically observers would find situations less real than active participants.

2. Adults will emphasize external correspondence more, and children will emphasize internal correspondence more. This could be true for either of two reasons. First, adults have more of both the experience and the mature faculties needed to extrapolate the consequences of their acts. They should thus be more likely to use whether or not behavior has important consequences as a criterion for judging whether or not a situation is real. Children are more likely to get involved in play and games that are not real in their consequences, whereas adults are more likely to insist upon the trappings of uniforms and stakes of money or lives before they play seriously. Second, as people get older, they learn to adapt to situations, lie, and play roles. Thus adults should be less likely to use whether or not behavior corresponds to important feelings as a criterion for judging whether or not a situation is real. Retaining a child's dedication to internal correspondence is likely to make a person either self-actualized — or hospitalized.

Children should be more likely to give a gift that expresses themselves but has no impact or an unintended impact on the recipient. Adults should be more likely to give a gift that has an impact on the recipient but does not express themselves. The ideal gift of course does both, at once saying "this is me" and "this is you," thus constituting a message of both sharing and understanding.

If concern for internal correspondence were the same as concern for the intentions of behavior rather than the consequences of behavior, our prediction would contradict the discovery that adults are more likely to judge behavior by its intentions and children by its consequences (Piaget, 1932/1948). However, attribution of reality is not a moral judgment, and there is evidence that the special weight attached by adults to intentions holds only for approval of moral behavior and not for approval of behavior in other domains. Weiner and Peter (1973), for example, found that although older children judged moral behavior more and more as a function of intention rather than outcome, they began to judge achievement behavior somewhat more on the basis of outcomes and somewhat less on the basis of intentions. Indeed, in nonmoral domains the thrust of socialization and the developmental shift may be the opposite from the trend in moral domains. Adults may tell children that it is the thought that counts and reward them for their trying rather than for their actual accomplishments. As children grow older, however, they learn that in the adult world, it is the results, not intentions, that count.

3. Males will emphasize external correspondence more, and females will emphasize internal correspondence more. This would seem to follow from the pervasive difference in sex-role socialization whereby males are taught to seek power and achievement and to suppress or ignore feelings (Pleck & Sawyer, 1974), whereas females are taught to pay attention to feelings and internal space (Erikson, 1964). Bakan (1966) sees males in general as following the principle of agency, emphasizing assertive control over the environment, and females as following the principle of communion, emphasizing feelings of closeness and commonality. The pursuit of external correspondence appears to embody Bakan's principle of agency, whereas the pursuit of internal correspondence appears to contain the principle of communion as a special case. Hoffman (1975) found that males were more concerned with being caught, or about the connection of their behavior to consequences, and that females were more concerned with guilt, or about the connection of their behavior to feelings. Sampson (no date) has shown that males are more likely to seek information about another person's background and situation and that females are more likely to seek information about what the person thinks and feels.

4. People self-conscious about the relationship between their behavior and environmental stimuli will emphasize external correspondence more, and people not especially conscious of the relationship between their behavior and environmental stimuli will emphasize internal correspondence more. The first category appears to include people who are high in objective self-awareness (Duval & Wicklund, 1972) — which means people who self-consciously think of themselves as objects rather than people who think objectively in the usual sense — and also people who are high in need for approval (Crowne & Marlowe, 1964) or externality (Schachter & Rodin, 1974). We would expect people in any of these states to use the relationship between their behavior and environmental stimuli as a stronger cue to what is real than people not in these states. Thus Wicklund and Duval (1971) found that individuals in a state of objective self-awareness were more likely to infer their attitudes from what they heard themselves say in a situation than individuals not in a state of objective self-awareness. Crowne and Marlowe (1964) reported a similar result for people high in need for approval, as did Norman and Watson (1976) for introverts, both groups being high in

self-conscious awareness of environmental stimuli. Schachter and his co-workers (Schachter & Rodin, 1974) found that obese individuals tend to be more sensitive to external cues and are more likely to infer their hunger from what they see, smell, and do in a situation than nonobese individuals (see also Younger & Pliner, 1976).

It is harder to categorize people high in self-monitoring (Snyder, 1974), since these individuals are described both as people who pay special attention to situational cues or interpersonal specifications of appropriateness and as people who pay relatively little attention to information about dispositions, attitudes, or other inner states of self. The heightened sensitivity to situational cues implies greater emphasis on external correspondence, and Ickes and Barnes (1977) found that subjects high in self-monitoring were particularly concerned to adapt their behavior in order to maintain a smooth flow of conversation. However, the reduced self-consciousness implies less readiness to make inferences from the observed relationship between behavior and consequences (Snyder & Tanke, 1976). One possibility is that external correspondence matters to both high and low self-monitors, but for different reasons — for highs, because it gives them clues as to their effects on others; for lows, because it gives them clues as to their own internal states.

Theater actors score much higher in monitoring the situational cues and consequences of their behavior than do psychiatric patients (Snyder, 1974). Thus people who emphasize internal correspondence and de-emphasize their monitoring of behavior and situations may pay the price of finding themselves not engaged in situations others define as real or overly engaged in situations others define as unreal. We sometimes say that people who play games — different games in different situations — are disturbed, but we judge still more harshly those who will not or cannot play games.

5. People intermediate in power will emphasize external correspondence more, and people either very low or very high in power will emphasize internal correspondence more. People who have a great deal of power have by definition the ability to produce events that correspond with their behavior. People who have no power have no hope of generating very much external correspondence. In the first case attempts to achieve or to influence others are likely to be un-necessary, whereas in the second case they are likely to be fruitless. Thus in both cases they are less likely to be observed (Atkinson, 1957; Gamson, 1968). At both extremes of power, people may place more empahsis on feelings and come to rely on the presence or absence of internal correspondence for their sense of reality. Among the powerless, the experience of suffering or being punished may generate sufficient correspondence between feelings and behavior to make situations real. Among both groups, drugs and alcohol may be alternative vehicles for producing an appropriate level of internal correspondence. Interestingly enough, we can now understand different chemical agents as alternatively in the service of deadening a level of internal correspondence or sense of reality that is too intense and painful, or heightening a level of internal correspondence or sense

of reality that is too flat and unstimulating. Nonusers think of drugs as primarily a device for deadening or escaping from reality, while users enjoy the idea that "Reality is just a crutch for those who can't handle drugs." In part seeking an optimal level of internal correspondence is simply seeking an optimal level of arousal, for which extroverts may need more stimulation than introverts (Eysenck, 1963). Alcohol, of course, decreases arousal, and coffee increases it.

6. People early in the process of involvement in a role will empahsize external correspondence more, and people late in the process of role involvement will emphasize internal correspondence more. This follows from the fact that it may be hard, early in a role, to know what the appropriate feelings or the appropriate behaviors are, let alone their relationship. Even if one is new at the role of fiancé, however, other people are familiar with how to treat fiancés; and the role may become real as one is treated as a fiancé with consistency and consensus by more and more people (cf. Kelley, 1967; Miller, Brickman, & Bolen, 1975). Dating couples who report that they are treated by others as a couple are more likely to continue and consolidate their relationship than couples not so treated (Lewis, 1973). Accounts are common of how a person did not feel like a champion, a president, a college student (cf. Brickman, 1964), or a teaching fellow until other people began treating them as one.

Early in a role when the behavior involved is still unfamiliar or perhaps even counterattitudinal, we would expect situations in which people infer their attitude from their behavior to be more important than late in the role. Such inferences require that people see their behavior as having had consequences, or that others have responded to their behavior, and that the actors see themselves as responsible for these consequences (Collins & Hoyt, 1972). Later in the role, however, the vigor and sincerity with which a person enacts the role may determine to what extent others are willing to respond to or follow him. Thus early in a role internal correspondence may depend to an extent on external correspondence, and late in a role external correspondence may depend to an extent on internal correspondence. We may also suspect that loss of external correspondence precedes loss of internal correspondence as a person withdraws from a role (others give up treating a person like an incumbent before the person gives up feeling like an incumbent), but there may be important instances in which people's feelings for the role die before they relinquish it.

PHENOMENOLOGICAL AND INFERENTIAL VALIDITY

Experiencing a situation as real means that the situation has a kind of validity for us. How does validity in this sense compare to validity of the sort we traditionally seek in our research? What would research that was valid in this sense look like? The present section seeks to answer these questions. Finally, in the following section, we will consider what social psychology as a whole would gain by seeking this additional kind of validity and how it might do so.

We evaluate our experiments by the criteria of internal and external validity (Campbell & Stanley, 1963). An experiment has internal validity if it permits a valid causal inference, or if change in the dependent variable can correctly be attributed to the causal influence of the independent variable. An experiment has external validity if it permits valid generalization, or if the causal relationship observed in the experiment can correctly be generalized to other settings, operations, and populations. Recently, efforts have been made to redefine, supplement, or challenge the analysis of internal and external validity. Cook and Campbell (1976) have suggested a closer look at the components of internal validity, including whether or not the independent variable was correctly labeled (construct validity) and whether or not the statistical analysis underlying the inference of causality or lack of causality was adequate (conclusion validity). Other authors have challenged the basic distinction between internal and external validity, asserting that external validity is the ultimate form of validity with internal validity being only conditions that must usually be met for external validity to be achieved.

The present analysis suggests that the categories of internal and external validity be supplemented in another way. We need terms to represent the components that determine whether an experience is seen as real or not, to represent whether an experiment has phenomenological validity or not. Let us say that an experiment has intrinsic validity if internal correspondence (correspondence between behavior and feelings) in the experiment is high. Let us say that an experiment has extrinsic validity if external correspondence (correspondence between behavior and consequences) in the experiment is high. An experiment that satisfies the usual demands of internal and external validity may be said to have inferential validity. An experiment that satisfies our demands of intrinsic and extrinsic validity may be said to have phenomenological validity.

Phenomenological validity could contribute to the inferential validity of an experiment. For example, if an experiment involved manipulating freedom, subjects' feelings that their experience of freedom (or the lack of it) in the experiment was real could increase our confidence that the experiment had construct validity or internal validity. In this sense ascertaining phenomenological validity is like conducting a manipulation check, although a check on something a little different (attributed reality) than a standard manipulation check. Similarly, if the phenomenological validity of an experiment manipulating freedom was high, we would have greater confidence that we could generalize the results of that experiment to other situations involving the experience of freedom. On the other hand, an experiment could be very high in phenomenological validity without allowing either any valid causal inference or any valid generalizations. Alternatively, an experiment could permit valid inference or generalization about a form of freedom that had no phenomenological validity for subjects. A situation could be high in inferential validity without observers of that situation finding it terribly real, involving, or consequential.

Our analysis of the phenomenological validity of experiments parallels in certain ways Aronson and Carlsmith's (1968) analysis of the realism of experiments. However, their analysis treats the question of realism as a methodological problem; ours looks at the question of phenomenological validity as a theoretical issue for attribution theory. Furthermore, Aronson and Carlsmith break realism down into two components quite different from ours. They distinguish between experimental realism and mundane realism. Experimental realism refers to how vividly subjects are involved in the procedures of the experiment. Mundane realism refers to how typical or representative the procedures of the experiment are of everyday life. Phenomenological validity in our sense could describe or contribute to both experimental realism and mundane realism. On the other hand, experimental realism and mundane realism do not distinguish between internal and external correspondence as determinants of the perception of reality.

Note that it is incorrect to say that phenomenological validity refers to subjective reality and inferential validity to objective reality. Modern philosophy has shown that the distinction between subjective reality and objective reality is hard to maintain. Both phenomenological validity and inferential validity are in the end determined by consensus. An experiment is high in phenomenological validity if qualified observers agree that it is high in phenomenological validity. An experiment is high in inferential validity if qualified observers agree that it is high in inferential validity. Who these qualified observers are, however, is different in the two cases. For phenomenological validity, the qualified observers are subjects or participants. For inferential validity, the qualified observers are experimenters or researchers.

A summary of a few of the major points that distinguish phenomenological and inferential validity is provided in Table 1.3. The first criterion for phenomenological validity is that outcomes be meaningful; all else is in the service of this end. "Meaningful" implies that things were seen as real by those involved, but it also implies something more. In Frankl's (1959/1963) sense, it implies that the feelings, behavior, and consequences are seen as mattering within the subject's larger frame of values. The first, familiar criterion for inferential validity is that outcomes be unambiguous; all else is in the service of this end.

TABLE 1.3
Elements of Phenomenological and Inferential Validity

	Phenomenological Validity	Inferential Validity
Outcomes must be:	Meaningful	Unambiguous
Evidence provided by:	Involvement	Calculation
Focus is on:	Diversity	Uniformity
Typified by:	Games	Experiments

The question of involvement versus calculation as a basis of knowing is an old one (cf. Merton, 1972). It is captured in the contrast between the pull of a case study and the power of a statistical sample. The problems with using a case study as the basis for inference are too well known to bear comment (Campbell & Stanley, 1963). We are surprised, and sometimes despairing, at the extent to which people allow a single case or several dramatic cases to override the evidence of a statistical sample (Nisbett & Borgida, 1975; Feldman, Higgins, Karlovac, & Ruble, 1976). Yet, the present analysis suggests they may not always be wrong to do so. Phenomenological validity can only be found in direct experience or in its next best form, vicarious experience gained through the vivid portrayal of someone else's direct experience. Thus, people may not be wrong in seeking direct experience or case studies if it is phenomenological rather than inferential validity that they seek. We would expect them to place more emphasis on the distinctiveness and consistency of individual cases than on the consensus of a representative sample (Kelley, 1967) when concern for phenomenological validity is high, and more emphasis on consensus than on case information when concern for inferential validity is high.

Culture, by definition, is information that can be transmitted to future generations without each generation's having to re-experience directly the events from which the information was derived. Universities are institutions for the transmission of culture. It should thus not be surprising that universities and their scholarly disciplines honor information that can be abstracted, summarized, and passed on by instruction more than information whose impact can only be felt by the direct involvement of participants. Yet, we also know that each older generation tries and fails to pass on to the young other lessons, which the young cannot hear, because the lessons do not make sense until people have tried and felt for themselves. These lessons are usually lessons of personal relevance, perhaps that alcohol or smoking is not worth the price or that fidelity to a vital relationship is more rewarding than its alternatives. These lessons make sense only if they have phenomenological validity, or only if people have had the emotionally involving experiences that convince them that these lessons apply to them and not just to their elders. People may change on the basis of phenomenological validity rather than inferential validity. Thus analytic therapists need to use patients' emotional transference as well as their objective insight into themselves to produce improvement.

Comedian Henny Youngman: "When a little girl says, 'I'm a girl and you're a boy,' and the boy says, 'I'll go ask my mother,' that's research. When he says, 'Let's see,' that's sex!", what we need is a balance of respect for phenomenological and inferential validity, so that we can focus on the question of moving back and forth between them. How should we use our firsthand experience to relate to an abstracted account of what other people have done, and how should we use accounts of other people's experiences (as summarized in research) to illuminate our understanding of our own lives? These are the questions we should be teaching – and studying. In trying to understand what it is like to be handicapped, when should people read the fine experiment by Clore and Jeffrey (1972) on the effects of role playing a handicapped person, and when should they follow around a handicapped person for a day or role play a handicapped person themselves? In trying to understand what it is like to be in a foreign country, or a politician, or divorced, or suicidal, when should people prefer an article describing the results of interviews with a hundred individuals in one of these states, and when should they prefer an article describing in detail the life of one individual in one of these states?

As indicated in Table 1.3, phenomenological validity values diversity while inferential validity values uniformity. The same situation may have different phenomenological validity for different participants. If some people find a situation real and others do not, it is hard to say that one group is right and the other is wrong. Each person is the best judge of whether the situation, has phenomenological validity for them. The inferential validity of a situation, however, should be the same for all observers. If people disagree about inferential validity, somebody is probably wrong, and future research will show who.

We have a special word for creations that emphasize inferential validity above all else. These are called experiments. Do we also have a word for creations that emphasize phenomenological validity above all else? I suggest that the word is games, and that this is what we mean — or should mean — when we talk about games. Chess, gambling, and hide-and-go-seek may to some extent simulate events in nature or events in society, but this is not why they are played. They are played because they create their own reality, which participants find engrossing. Goffman (1961) points out that a game is made real for players, yet still kept a game, by a delicate calculation that sets the tangible and symbolic stakes of the game at just the right level. Curiously enough, players' involvement in a game can be threatened if the stakes are either too small or too large. In our framework, stakes that are too small threaten the external correspondence of the game. Players can no longer feel that their behavior in the game has sufficient conse-quences for others, or for themselves, to make the game real. Stakes that are too large threaten the internal correspondence of the game. Players must be so con-cerned with the outcome that they can no longer allow their behavior in the game to reflect their moods, their feelings, or their personalities. Since phenomeno-logical validity is the goal of games, it should not be surprising that games and sports at their best can be more real to players than anything outside them. The experience of total physical and mental involvement in play is called flow by Csikszentmihalyi (1975), who reports that participants find it to be an emotional peak in their lives.

It may seem paradoxical that an artifically created game environment can be more real for participants than activities outside the game. The paradox vanishes, however, when we realize that internal correspondence is every bit as important for the attribution of reality as external correspondence. Stakes and consequences may be limited in a game, making it unreal in that sense — but only in a manner designed to heighten expressiveness and internal correspondence beyond what would obtain without the appropriate setting of the stakes. It is the joint product of internal and external correspondence, not either one alone, that determines the experiencing of events as real; and it is only games that specifically set out to maximize this joint product.

An example may make these ideas clearer. DeKoven (1973) describes a children's game called "The Belt" and his efforts to modify it. The game begins with one player hiding an ordinary man's belt. The other players then leave the base (a safety zone) and begin to search for the belt. The player who finds the belt is entitled to whip any people he can catch before they succeed in getting back to the base. The finder then hides the belt to start a new round of the game. The game has a nice balance of tension in that timid players can avoid being hit by staying close to the base, but they also thereby give up much chance of finding the belt. DeKoven tried to get children to play the game with a rolled-up newspaper instead of a belt, which seemed dangerous to him. The children refused. The belt was required to make the game real. On the other hand, they might also have refused to play with a baseball bat or a sword; in any case they did not beat each other with the same consequences that a mother or father wielding the belt might produce. DeKoven noted that children who did mete out consequences seen by other players as too severe were beaten in retaliation until they had learned their lesson. Thus, the players insisted on regulating the stakes of the game in a way that kept the game as real and as engaging for them as possible.

We need to differentiate our speaking of games from two other usages of the word *games* in the social sciences, games in the sense of game theory (Rapoport, 1960; Shubik, 1964) and games as a form of pathology (Berne, 1964). Game theory is essentially a mathematical language for describing and classifying interaction situations, and especially conflict situations. It is concerned with inferential validity, not phenomenological validity, and deals with games that are clear and simple but not necessarily very meaningful to participants. Perhaps the most famous of these is the trust dilemma or Prisoner's Dilemma, in which it is rational for each player to make a selfish choice, but everyone is worse off if they all make selfish choices than if they all make unselfish choices. I believe we could show that this game does have some phenomenological validity, but even so it is often presented in a way that has little meaning for participants (Wrightsman, O'Connor, & Baker, 1972) and is in any case hardly a game that anyone would want to play voluntarily or for very long. Berne (1964) speaks of games in a pejorative sense. Playing games is pathological and neurotic. But Berne is talking about a special kind of game in which the player ostensibly has one purpose but secretly, at another level of their personality, seeks something else. For example, a person is playing a game in Berne's sense if they appear to be asking their mate for reassurance but are really engaged in trapping their mate into saying something that they can either attack or deny. A game in this sense is simply a dishonest interaction. In one sense the game is not valid or the interaction is not real, since the player's overt behavior does not reflect his inner feelings, or internal correspondence is low. It is not inappropriate to use the word *games* to describe what game theory is interested in or what transactional analysis is interested in, but it should be understood that the games they are interested in do not exemplify our notion of phenomenological validity.

Phenomenological and inferential validity are not incompatible, but nor are they easy to pursue simultaneously. The problem is a familiar one. Phenomenological validity is associated with complexity, multiple meanings, and many plausible interpretations. Inferential validity is associated with simplicity, clarity, and control. So far social psychology has tended to seek inferential validity at the expense of phenomenological validity. There are two possible ways we might increase concern for phenomenological validity and thus produce a more even balance of respect for the two forms of validity. The first is to show, as we have tried to do, that the question of phenomenological validity is of theoretical interest in its own right. As such, we have tried to derive a theoretical framework for predicting when people will see things as real and to place this question within the context of attribution theory. The second is to show that there are practical reasons for social psychology to seek phenomenological validity through games, just as it seeks inferential validity through experiments. This is the purpose of the next section.

GAMES AND EXPERIMENTS FOR SOCIAL PSYCHOLOGY

Experiments are sometimes criticized, mistakenly, because they are artificial. In fact, they derive their power from creating artificial environments that provide an unusual opportunity for valid causal inference. Vacuums on earth may exist only in artificial laboratory environments, and praise for people may be un-correlated with their actual performance only in artificial laboratory environments. If we are then interested in whether the causal effects observed in the laboratory account for important events outside the laboratory, we must go and test for their presence outside the laboratory. But the artificial environment of the laboratory is a critical asset, not a drawback, to our search for inferential validity. Similarly, games, too, as vehicles for inquiry, are often criticized because they are artificial. This criticism is much more tenacious for games than for experiments because somehow we have gotten it into our minds that games are valid only when they are accurate simulations or imitations of something outside themselves. The analysis of this paper suggests that this criticism is mistaken for games, too, and indeed has seriously hindered their development. The fact that games are artificial environments is their strength, not their weakness, if we focus on the unusual opportunity they provide for seeking phenomenological validity and for testing what people will see as real and why. Secondarily, and only secondarily, if we are interested in whether the processes observed in the games resemble important processes outside the games, we can ask if they occur outside the games.

Curiously enough, two of the most hardheaded areas of society — business and the military — have made the most extensive use of games. Business games do not use real money, and the models of the marketplace to which they expose players are highly simplified. The bullets in war games are not lethally aimed, if real, and the tactics to which players are exposed may or may not resemble those actually used by the enemy in the next war. It could be that business and the military have made more use of games than social psychology because their criteria for validity are less demanding, but I supect this is not the case. Rather the practitioners are as usual a little ahead of the theoreticians, in this case in recognizing that games have a validity — and a training value — separable from whether or not they are perfect imitations of outside events. Players will get involved, even if the games seem unreal at first, and they will learn things from their behavior when they are involved that will stand them in good stead when they get involved even in somewhat different situations in the future. No textbook learning, whatever its inferential validity, can substitute for this experience.

Games should be a natural concern for social psychology as science, as history, and as a vehicle for personal growth. Social psychology has recently seen the growth of interest in the scientific study of environmental systems (Altman, 1975). Games essentially involve creating entire environments and exploring people's response to them (Duke, 1974). The field has also witnessed a growing sense that its knowledge is historically conditioned (Gergen, 1973), that the tendency of people to trust one another or to seek power or to be attracted to certain types of others may all change over time. Most parties to the debate, whether they accept this idea or deny it, have found it rather gloomy to contemplate the prospect that social psychology may not be able to accumulate a body of facts of the same enduring validity as those in the natural sciences. However, the idea that each generation may to a degree need to rediscover social psychology, to discover for themselves what principles govern their own behavior, is not entirely unappealing. It may be that the appropriate task for social psychology is to help people in the future discover themselves more easily and more effectively than is possible today, rather than to discover now what people in the future will be like. Games would certainly be at least as much a part of this rediscovery process as experiments. The very word *game* conveys the idea of an enterprise in which a variety of outcomes are possible and reasonable. If more than one outcome occurs or the typical outcome changes over time, this in no way implies that the game was no good or invalid. The idea of an experiment, however, implies a particular and fixed set of outcomes. If more than one set of outcomes occur or the typical outcome changes over time, we do indeed take this as indicating that the experiment was not especially good or valid.

A larger role for games would also bring us closer to a social psychology of human concerns. Experiments must hold the interest of researchers. Games must hold the interest of participants. Games may be thought of as relatively safe environments in which -people can have access to experiences that would otherwise be inaccessible to them (sometimes traumatic, sometimes ecstatic) and can observe how they and other people handle these experiences. An experiment on conformity or freedom is something hardly anyone would choose to be in once they knew what it was about. A game on conformity or freedom is something people might well choose to be in, because they can see it as capturing an important phenomenological concern of theirs and as a learning experience they might profit from. A social psychology that devoted one part of its work to designing such games would be a discipline working closer to the central concerns of its subjects.

The games that would be most appropriate for social psychology would be different from the kinds of games currently available. Some current games are too complex or too loosely structured. These include the various international simulations and business market games, which are very elaborate and may take days or weeks to play, and also many encounter group games, which are primarily devices to get people to express their feelings. Both these kinds of games may engage people; but they do not focus on particular concepts of interest to social psychology, and they do not make it easy for people to observe fundamental patterns in their own or other people's behavior. Other current games are too simple or too tightly structured. A good example is the Prisoner's Dilemma or trust dilemma, in which each player without communication repeatedly chooses between two options, a cooperative one and a competitive one. In its usual laboratory form the game is quick and certainly focuses on a concept of interest (trust), but permits only rather limited involvement by players and taps only a small part of their concerns and capacities. The ideal game is one with a simple focus and structure that engages a rich and complex involvement by players.

The learning that takes place through games is akin to the learning that takes place through living theater or educational drama. Highly skilled practitioners of this art, like Dorothy Heathcote, produce results that are breathtaking to observers: "Children are seen to be working very hard and achieving levels of understanding both of ideas and emotions and in their range of language that are extraordinary for children of the age group being observed [Fines & Verrier, 1976, p. 12]." It is illuminating to look at the subtle pressures such teachers use to make the dramatic situations real for the children. These typically involve asking them to justify their actions either by elaborating the feelings involved (heightening internal correspondence) or to consider consequences they may not have thought of (heightening external correspondence). More fundamentally, however, let us consider the description Fines and Verrier (1976) offer of educational drama:

Educational drama takes what the pupils bring with them to the classroom — their knowledge and understanding of the world and how it works, their feelings and emotions, their enthusiasms and above all their attitudes. These are the basic ingredients with which they will start to explore reality.

> Drama is a three-dimensional moving picture of life, lived at life rate and not preplanned so that pupils know the ending beforehand. This they will discover in action as we discover what happens by living our own lives. Drama is never coldly intellectual, nor is it submissive to undirected emotions; rather it is a making real and concrete of abstractions so they can be tested out and later on refined in the light of new understanding [p. 6].

Note that educational drama differs from ordinary drama in two ways of critical importance to us. First, it is built of what participants bring with them, rather than scripts provided by a playwright. It is not preplanned with an ending known in advance, since the object is that the participants create and explore their own reality. Second, most remarkably, it is defined as a making real and concrete of abstractions so that they can be tested out and refined in the light of new understanding. Making real for participants important concepts in social psychology and allowing them to test and challenge their meaning is the purpose we see for games in social psychology.

We have been working on a number of games. In the power game, described earlier, the central rule is that players who vote on the winning side of an issue gain somewhat more power to influence subsequent issues than players who vote on the losing side. The purpose of the game is to make people think about whether this is a reasonable assumption, about whether equal power is easy or hard to maintain, and about how they and other players acted in the situation. In another game, players choose a discussion partner each round. The central rule is that players must decide at the start of each round whether they wish to remain free to seek other partners in future rounds or to make a commitment to return to their partner of the current round. The purpose of the game is to make people think about what they gain and what they lose by choosing freedom or choosing commitment, and to compare their own behavior and reasons with those of other participants.

By varying the parameters of these situations we could do experiments on power or freedom that should have inferential validity. But our first concern for them as games is phenomenological validity. We want people to want to play them and to feel that they get something out of them. The games work only if people feel they are real. This probably requires allowing them sufficient freedom in the game to express their feelings about a variety of matters, so that internal correspondence is high. This probably also requires having their choices in the game carry sufficient consequences for other people so that others react with feeling, making external correspondence high. In addition, it requires future research on the starting point and central question of this paper, namely, the conditions under which people will see things as real. Games should be a rich site for testing this central question in attribution theory. Research on the attribution of reality should in turn be a vital input both into the future development of games and into the general development of social psychology.

ACKNOWLEDGMENTS

I am especially grateful to Bill Ickes for an enthusiastic and provocative reading of an earlier version of this chapter, and to Ann Gordon for her longstanding dedication to the development of the games described herein. I have also profited from the thoughtful comments made by Ellen Cohn, Bob Kidd, John Harvey, and Camille Wortman.

REFERENCES

Altman, I. *The environment and social behavior: Privacy, personal space, territory, crowding.* Monterey, Calif: Brooks/Cole, 1975.

Aronson, E., & Carlsmith, J. M. Experimentation in social psychology. In G. Lindzey & E. Aronson (Eds.), *Handbook of social psychology* (2nd ed., Vol. 2). Reading, Mass.: Addison–Wesley, 1968.

Atkinson, J. W. Motivational determinants of risk-taking behavior. *Psychological Review,* 1957, *64,* 359–372.

Bakan, D. *The duality of human existence.* Chicago: Rand–McNally, 1966.

Bandura, A. *Social learning theory.* Englewood Cliffs, N.J.: Prentice–Hall, 1977.

Bandura, A., & Walters, R. H. *Social learning and personality development.* New York: Holt, Rinehart & Winston, 1963.

Bem, D. J. Self-perception theory. In L. Berkowitz (Ed.), *Advances in experimental social psychology* (Vol. 6). New York: Academic Press, 1972.

Berger, P. L., & Luckmann, T. *The social construction of reality.* Garden City, N.Y.: Doubleday Anchor, 1966.

Berne, E. *Games people play.* New York: Grove Press, 1964.

Brewer, M. Erhard Seminars Training: "We're gonna tear you down and put you back together." *Psychology Today,* August 1975, pp. 35–40; 82–89.

Brickman, P. Attitudes out of context: Harvard students go home. Undergraduate honors thesis, Harvard University, 1964. Described in K. J. Gergen, *The psychology of behavior exchange.* Reading, Mass.: Addison–Wesley, 1969.

Campbell, D. T., & Stanley, J. C. *Experimental and quasi-experimental designs for research.* Chicago: Rand–McNally, 1963.

Clore, G. L., & Jeffrey, K. M. Emotional role playing, attitude change, and attraction toward a disabled person. *Journal of Personality and Social Psychology,* 1972, *23,* 105–111.

Collins, B. E., & Hoyt, M. F. Personal responsibility-for-consequences: An integration and extension of the "forced compliance" literature. *Journal of Experimental Social Psychology,* 1972, *8,* 558–593.

Cook, T. D., & Campbell, D. T. The design and conduct of quasi-experiments and true experiments in field settings. In M. D. Dunnette (Ed.), *Handbook of industrial and organizational psychology.* Chicago: Rand–McNally, 1976.

Cooper, J. Deception and role playing: On telling the good guys from the bad guys. *American Psychologist,* 1976, *31,* 605–610.

Crowne, D. P., & Marlowe, D. *The approval motive.* New York: Wiley, 1964.

Csikszentmihalyi, M. *Beyond freedom and anxiety.* San Francisco, Calif.: Jossey–Bass, 1975.

Deci, E. L. *Intrinsic motivation.* New York: Plenum Press, 1975.

DeKoven, B. L. The belt. *Simulation/Gaming/News,* January, 1973, pp. 1; 5.

Duke, R. D. *Gaming: The future's language.* Beverly Hills, Calif.: Sage Publications, 1974.

Duval, S., & Wicklund, R. A. *A theory of objective self-awareness.* New York: Academic Press, 1972.

Erikson, E. H. Inner and outer space: Reflections on womanhood. *Daedalus,* 1964, *93,* 582–606.

Etzioni, A. Basic human needs, alienation and inauthenticity. *American Sociological Review,* 1968, *33,* 870–885.

Eysenck, H. Biological basis of personality. *Nature,* 1963, *119,* 1031–1034.

Feldman, N. S., Higgins, E. J., Karlovac, M., & Ruble, D. M. Use of consensus information in causal attributions as a function of temporal presentation and availability of direct information. *Journal of Personality and Social Psychology,* 1976, *34,* 694–698.

Fines, J., & Verrier, R. The work of Dorothy Heathcote. *Young Drama,* 1976, *4,* 3–12.

Forward, J., Canter, R., & Kirsch, N. Role-enactment and deception methodologies: Alternative paradigms? *American Psychologist,* 1976, *31,* 595–604.

Frankl, V. E. *Man's search for meaning.* New York: Washington Square Press, 1963. (Originally published, 1959.)

Gamson, W. A. *Power and discontent.* Homewood, Ill.: Dorsey Press, 1968.

Gergen, K. J. Multiple identity. *Psychology Today,* May 1972, *5,* pp. 31–36, 64–66.

Gergen, K. J. Social psychology as history. *Journal of Personality and Social Psychology,* 1973, *26,* 309–320.

Goethals, G. R., & Reckman, R. F. The perception of consistency in attitudes. *Journal of Experimental Social Psychology,* 1973, *9,* 491–501.

Goffman, E. *The presentation of self in everyday life.* New York: Doubleday Anchor, 1959.

Goffman, E. Fun in games. In E. Goffman, *Encounters.* Indianapolis, Ind.: Bobbs–Merrill, 1961.

Goffman, E. *Frame analysis.* Cambridge, Mass.: Harvard University Press, 1974.

Hamilton, V. L. Role play and deception: A re-examination of the controversy. *Journal for the Theory of Social Behaviour.* 1976, *6,* 233–252.

Hampden–Turner, C. Dramas of Delancey Street. *Journal of Humanistic Psychology,* 1976, *16,* 5–54.

Harré, R., & Secord, P. F. *The explanation of social behaviour.* Oxford, Great Britain: Basic Blackwell, 1972.

Harvey, J. H., Ickes, W. J., & Kidd, R. F. (Eds.), *New directions in attribution research* (Vol. 1). Hillsdale, N.J.: Lawrence Erlbaum Associates, 1976.

Hoffman, M. L. Sex differences in moral internalization and values. *Journal of Personality and Social Psychology,* 1975, *32,* 720–729.

Hunt, M. M. *The natural history of love.* New York: Grove Press, 1959.

Ickes, W., & Barnes, R. D. The role of sex and self-monitoring in unstructured dyadic interactions. *Journal of Personality and Social Psychology,* 1977, *35,* 315–330.

James, W. J. The perception of reality. In W. James, *Principles of psychology* (Vol. 2). New York: Dover Publications, 1950. (Originally published, 1869.)

Jones, E. E., & Nisbett, R. E. *The actor and the observer: Divergent perceptions of the causes of behavior.* Morristown, N.J.: General Learning Press, 1971.

Jones, E. E., & Wortman, C. B. *Ingratiation: An attributional approach.* Morristown, N.J.: General Learning Press, 1973.

Kelley, H. H. Attribution theory in social psychology. In D. Levine (Ed.), *Nebraska Symposium on Motivation* (Vol. 15). Lincoln, Neb.: University of Nebraska Press, 1967.

Kundera, M. *Laughable loves.* New York: Knopf, 1974.

Laing, R. D. *The politics of experience.* New York: Penguin, 1967.

Lewis, R. A. Social reaction and the formation of dyads: An interactionist approach to mate selection. *Sociometry,* 1973, *36,* 409–418.

Manis, M., Cornell, S. D., & Moore, J. C. The transmission of attitude-relevant information through a communication chain. *Journal of Personality and Social Psychology*, 1974, *30*, 81–94.

McGuire, W. J. Suspiciousness of experimenter's intent. In R. Rosenthal & R. L. Rosnow (Eds.), *Artifact in behavioral research*. New York: Academic Press, 1969.

Merton, R. K. The self-fulfilling prophecy. *The Antioch Review*, 1948, *8*, 193–210.

Merton, R. K. Insiders and outsiders: A chapter in the sociology of knowledge. *American Journal of Sociology*, 1972, *78*, 9–47.

Milgram, S. Some conditions of obedience and disobedience to authority. *Human Relations*, 1965, *18*, 57-76.

Miller, R. L., Brickman, P., & Bolen, D. Attribution versus persuasion as a means for modifying behavior. *Journal of Personality and Social Psychology*, 1975, *31*, 430–441.

Mixon, D. Instead of deception. *Journal for the Theory of Social Behaviour*, 1972, *2*, 146–177.

Mixon, D. Studying feignable behavior. *Representative Research in Social Psychology*, 1976, *7*, 89–104.

Mixon, D. Temporary false belief. *Personality and Social Psychology Bulletin*, 1977, *3*, 479–488.

Monson, T. C., & Snyder, M. Actors, observers, and the attribution process: Toward a reconceptualization. *Journal of Experimental Social Psychology*, 1977, *13*, 89–111.

Nisbett, R. E., & Borgida, E. Attribution and the psychology of prediction. *Journal of Personality and Social Psychology*, 1975, *32*, 932–943.

Nisbett, R. E., & Wilson, T. D. Telling more than we can know: Verbal reports on mental processes. *Psychological Review*, 1977, *84*, 231–259.

Norman, R. M. G., & Watson, L. D. Extraversion and reactions to cognitive inconsistency. *Journal of Research in Personality*, 1976, *10*, 446–456.

Petrakis, H. M. Rosemary. *Mademoiselle*, February 1968, *66*, pp. 134–135.

Piaget, J. *The moral judgment of the child*. Glencoe, Ill.: Free Press, 1948. (Originally published, 1932.)

Pleck, J. H., & Sawyer, J. (Eds.). *Men and masculinity*. Englewood Cliffs, N.J.: Prentice–Hall, 1974.

Rapoport, A. *Fights, games, and debates*. Ann Arbor, Mich.: University of Michigan Press, 1960.

Rosenthal, R. *Experimenter effects in behavioral research*. New York: Appleton–Century–Crofts, 1966.

Ross, M. The self-perception of intrinsic motivation. In J. H. Harvey, W. J. Ickes, & R. F. Kidd (Eds.), *New directions in attribution research* (Vol. 1). Hillsdale, N.J.: Lawrence Erlbaum Associates, 1976.

Rytina, S., Cohn R., Fireman, B., Gamson, W., Morgan, D., & Taylor, B. *An introduction to the Mobilization in Miniature Project: A dramaturgical perspective*. Ann Arbor, Mich.: University of Michigan, Center for Research on Social Organization, no date.

Sampson, E. E. *Agency, communion and identity mastery: Theory and preliminary research data*. Unpublished paper, Clark University, no date.

Schachter, S., & Rodin, J. *Obese humans and rats*. Washington, D.C.: Erlbaum/Halsted, 1974.

Seligman, M. E. P. *Helplessness*. San Francisco, Calif.: Freeman, 1975.

Shubik, M., (Ed.). *Game theory and related approaches to social behavior*. New York: Wiley, 1964.

Snyder, M. The self-monitoring of expressive behavior. *Journal of Personality and Social Psychology*, 1974, *30*, 526–537.

Snyder, M., & Tanke, E. D. Behavior and attitude: Some people are more consistent than others. *Journal of Personality*, 1976, *44*, 501—517.

Snyder, M. L., & Jones, E. E. Attitude attribution when behavior is constrained. *Journal of Experimental Social Psychology*, 1974, *10*, 585—600.

Watzlawick, P., Beavin, J. H., & Jackson, D. *Pragmatics of human communication.* New York: Norton, 1967.

Weiner, B., & Peter, N. A cognitive-developmental analysis of achievement and moral judgments. *Developmental Psychology*, 1973, *9*, 290—309.

Wicklund, R. A., & Brehm, J. W. *Perspectives on cognitive dissonance.* Hillsdale, N.J.: Lawrence Erlbaum Associates, 1976.

Wicklund, R. A., & Duval, S. Opinion change and performance facilitation as a result of objective self-awareness. *Journal of Experimental Social Psychology*, 1971, *7*, 319—342.

Windeler, R. David Carradine barely flew past the cuckoo's nest, but now says he is glorybound. *People,* March 21, 1977, pp. 50—54.

Wrightsman, L. S., Jr., O'Connor, J., & Baker, N. J. (Eds.). *Cooperation and competition: Readings on mixed-motive games.* Belmont, Calif.: Brooks/Cole, 1972.

Younger, J. C., & Pliner, P. Obese—normal differences in the self-monitoring of expressive behavior. *Journal of Research in Personality*, 1976, *10*, 112—115.

Zajonc, R. B. The attitudinal effects of mere exposure. *Journal of Personality and Social Psychology*, 1968, *9*(No. 2 Part 2), 1—27.

Zimbardo, P., Haney, C., & Banks, W. C. A Pirandellian prison. *New York Times Magazine,* April 8, 1973, pp. 38—60.

2 Rethinking the Role of Thought in Social Interaction

Ellen J. Langer
Harvard University

Over the past decade there have been numerous articles devoted to the study of attribution theory. Each of these has attempted to uncover in some way the rules that people use in giving meaning to the world about them. Each sought to understand the various factors involved in consciously assigning causality, a process assumed to be essential if one is to make sense of one's world. According to this view, people are pictured primarily as information processors who continually and consciously ask what and why questions and then behave in ways that correspond to their answers.

Though much attention has been given in social psychology to the attribution process and its attendant behavior, there has been little concern with just how pervasive the attribution phenomenon actually is. That is, we know that people are capable of perceiving their world in cause-and-effect terms, but how often and under what circumstances do they actually do this? In fact, rather than just ask how often people make attributions — perhaps the more general issue that needs to be considered is how much time is actually spent in *any* kind of thoughtful action. Much psychological research relies on a theoretical model that depicts the individual as one who is cognitively aware most of the time, and who consciously, constantly, and systematically applies "rules" to incoming information about the environment in order to formulate interpretations and courses of action. Attribution theorists rely on this model in attempting to uncover the sources of regularities in human behavior. But if in fact it can be demonstrated that much complex human behavior can and does occur without these assumed cognitive assessments, then we must question both the pervasiveness of attribution making as a cognitive process and the assumptions made by most social psychologists.

I do not propose to conclusively resolve this issue in this chapter. Rather, by discussing a number of diverse research findings, largely from my own work, I hope to illustrate the ramifications of the question and suggest some alternate assumptions that might be more useful. The approach to be taken is twofold. First after looking briefly at recent historical material dealing with these issues, research findings are offered to demonstrate the apparent mindlessness of ostensibly thoughtful action. Then, research is presented to show some of the potentially positive effects of actually engaging subjects in mindful action.

Throughout the chapter reference is made to thinking, thoughtful action, awareness, and mindfulness. These words are used interchangeably to mean that process of consciously making use of information relevant to the situation. The major premise implicit in the argument that follows is that when individuals consciously process information, subsequent action very often may take a different form from that action that may be guided by cues that result from overlearning. This is because in the latter instance only that information of which the individual was conscious that has recurred repeatedly becomes overlearned. Upon each repetition, less information is processed. Thus, those actions guided by these cues are based eventually only on minimal information. When a social situation incorporates enough information that is redundant with the past to suggest a course of action, i.e., when these cues are present, new goal-relevant information may be ignored. Therefore, an individual who is "thinking" about a course of action in a situation may respond quite differently from a "nonthinking" individual.

HISTORICAL BACKGROUND

Recent social psychology has not paid much attention to the degree to which complex behavior may be performed automatically, but as early as 1896 experimental psychologists Leon Solomons and Gertrude Stein[1] addressed this issue directly. In arguing against the idea that the mindless performance of the "second personality" of people with so-called double personalities was essentially different from that of normal people, they proposed that normal people also could in fact enact a great deal of complex behavior without conscious awareness. They conducted several experiments, serving as their own subjects, to demonstrate that both writing and reading could be performed automatically. They were successful in writing English words without awareness while they were otherwise engaged in reading an absorbing story. The stimuli for the behavior were sensory in nature. With practice, they were able to take dictation automatically while reading. They were completely unable to recall the words they had written but were nevertheless quite certain they had written

---

[1]A historical note for the curious reader: This is the same Gertrude Stein who most people today remember as a famous poet and author. From 1893 to 1898 she was a graduate student in experimental psychology at Harvard University, working under the guidance of William James.

something. To demonstrate that reading could take place automatically, the subject read aloud from a text while an interesting story was read to him. Again, they found that, with practice, oral reading could proceed unhampered while full attention is given to the story being read to the subject. In discussing the difficulties encountered in their experiments, Solomons and Stein (1986) state "It must be remembered that our training was purely a training of attention. Our trouble never came from a *failure of reaction,* but from a *functioning of attention. It was our inability to take our minds off the experiment that interfered.* [Italics mine.] From the start, whenever, by good luck, this did happen, the reaction went on automatically [p. 502]."

Solomons and Stein (1896) sum up their experiments by concluding that a vast number of actions ordinarily called intelligent, such as reading and writing, can go on quite automatically in normal people:

We have shown a general tendency, on the part of normal people, to *act,* without any express desire or conscious volition, in a manner in general accord with the *previous habits* of the person, and showing a full possession of the faculty of *memory;* and that these acts may go on just as well outside the field of consciousness; that for them, not only volition is unnecessary, but that consciousness as well is entirely superfluous and plays a purely cognitive part, when present [p. 509].

Their methodology, which would not meet present-day standards, may not allow for full acceptance of their results. Nevertheless, their conclusions are provocative.

It clearly would be inaccurate to give the impression that this topic has been ignored by researchers since the appearance of Solomons and Stein's article. Social psychologists have not paid much attention to it, but communication theorists, linguists, and cognitive psyhologists have devoted much time and energy to the study of information processing, discovering very early on (cf. Broadbent, 1958; Shannon & Weaver, 1949) that persons have limited channel capacities for processing information and that as a result of this, people process information selectively — that is, they selectively ignore information. So, for example, spoken messages may be acted upon although the content of the messages is frequently not accessible to consciousness (cf. Deutsch & Deutsch, 1963; Moray, 1969; Treisman, 1964a, 1964b). Psychologists studying subliminal perception, hypnosis, forgetting, and verbal conditioning, among other topics, have all addressed the issues of responding without complete awareness. To give a review of all the work psychologists have done in the area of "non-thinking" is well beyond the scope of this chapter; however, such a review would reveal an obvious omission of social psychologists and social psychological topics.

Most theories in social psychology (e.g., attribution theory, consistency theories, social comparison theory, equity theory) take as a given that, for the most part, people move through their environments either thoughtful of people, objects, events, and relationships among these, or at least they assume that the person can be treated "as if" such thought is present. That is, when unaware, the individual is assumed to be behaving as if (s)he is making the *same* calculations and assessments to arrive at a course of action that is equitable or consistent, for example, as when (s)he is aware. It is odd that this cognizant cognitive view seems more vitally presumed in the research of social psychology than in cognitive psychology. There are, of course, social psychological theories like Bem's Self-Perception Theory (Bem, 1972) that do not assume that people typically behave with conscious attention to the particulars of the situation. Rather, the issue is ignored as in the Skinnerian tradition, with the only principle being that *when* (not *if*) people engage in conscious thought, the likelihood of any particular thought occurring is dependent on the individual's past reinforcement history, just as is the case with overt behavior. Thus, while Bem's work was a move in the right direction, we are still left with a void of theory and research concerning the degree to which behavior studied in the domain of social psychology is performed automatically and the degree to which automatic behavior corresponds to nonautomatic behavior. As the environment changes, such correspondence should diminish. Some might argue that sustained social interaction cannot, because of its complexity, be enacted without thinking in the way smaller units of behavior are (i.e., those typically studied by cognitive psychologists). An individual in a dichotic listening experiment in the laboratory may be able to switch ears to pick up a simple message without knowing that (s)he switched attention from one ear to the other to do it. But that seems to be a long way from expecting two or more individuals in a rich environment to carry out a dialogue with each other while possibly engaging in secondary motor acts without conscious attention to what they are doing. Nevertheless, this is what is presently being suggested: *Most* behavior may be enacted without paying attention to it, even complex social interaction.[2]

There is a continuum of awareness along the thought dimension that accompanies action that ranges from consciousness of such awareness to nonconsciousness. In between consciousness of awareness (the class of thoughts for which there is an almost one-to-one relationship between action and awareness of action) and nonconsciousness, lies the category of mindlessness of ostensibly thoughtful action. Here, I suggest, lies the most interesting and pervasive feature of social phenomena; the instances where people believe they had been thinking, but where, in fact, they were behaving according to well-learned

---

[2]An article by Nisbett and Wilson appeared in *Psychological Review* (1977, *84*, 231–259) shortly after this chapter was written. A point of clarification is in order to distinguish my position from theirs. Their claim is that people are basically mindful but nevertheless do not have access to their cognitive processes. In contrast, the position advanced in this chapter is that most of the time people are not consciously seeking explanations or trying to assess their cognitive processes.

and general scripts, rather than on the basis of new, incoming information. How do these instances come about? When one first experiences an event, conscious attention is paid to the particulars of the situation to guide behavior. With repeated experiences of the same event there is diminished need for such conscious attention. Sooner or later the behavior will become overlearned; one will enact the behavior with minimal (if any) awareness. Suppose, now, that one finds it necessary to remember (or think about) the event. This will most likely occur without the realization that in remembering an event, it first must have been forgotten. But because of the phenomenological similarity between the attention given upon first learning something and the attention necessary for rethinking it, it is easy to mistake other enactments of that behavior as having been equally thoughtful. That is, we may be able to describe an event just engaged in, not because of awareness of the particulars of that event but rather because of earlier attention to the same characteristics present when first engaging in the event — when first learning the script.

A continuum of awareness varies directly with the degree of repeated experience that we have with the activity. The more often we have engaged in the activity the more likely it is that we will rely on scripts for the completion of the activity and the less likely it is that there will be any correspondence between our actions and those thoughts of ours that occur simultaneously. This would suggest that children often may actually engage in more thinking relevant to the situation than adults, since they have not yet had the opportunity for such overlearning to occur. This may account for some of the differences in behavior previously ascribed to their undeveloped intelligence. Similarly, the elderly often may represent the opposite extreme, because they have had the opportunity to overlearn more behavior than the majority of the population. This is not a necessary consequence of aging but rather is more a function of restricted mobility. If the same environment is repeatedly experienced, there is little new to think about.

When we think about other people, since we ourselves are then *thinking*, we are not likely to assume their behavior to be automatic. If that behavior is not correspondent with the information the situation presents to us, we may be led to erroneous conclusions about why they behaved the way they behaved. It is probably because of this — that people think they, or others were thinking; that we, when behaving as people, think we are thinking; and when, we behaving as psychologists studying people, may actually *be* thinking — that we have overlooked all the nonthinking that takes place.

Let me illustrate the above with a rather mundane example. Do you flush the toilet after you use it? You probably, confidently (so confidently as to think me absurd for asking) answered "yes." What is your response to the question, "Did you flush the toilet the *last time* you went to the bathroom?" Because of the general rule you follow, you'll answer, "yes"; but after thinking about it, how certain in this particular case are you? The difference in certainty, I suggest, is a function of the different states of awareness attendant in the initial and present enactments of the behavior. To be more convinced, after the next person leaves the closest bathroom to you, ask him or her the question and ask for their degree of confidence in the answer. People may be startled at not *really* being sure of the answer. They think they should know, but should they? It is often only when we forget things we were expected to remember, like mailing a letter, or misplacing things like keys and bankbooks, that we question "where our minds were" at the time. It usually comes as a jolt, because people don't realize how much they do without paying attention to it. Often the problem is that people think they *should know,* rather than that they *don't know.*

We typically have assumed that virtually all behavior other than overlearned motor acts are performed with conscious awareness. Perhaps a more efficacious strategy is one that assumes that by the time a person reaches adulthood, (s)he has *achieved* a state of "ignorance" whereby virtually all behavior may be performed without awareness and tends more often than not to be performed in this manner unless special circumstances are invoked. That is, unless forced to engage in conscious thought, one prefers the mode of interacting with one's environment in a state of relative mindlessness, at least with regard to the situation at hand. This may be the case, because thinking is effortful and is often just not necessary. It is also not unreasonable to postulate that beyond some as yet undefined point, the quantity and quality of thought over possible strategies or over occasions may be inversely related. That is,

> By relieving the brain of all unnecessary work, a good rotation sets it free to concentrate on more advanced problems, and in effect increases the mental powers of the race. . . . It is a profoundly erroneous truism, repeated by all copy-books and by eminent people making speeches, that we should cultivate the habit of thinking of what we are doing. The precise opposite is the case. Civilization advances by extending the number of operations which we can perform without thinking about them. Operations of thought are like cavalry charges in a battle — they are strictly limited in number, they require fresh horses, and must only be made at decisive moments.
>
> A. N. Whitehead

Philosophers and cognitive psychologists have been concerned with the process of shifting attention from something to something else. However, it may also be the case that in many waking instances we shift attention from something to nothing else. At those times, we process a minimal amount of information to get us through whatever activity engages us. But by and large, our "minds" are virtually at rest. This is not to say that an achieved state of ignorance is preferred because it is restful or because it may enable more productive thought at a later time — both of which may be true — but rather because it is the only way people can get things done.

Thinking about behavior as we are enacting it often destroys the spontaneous continuity of that action. If we are talking, for example, then thinking of every word we will say before we say it will disrupt the flow of our speech. Moreover, when behavior is accomplished without premeditation to particulars, the individual may be in a more, rather than less, positive affective state. Imagine that you are a spectator or participant at a sports event, a concert, the theater, a sexual encounter, or even engaged in writing a paper. I would hypothesize that the positivity of the post-activity evaluation varies inversely with the degree of conscious awareness of the activity. That is, the more one is involved in the event, the more one enjoys the event; and increased levels of involvement are achieved by not paying attention to the particulars of the situation (by not thinking).

To say that individuals can conduct complex interactions automatically relies on the understanding that large units of varied behavior can be chunked together to form fewer coherent cognitive units that are capable of being overlearned. These units, the abstracted essences of social events, have been called *scripts* by Abelson (1976) and form the basis of his elaborate computer simulation of belief systems. To Abelson, a script is "a coherent sequence of events expected by the individual, involving him either as a participant or as an observer [p. 33]." He proposes that there are features abstracted from many single vignettes, the basic ingredient of scripts, that help group similar experiences and also help differentiate contrasting ones. With experience, these features come to be processed instead of the original vignette or vignettes. Although Abelson acknowledges the problems inherent in arriving at a taxonomy of scripts (Schank & Abelson, 1975), he is, nevertheless, one of the few social psychologists to address the importance of understanding scripted behavior and the importance of de-emphasizing our "overly rationalistic orientation in social cognition [Abelson, 1976, p. 45]." Abelson's formulation allows for a lesser amount of conscious cognitive work on the actor's part in order to get the behavior done. His alternative, in contrast to that of attribution theory, would be congenial to data that reveal the relative lack of awareness in social behavior.

INDIRECT RESEARCH SUPPORT FOR THE NONTHINKING HYPOTHESIS

Some earlier research that Robert Abelson and I conducted (Langer & Abelson, 1972) provides some indirect evidence of scripted behavior in social interactions. Groups of subjects were presented with requests for help where only the order of the words spoken was varied. Nevertheless, depending on the legitimacy of the request made, the appeals were differentially effective in receiving compliance. Had subjects processed the full request, rather than used the few opening words to cue them into some script, there would have been no difference between the groups, because the semantic content was identical for all subjects. Specifically, what we did in two field studies was to vary the opening words of a plea for help and to give the request either a victim or a target orientation. (Abelson, 1976, refers to these as an empathy script and a social obligation script.) In the former, the request begins with a statement of the victim's need, whereas the latter begins with a statement that shifts attention immediately to the target of the request, asking implicitly if (s)he's the kind of person who helps people. A woman who feigned a minor knee injury limped over to people in a shopping mall and asked them a favor in either a victim-oriented manner ("My knee is killing me. I think I sprained it. Would you do something for me? Please do me a favor and call my husband and ask him to pick me up.") or using a target-oriented approach ("Would you do something for me? Please do me a favor and call my husband and ask him to pick me up. My knee is killing me. I think I sprained it.") The legitimacy of the favor also was varied by requesting subjects either to call her husband to pick her up (legitimate) or to call her employer and tell him she'll be late (less legitimate or credible — primarily because of its occurrence in a shopping center, a peculiar place to call one's boss). Thus, the study employed a 2 × 2 design with 20 subjects per cell. Our understanding at that time was that if the victim merits legitimate concern, an appeal that focuses attention on her plight would be more effective than a call to duty in getting help but that cueing a potential helper into an empathetic role would not be as effective when the favor asked was illegitimate, since empathy is then less appropriate.

A conceptual replication of this study was conducted wherein a victim asked someone to mail a package for her, using either a victim or a target-oriented approach because she either had to catch a train or go shopping. The data from the two studies are summarized in Table 2.1. There was clear support for the hypothesis that for a legitimate favor a victim-oriented approach that sets in motion an empathy script is clearly superior to a target-oriented approach, but that for an illegitimate favor, a target-oriented approach that sets in motion a social obligation script is more effective. Thus, it would seem that subjects who were engaged in the social situation relied on its scripted aspects for behavioral cues rather than cognitively processing all of the information that might have seemed relevant to an outside observer or to the speaker. In more recent research (Langer, Blank, & Chanowitz, in press) on this topic of minimal information processing, discussed later, this legitimacy variable has been reinterpreted in terms of effort.

In a series of studies designed to elucidate a phenomenon referred to as the "Illusion of Control" (Langer, 1975; Langer, in press; Langer & Roth, 1975), I tried to demonstrate the importance of the perception of control by showing that with little provocation, people participating in chance events will behave as if they were engaged in a skill situation and attempt to exert an influence over

TABLE 2.1
Frequency of Helping Behavior — Two Experiments Combined [a, b]

| | | Appeal | |
		Victim-oriented	Target-oriented
Favor	Legitimate	30 (75%)	17 (42%)
	Illegitimate	11 (27%)	19 (47%)

[a]From Langer & Abelson (1972).
[b]N = 40 per cell.

the outcome. This perspective asserts that by encouraging or allowing participants in a chance event to engage in behavior that they typically engage in when participating in a skill event, the person is initially led into perceiving that (s)he is involved in a skill situation. Continued behavior that implies a skill orientation demonstrates a reliance on the "skill" *script* and a lack of ongoing cognitive processing of the incoming information that would reveal the "chance" quality of the situation. Skill-relevant behaviors that induce this illusion of control include making choices, thinking about the possible strategies that may be employed, exerting effort while actively engaged in the task, familiarizing oneself with the materials to be used and the responses to be made, and competing with other people as a way to assess one's skill. By introducing any of these aspects of a skill situation (choice, passive and active involvement, stimulus and response familiarity, competition) into a chance situation, where they do not objectively influence the outcome, one induces an illusion of control. When the illusion is operative, people are more likely to take risks. When there is an *intrusion of reality* (italics not in original) such that the focus of attention is shifted to the chance elements in the situation and away from the skill characteristics that were predominating, the illusion will dissipate.

Several experiments were conducted to test these propositions. In each study, one or more of these skill-related factors was introduced into a chance setting and was found to occasion behavior more appropriate to a skill than to a chance activity. In the language of the present chapter, the presence of the skill-relevant variable cued a skill script that subjects then "unthinkingly" enacted. Had subjects been processing all the information in the situation, rather than just relying on a few cues, there probably would have been no differences between conditions where these "skill" cues were present and where they were absent, since their presence was objectively irrelevant to the outcome. This is why I previously emphasized the proposition regarding the "intrusion of reality." What this means in the present context is that if people were made to attend to the chance elements in the situation, that is, if people were made to think about the event while engaging in it, the illusion of a skill situation and consequently the illusion of control would dissipate. Because they would be processing more of the information in the situation, people would behave in a way that to "thinking" others would appear rational. When the illusion is present, people are not behaving irrationally but simply automatically.

Examples from a few of these studies may help to illustrate the point. In one study, subjects competed in a completely chance-determined card game (drawing for high card) against either an attractive, confident confederate or an awkward, nervous confederate. If the subjects were thinking about the situation rather than responding on the basis of minimal cues, then one would expect them to either bet equally against both these confederates or perhaps to bet more against the confident confederate to appear similar to him because he would be expected to bet more or because risk is a value in our society. However, if competition invokes a skill script and implies the possibility of control, then subjects would try to maximize their winnings by betting more against the apparently less skilled confederate. This is just what they did.

In another study, office workers participated in a lottery where they had choice or no choice of a familiar or unfamiliar lottery ticket. If one could exert control over a lottery ticket drawing, then one would do so by increasing the likelihood of one's ticket being selected. One's chances of winning would then seem better than the chance of someone without this control. All the ticket holders were given the opportunity to keep their original ticket or to trade it for a ticket in a lottery where the total number of tickets to be drawn from was smaller. Both choice and familiarity were effective in increasing the proportion of participants who rejected this opportunity to improve their chances of winning. When one feels (s)he is already going to win, there is no reason to make a change. Again, if subjects were thinking about what they were doing, all would have traded in their tickets to take advantage of objectively better odds. In all of the studies of this series, whether dependent measures were the subjects' overt behavior or the subjects' report of his confidence in winning, people from a wide range of backgrounds rather consistently behaved as if they were following some skill script, when, because of the totally chance-determined nature of the situation, skill was inappropriate. Wortman (1975) and Ayeroff and Abelson (1976), employing different methodologies, found virtually the same results.

When people are engaged in skill tasks, they seem to follow some general rule that says, "When *asked* about the responsibility for outcomes, attribute positive but not negative outcomes to self." Cohen (1960), Kelley (1967); Streufert and Streufert (1969); Weiner, Frieze, Kukla, Reed, Rest, and Rosenbaum (1972); Wortman, Constanzo, and Witt (1973); and Shaver (1970); among others, have all discussed this tendency for people in skill situations to make "defensive attributions." In a study dealing with the effects of the sequence of outcomes on the illusion of control (Langer & Roth, 1975), we were able to generate this effect in a chance situation whose scripted aspects suggested once again that it was a skill situation. We found that when people who are engaged in predicting coin tosses are presented with a sequence of outcomes in which there are several "wins" early in the sequence, they see themselves as good at the task and are confident of future success. In addition, 25% of these subjects reported that their performances would be hampered by distraction, and 40% felt their performance would improve with practice. Thus the descending sequence of outcomes employed seems to have induced a skill orientation toward this purely chance-determined task. Again, had subjects been thinking about what they were doing when engaging in the task or when reporting their sentiments to us, it is hard to imagine that they (Yale undergraduates) would insist that anyone could be either good or bad at coin tossing or that skill-relevant factors like distraction and practice could matter.

A very different kind of study from those reported so far, which Jeremy Miransky and I conducted (Miransky & Langer, 1977), also provides some indirect evidence that people don't spend much time engaged in thoughtful action. This study compared the attributions of responsibility for crime prevention and the actual steps taken to prevent a particular crime, burglary, by people who had been burglarized with people who had not had this experience. A pilot study had revealed, to our surprise, that, at least for New Yorkers, past victimization did *not* result in the taking of greater precautionary measures, such as locking doors either when present or absent from one's home. Although 66% of the total sample of burglarized and nonburglarized subjects reported that they believed burglary could be prevented, 65% of these subjects used fewer locks than they had on their doors, the clearest measure for preventing the crime. With these results in mind, a study was conducted to assess the reason for finding no difference between burglarized and nonburglarized people. To do this, we tried to push the two groups apart by making burglarized individuals attend to their past burglary. Communications were presented to individuals in their own apartments that increased the vividness of the burglary experience either through a graphic description of the disarray resulting from a burglary of a home nearby (property condition), or through a communication that increased the similarity ot the subject to a recent burglary victim (similarity condition), or through a communication that stressed both of these (combined condition). (The uncertainty of the relative effectiveness of each of these dictated the necessity for the three distinct experimental treatments.) Both behavioral and attitudinal measures of reactions to the crimes were obtained. The behavioral measures consisted of the difference between the number of locks subjects had on their doors and the number they used, and the number of people who called the police to make use of a free security service we advertised in the communication. The questionnaire dealt with recalled and anticipated reactions to burglary such as feelings of helplessness.

As in the pilot study, the large majority of subjects prior to the experimental treatment used fewer locks than they had on their doors (84%), although 88% of the people reported that burglary could be prevented. Once again, there were no differences on these measures between burglarized and nonburglarized people. Group differences did emerge as a function of the experimental treatments in regard to questionnaire responses. When reading a communication designed to make you think about an event and then filling out a questionnaire that is in itself thought-provoking because of its novelty, it is not surprising that people with different experiences, when thinking about those experiences, should respond differently. Thus, subjects who had been burglarized reported greater feelings of helplessness than subjects who had never been burglarized. An interaction obtained such that the property manipulation made nonburglarized subjects feel most helpless, and the similarity manipulation made the burglarized subjects feel most helpless. It is unnecessary for the present argument to go into a detailed explanation of these differences. What is important to note is that although the communications were differentially effective in making people remember or imagine the negative event, none of the communications had the continued effect of resulting in greater lock use on doors or windows or calls to the police department for use of their free security service. Both burglarized and nonburglarized subjects were equally as unlikely to take preventive measures against burglary after our intervention as before it, although, again, when made to think about burglary (i.e., when asked) both groups attributed responsibility for burglary prevention to nonchance factors and responded differentially. This is, of course, reminiscent of the attitude-behavior discrepancy long acknowledged by social psychologists studying attitude change (Wicker, 1969). To expect people who report beliefs and opinions on issues, when made to think about those issues, to later behave in a manner consistent with those beliefs and opinions is to assume that they are thinking while they are behaving.

DIRECT RESEARCH SUPPORT FOR
THE NONTHINKING HYPOTHESIS

This assumption, that people typically engage in thoughtful action or behave "as if" they are engaging in thoughtful action, has been questioned throughout this chapter. Indirect evidence against this notion has been advanced. At this point, I would like to discuss some work that Arthur Blank, Buzzy Chanowitz, and I (Langer, Blank, & Chanowitz, in press) recently have completed that directly tests some of the ideas only alluded to up until now. We conducted three field experiments to test the mindlessness of ostensibly thoughtful action in the domain of spoken and written communication. It was hypothesized that thoughtful behavior will only occur when habit is inadequate for the occasion. This will be the case when either of two conditions is met: (1) when the message transmitted is structurally (rather than semantically) novel; (2) when the interaction between the participants requires an effortful response. Scripts, we asserted, are used by the individual, since they reduce the cognitive work necessary to engage in interactional activity. The recurrence of certain typical activities in the person's life provides for the possibility of and encourages the perception of a typical structure to this everday activity. As the person is repeatedly exposed to the activity, a structure emerges. With repeated exposure and emerging structure, the person pays less and less attention to the semantics of the activity. Rather, (s)he cues into the scripted structure of the activity that signifies, in its felicitous repetition, that the typical activity is as it was before. So long as the structures satisfy the script, the person takes it for granted that the unexamined semantics of the (repeated) activity are similar to the semantics of his/her earliest encounters with the activity when (s)he fully examined the semantics of the particulars of the situation and when the scripted structure was first emerging. If this phenomenon is a feature of recurring, typical activities in people's lives, then the more people engage in particular activities, the more they rely on scripts in order to get these activities done. The most potent advantage of this feature is that it allows the person to engage in less "mental activity" and effort for those interactions that (s)he has learned are invariably repeated in his/her life. There is no longer any necessity to repeatedly and laboriously re-examine the semantics of a typical interaction. As mentioned earlier, this scripted feature of activities will not be used by the individual under two circumstances. If the structure of the situation is novel, the indications are that this activity is not representative of earlier occurrences of this activity. The novel structure indicates a novel semantics. If the response required by this activity is an effortful one, then the script has failed in its purpose of reducing effort. The features of novelty or effort indicate a need to re-examine the semantics of the situation (i.e., thoughtfully process as much of the incoming information that seems relevant as possible). Under "normal" circumstances, however, minimal information (that which invokes a satisfactory script and therefore a typical activity) will be processed, and there will be a reduced level of mental activity. We suggest that the essence of a script may not even lie in recurring semantics, but rather in more general paralinguistic features of the message.

These ideas were tested in the context of a compliance paradigm, where, depending on the study, either a meaningless reason was offered for compliance with a request or the request itself was meaningless. These were either structurally congruent with subjects' past experience or not. Generally, we found that when the structure of a communication (either oral or written, either semantically sound or sematically senseless) is congruent with subjects' experiences, it occasions behavior that appears thoughtful but that in fact may be thoughtless.

The studies are simple in design but nevertheless are quite revealing. In the first study, people about to use a copying machine in the library at the Graduate Center of the City University of New York were approached and asked in one of several ways to let another person use it first. Either just a request, a request plus "placebic" information, or a request plus real information was made for compliance with a favor that entailed either more or less effort. Thus, the study was a 3 x 2 factorial design. The specific request made by the experimenter to use the machine was one of the following:

1. *Request Only:* Excuse me, I have 5 (20) pages. May I use the Xerox machine?
2. *Placebic Information:* Excuse me, I have 5 (20) pages. May I use the Xerox machine because I have to make copies?
3. *Real Information:* Excuse me, I have 5 (20) pages. May I use the Xerox machine because I'm in a rush?

Condition 2 is called placebic information, because the reason offered is entirely redundant with the request. What else would one do with a copying machine except make copies of something? Thus, if subjects were processing the information communicated by the experimenter, then the rate of compliance for subjects in the request-only condition and in the placebic information condition should be the same, since the same information is conveyed. Both may be different from Condition 3, since here additional information is given. If, on the other hand, people are only processing a minimal amount of information and are responding to a script that goes something like "favor X + reason Y → comply," then the rate of compliance should be the same for the placebic and real information conditions and different from the request-only group. If the latter case obtained, then the placebic information would be redundant only in an information theory sense (Shannon & Weaver, 1949) and not in a script sense. Although we maintain that people proceed mindlessly as long as script requirements are met (in this case that means when request + reason are given), it is also the case that a potentially very effortful response would be sufficient to shift attention from the simple physical characteristics of the message to its semantic content. (As long as it's easier to ignore what is being said than to process it, ignoring will be the case. However, when ignoring is potentially effortful, it pays to think, in which case thoughtful action will ensue.) Thus we predicted that when the favor asked was effortful (determined by the number of pages both subject and experimenter had to copy), subjects would process the experimenter's communciation, which would result in a similar rate of compliance for the request-only group and the placebic information group. On the other hand, a similar rate for the placebic and real information groups was predicted to obtain when little effort was involved. A look at Table 2.2 shows that this is just what we found. There was a main effect for small/big favor, as one would expect; but note that in the Small Favor condition, the placebic information group looks like the real information group, and in the Big Favor condition, the placebic group looks like the no information (request-only) group.

One would probably have assumed that an interaction between two strangers, where one was making a request of the other, would proceed mindfully. This would seem especially likely when the people involved were spending their time

TABLE 2.2
The Proportion of Subjects Who Agreed to Let
the Experimenter Use the Xerox[a]

		Reason		
		No Info	Placebic Info	Real Info
Favor	Small	.60 (15)[b]	.93 (15)	.94 (16)
	Big	.24 (25)	.24 (25	.42 (24)

[a]From Langer, Blank, & Chanowitz (in press).
[b]N per cell = number in parentheses.

in as thought-provoking an atmosphere as a library. Nevertheless, this appears not to have been the case. Even in this situation when minimal requirements were met, people proceeded mindlessly.

The next study tried to determine whether the processing of only minimal information was also the rule for responding to written communication, another ostensibly thoughtful action. In this study, subjects were randomly selected from the New York City telephone directory and sent a written communication that varied in its adherence to structural requirements for a script through mail. A meaningless five-item questionnaire was mailed with a cover letter that either demanded or requested the return of the questionnaire and was either signed or unsigned at the bottom of the letter. We assumed that signed requests and unsigned demands were more congruent with the structure of written communications than unsigned requests or signed demands. "Thoughtful" processing of the cover letter would not reveal any rational reason for returning the questionnaire (in stamped envelopes provided addressed to a post office box number). We expected greater compliance for the congruent communications, signed requests and unsigned demands, than for unsigned requests and signed demands. The subjects of the study comprised a high-status group made up of physicians and a random-status group. It was predicted that because of their experience with greater volumes of mail, high-status subjects would return more questionnaires than low-status subjects and that this would be more the case for congruently phrased communications than for those that were structurally incongruent. There were an equal number of returns for the high- and random-status groups. However, while congruency did not seem to affect returns for the random-status group, there were significantly more (nonthinking) responses for the congruent communications than for the incongruent communications for the high-status group. Rather than assume that only physicians were non-thinking, we instead assumed that we had not determined the appropriate script for the random-status group. Therefore, we conducted another experiment to test more rigorously the potential mindlessness of responding to written communications.

Memoranda were collected from the trash baskets of secretaries of various departments at the Graduate Center of the City University of New York so that the structure of a communication could be tailor-made to the experience of our next group of subjects. Though varying in content, most of the memos we found requested rather than demanded that the secretary do something, and none were signed at the bottom. Thus, incongruent (and thereby thought-provoking) communications for these subjects would be memoranda that contained a signed request, or a signed or unsigned demand. We sent secretaries either structurally congruent or structurally incongruent memoranda that asked them to return the memo by interoffice mail to a room that did not, in fact, exist in the building. The workers in the mailroom collected the memos for us. Determined to make our point most convincingly, the memos contained nothing else. The *only* thing that each memo asked was that the memo be returned ("This paper is to be returned immediately to Room 238 through interoffice mail"; or, "I would appreciate it if you would return this paper immediately to Room 238 through interoffice mail. The memo was either signed, "Sincerely, John Lewis" or unsigned. Any thoughtful processing of the communication should result in the secretary *not* returning the memo. Why return it? If the sender wanted the piece of paper, he shouldn't have sent it in the first place! Nevertheless, 68% of the subjects did in fact comply with the request. One could argue that these subjects thoughtfully, rather than mindlessly, complied with the request to avoid negative repercussions from someone who possibly had more authority than they did. However, this alternative becomes less viable in light of the fact that there was significantly greater compliance when the memo was structurally congruent with subjects' past experience than when it was not.

These studies seem to support the idea that much of the behavior we assume to be performed mindfully instead is enacted rather mindlessly; that unless there is no well-learned script to follow or effortful response to make, people may process only a minimal amount of information to get them through their day.

CONCLUSIONS

Where should we go from here? Attribution theorists like Heider, Kelley, Jones, Davis, Nisbett, and Weiner, to name but a few, have taught us much about the rules people follow when they make attributions. However, as the research that has been presented suggests, attribution theorists, by and large, may have presumed too much mental activity on the part of individuals engaging in many of their everyday activities. Much of the interaction that people mundanely enact in the everyday world would seem to rely on the scripted structure of typical activities rather than on the active processing of incoming information that attribution making relies on. If in fact most of human behavior is scripted in character, then two problems arise that warrant consideration: (1) the misleading inferences drawn by the scientist in the experimental situation; and (2) the misleading occurrence of an atypical semantics despite the appearance of a typical script.

First, looking back, many of the results of our laboratory research efforts may have been mistaken in their application to the everyday world, since with the possible exception of "professional" subjects, people in the laboratory do not have "subject" scripts to follow. When they come to the novel laboratory situation, they are self-conscious and unsure of what is expected of them. They are in a state of heightened arousal that leads to a focusing of attention (Berlyne, 1960; Kahneman, 1973) to the particulars of the situation (see quote on p. 37). The fact that they are in this heightened state of awareness leaves us in the possible position of generalizing our results from thinking subjects in the laboratory who are actively processing incoming information to nonthinking people in the everyday world who rely on scripts to accomplish their mundane activities. Ignoring the distinction raised here may lead to the building of erroneous theoretical models and to the misleading application by social psychologists of laboratory results to the "real" world.

A case in point would be the research of Rotter and his colleagues (see Rotter, Chance, & Phares, 1972) and my own work on the illusion of control described earlier. In several experiments, Rotter and his colleagues attempt to verify the distinction between an internal and an external locus of control by showing performance differences when people are given skill instructions (which occasion internal responding) versus chance instructions (which occasion external responding) for some ambiguous task. They find, for example, that people engaged in a chance task follow a lose–stay/win–switch strategy, whereas people engaged in a skill task follow a lose–switch/win–stay strategy. However, I have found that people typically do not make chance/skill discriminations. The different results may be understood in light of the different scripts called into play in the different situations. When instructed to attend to the chance elements in the situation, people can and will readily follow their chance scripts. However, without this "intrusion of reality" (i.e., instructions as to how to think about the situation), they typically respond to chance situations with an illusion of control.

This is not to say that attribution research in general or attribution research on the locus of control in particular is not valuable. However, we must be cognizant of the situations in the everyday world where it might be meaningfully applied and where it may be misapplied.

Second, looking forward, if much of human behavior is scripted in character, then a good deal of research in social psychology must be redirected. A problem, as mentioned earlier, in the use of scripts by individuals is that in many cases the emergence of a regular script structure in an activity is taken as a signal that attention need not be paid to the semantics of the situation. That is, if the *structure* is congruent with the earlier encounter of the activity, the individual assumes the *semantics* are congruent with the earlier encounters and will consequently process a minimal amount of the information available. Acting upon these assumptions, however, may lead to trouble for the individual if structures of activities seem "normal" and regular but in fact are distinctive in ways unperceived by the individual. Seemingly similar scripts may contain distinctive sets of semantics. The individual, however, is not privy to these distinctions and assumes that (s)he can act on the basis of a minimal amount of information. In some instances, then, this inevitably leads to unanticipated and seemingly inexplicable outcomes. In addition, if one assumes that not thinking is less arousing than thinking and that an optimum level of arousal is necessary for survival, then people who are overaroused prosper from their scripts, but people like the institutionalized elderly who are probably behaving in a maximally scripted way are likely to be underaroused and thereby deteriorating prematurely.

While the enactment of much behavior is scripted in quality, the original scripts learned do not necessarily provide the actor with the most efficacious role to play. Learning a new script may be profitable. Attention to a previously unattended scenario will momentarily disrupt the flow of behavior and cause some difficulty. However, careful attention to important but previously unincorporated particulars of the situation could prove worthwhile and of course, once well learned, should proceed just as smoothly as the enactment of the earlier less adaptive script. With this in mind, social psychology must take as part of its task the demonstration of the scripted quality of human behavior in many typical activities. That experimental demonstration, however, must also include an examination of the particulars in order to assess whether or not the scripts in use, as representatives of the underlying semantics, are the most adaptive ones. Social psychologists, using mindful rationality, should take a more active role in constructing the scripts people use, rather than cataloguing the scripts that people mindlessly construct in an ad hoc manner.

It is interesting to consider that once we find the means of determining what "better" scripts are and how to incorporate them into the repertoires of different groups of individuals, we might find ourselves in the position of enabling *more* mindless activity. The feelings of stress, guilt, and incompetence, for example, are all thought engaging. By focusing attention on previously overlooked elements in the situation, for example, on effort rather than incompetence as an explanation for a bad outcome, the person may be freed from concern over past and possibly future failures. While not thinking often may serve an adaptive function, there are clearly instances where people should be taught to make attributions and to think (but to think differently) for their own peace of mind.

AN EXPERIMENTAL POSTSCRIPT

What follows are a number of experimental demonstrations and manipulations that proved useful in inducing people to think about previously unexamined aspects of the activities in which they were involved. These strategies helped to develop new ways to process information that were always actually available to the subjects but not attended to. They involved modifying the script structures people had been using in these situations.

In a study that Irving Janis, John Wolfer, and I conducted (Langer, Janis, & Wolfer, 1975), we assessed the relative effectiveness of a cognitive coping strategy for stress reduction based on the use of preparatory information. Patients about to undergo major surgery were given one of several communications. The most effective of these employed the reappraisal of anxiety-provoking events, calming self-talk, and cognitive control through selective attention. These subjects were told that they could control much of the stress they were experiencing and that it was rarely the events themselves that caused stress, but rather the views people take of them and the attention they give those views. Alternative views of particular stressful events were then given. Thus, subjects who previously were spending much of their time engaged in thoughtful worrying were taught to think differently about the situation and to attribute responsibility to themselves for such thought. The result was that these people became less stressed on both pre- and post operative measures. They were rated by nurses as being less anxious, requiring fewer pain relievers and sedatives, and tending to leave the hospital earlier than the comparison groups.

Susan Saegert and I carried out an experiment on crowding and cognitive control that also is relevant to this issue (Langer & Saegert, 1977). Our subjects were New Yorkers doing their grocery shopping when the store was either crowded or uncrowded. Their participation in the study engaged them in the complex cognitive task of quickly determining the most economical way to purchase various products. Subjects who were given the information that many people feel somewhat anxious and aroused when supermarkets became crowded apparently reattributed their arousal to the environment. In this way, the information resulted in increased positive affect and test performance. Although several studies have shown that attributions can affect physiological processes, comfort, and poststress task performance (e.g., Ross, Rodin, & Zimbardo, 1969; Calbert—Boyanowski & Leventhal, 1975; Rodin, 1976), we were interested in showing that complex task performance can also be affected in this way. The results showed that crowding was affectively negative and interfered with task performance but that both affect and performance could be ameliorated with the provision of information about the nature of the situation.

In an experiment concerned with postdivorce adaptation, Helen Newman and I (Newman & Langer, 1977) tested the idea that making situational attributions for divorce would have beneficial effects. In pilot work, we found that when we asked people to think about the reasons for their divorce, those who gave primarily situational explanations showed a tendency to be happier than those who blamed the failure of the marriage on characteristics of self or spouse. Hence, it seemed that if we could impart information about the importance of external factors in guiding behavior, then subjects, when thinking about the divorce, would feel more positive. In addition, we believed that making situational rather than dispositional attributions for a past relationship would result in greater predictions for success in new relationships.

These ideas were tested by contrasting the effects of two communications delivered to divorced subjects after ascertaining the attribution they made for their divorces. One communication explained and stressed the importance of understanding the failure of the marriage through situational as opposed to dispositional attributions. The other communication basically stressed that it was important to think of several attributions (dispositional and situational were explained) for their divorce, so as to better identify why the relationship had failed and to prevent these problems from recurring. Both groups were given a pretreatment questionnaire and diaries to keep for two weeks, at the end of which time another questionnaire was administered. A third group of divorced subjects heard no communication but kept the diary. Though certain circumstances mitigated against the communication's effectiveness in teaching people to make situational attributions, those subjects who originally did make them were found to be happier, more confident, more active, and more optimistic about future relationships than were the other groups. This suggests the possible overall advantage of making situational attributions when thinking about negative events.

The last three studies described demonstrate some of the positive effects of encouraging people who were stressed thinking about an event one way, to think about it in another way (cf. Valins & Nisbett, 1972). In this next study, we tried to get people who may be spending too little time in any kind of thoughtful action to think. This final experiment is also the most dramatic one. It was conducted by Judith Rodin and myself (Langer & Rodin, 1976) in a nursing home in Hamden, Connecticut, with a population of subjects whose average age was in the late 70s. This environment is characterized by almost a complete lack of novelty and hence offers very little to think about. There were two floors in the nursing home that housed residents of equivalent socioeconomic backgrounds and psychological as well as physical health. One floor was randomly selected as the experimental floor, and the other served as the comparison floor. Residents in the experimental group were brought together in the lounge on their floor to hear a communication delivered by the nursing home's director. The message he delivered stressed the residents' responsibility for themselves and their responsibility to make decisions concerning how they want to spend their days — decisions, they were told, that they used to make and still were capable of making. He then told the residents that he had a gift for them. At that point, a nurse passed around a box of small plants from which each resident could select one if (s)he so desired. They were told that the plants were theirs to keep and to take care of. Finally, an announcement was made regarding a movie to be shown in the home. Residents were told to decide whether they wanted to see it on Thursday or Friday, if they chose to see it at all. This communication was intended to induce a sense of responsibility for caring for themselves in those residents who heard it. Residents in the comparison group were also brought together in their lounge to hear a communication delivered by the home's director. However, in this instance, the communication stressed the staff's responsibility to take care of the residents rather than the residents' responsibility for themselves. Instead of being given the opportunity to select a plant to take care of, these residents were given a plant to be cared for by the staff members. And finally, they were assigned a night to see the movie, as was typically the case in the home, rather than given a choice of nights. Rather than view the experimental group as a "responsibility-induced" group as was originally conceived, it may also be viewed as a "thought-encouraged" group.

Some of the dependent measures included self-reported ratings, ratings by an interviewer blind to the experimental manipulations, and nurses' ratings.

On these measures, the thought-encouraged (responsibility-induced) group showed a significant improvement over the comparison group on alertness, happiness, active participation, and a general sense of well-being. This was also the case for the behavioral measures taken. Significantly more people in the thought-encouraged group than in the comparison group participated in a contest we organized, and significantly more of these residents showed up for the movie. However, what was most satisfying was the result of a follow-up study (Rodin & Langer, 1976) 18 months later, when we found that approximately half as many people in the experimental group (7 out of 47) had died as in the control group (13 out of 44). It thus appears that helping people to think about events in a meaningful, more productive manner may indeed be important.

Throughout this chapter speculations regarding when thinking will or will not occur were "quietly" suggested. To have spelled them out earlier would have necessitated more justification and explanation than I am now prepared to offer or than space permits. Nevertheless, it might prove useful to list my conjectures. People will engage in thought primarily:

1. When encountering a novel situation, for which, by definition, they have no script.
2. When enacting scripted behavior becomes effortful, i.e., when significantly more of the same kind of scripted behavior is demanded by the situation than was demanded by the original script.
3. When enacting scripted behavior is interrupted by external factors that do not allow for its completion.

4. When experiencing a negative or positive consequence that is sufficiently discrepant with the consequences of prior enactments of the same behavior.
5. When the situation does not allow for sufficient involvement.

When thinking *is* taking place:

1. People will make attributions to others if the other's behavior is *both* automatic and based on fewer cues than are appropriate for that situation (i.e., when there is a discrepancy between the observer's hypothetical thoughtful action and the actor's nonthoughtful action).
2. No attribution to others will be made if the actor's scripted behavior and the observer's nonscripted behavior correspond.
3. Attributions to self will be made if one experiences a negative or positive consequence that is sufficiently discrepant with the consequences of prior enactments of that same behavior.
4. Self-relevant attributions will not be made if the consequences are not discrepant with prior scenarios.

If these speculations are correct, then we must agree with Virginia Woolf when she said, "Every day includes much more nonbeing than being." And in line with the perspective taken in this chapter, we might conclude by saying that every day includes much more nonbeing than yesterday.

ACKNOWLEDGMENTS

Preparation of this chapter was facilitated by a grant from N.S.F. (BNS 76–10939) to the author. I am especially grateful to the following people for their thoughtful comments on an early draft of this chapter: Robert Abelson, Paul Secord, Benzion Chanowitz, Helen Newman, and Phyllis Katz. I would also like to thank Susan Saegert, Irwin Katz, Shelley Taylor, and Stanley Milgram for a careful reading of a later draft.

REFERENCES

Abelson, R. P. A script theory of understanding, attitude, and behavior. In J. Carroll & T. Payne (Eds.), *Cognition and social behavior*. Hillsdale, N.J.: Lawrence Erlbaum Associates, 1976.

Ayeroff, F., & Abelson, R. P. E.S.P. & E.S.B.: Belief in personal success at mental telepathy. *Journal of Personality and Social Psychology*, 1976, *34*, 240–247.

Bem, D. J. Self-perception theory. In L. Berkowitz (Ed.), *Advances in experimental social Psychology* (Vol. 6). New York: Academic Press, 1972.

Berlyne, D. E. *Conflict arousal and curiosity*, New York: McGraw–Hill, 1960.

Broadbent, D. *Perception and communication*. New York: Pergamon, 1958.

Calvert–Boyanowski, I., & Leventhal, H. The role of information in attenuating behavioral responses to stress: A reinterpretation of the misattribution phenomenon. *Journal of Personality and Social Psychology*, 1975, *32*, 214–221.

Cohen, J. *Chance, skill, and luck: The psychology of guessing and gambling*. Baltimore: Penguin Books, 1960.

Deutsch, J., & Deutsch, D. Attention: Some theoretical considerations. *Psychological Review*, 1963, *70*, 80–90.

Heider, F. *The psychology of interpersonal relations*. New York: Wiley, 1958.

Kahneman, D. *Attention and effort*. Englewood Cliffs, N.J.: Prentice-Hall, 1973.

Kelley, H. H. Attribution theory in social psychology. In D. Levine (Ed.), *Nebraska Symposium on Motivation*. Lincoln, Neb.: University of Nebraska Press, 1967.

Langer, E. J. The illusion of control. *Journal of Personality and Social Psychology*, 1975, *32*, 311–328.

Langer, E. J. The psychology of chance. *Journal for the Theory of Social Behavior*, in press.

Langer, E. J., & Abelson, R. P. The semantics of asking a favor: How to succeed in getting help without really dying. *Journal of Personality and Social Psychology*, 1972, *24*, 26–32.

Langer, E. J., Blank, A., & Chanowitz, B. Minimal information processing. *Journal of Personality and Social Psychology*, in press.

Langer, E. J., Janis, I. L., Wolfer, J. Reduction of psychological stress in surgical patients. *Journal of Experimental Social Psychology*, 1975, *11*, 155–165.

Langer, E. J., & Rodin, J. The effects of choice and enhanced personal responsibility: A field experiment in an institutional setting. *Journal of Personality and Social Psychology*, 1976, *34*, 191–198.

Langer, E. J., & Roth, J. Heads I win, tails, it's chance: The illusion of control as a function of the sequence of outcomes in a purely chance task. *Journal of Personality and Social Psychology*, 1975, *32*, 951–955.

Langer, E. J., & Saegert, S. Crowding and cognitive control. *Journal of Personality and Social Psychology*, 1977, *35*, 175–182.

Miransky, J., & Langer, E. J. Burglary (non)prevention: An instance of relinquishing control. Mimeograph. Harvard University, 1977.

Moray, N. *Attention: Selective processes in vision and hearing*. London: Hutchinson, 1969.

Newman, H., & Langer, E. J. Post-divorce adaptation as a function of the attribution of responsibility for the divorce. Mimeograph. Harvard University, 1977.

Rodin, J. Menstruation, reattribution and competence. *Journal of Personality and Social Psychology*, 1976, *33*, 345–353.

Rodin, J., & Langer, E. J. Long-term effect of a control-relevant intervention. *Journal of Personality and Social Psychology*, 1977, *35*, 897–902.

Ross, L., Rodin, J., & Zimbardo, P. Toward an attribution therapy: The reduction of fear through induced cognitive emotional misattribution. *Journal of Personality and Social Psychology*, 1969, *12*, 279–288.

Rotter, J. B., Chance, J. E., & Phares, E. J. *Application of a social learning theory of personality*, New York: Holt, Rinehart & Winston, Inc., 1972.

Schank, R., & Abelson, R. P. *Scripts, plans, and knowledge*. Prepared for presentation at the Fourth International Joint Conference on Artificial Intelligence, Tbilisi USSR, 1975.

Shannon, C., & Weaver, W. *The mathematical theory of communication*. Urbana, Ill.: Illinois University Press, 1949.

Shaver, K. G. Defensive attribution: Effects of severity and relevance on the responsibility assigned for an accident. *Journal of Personality and Social Psychology*, 1970, *14*, 101–113.

Solomons, L. M., & Stein, G. Normal motor automatism. *Psychological Review*, 1896, *3*, 492–512.

Streufert, S., & Streufert, S. C. The effects of conceptual structure, failure, and success on attribution of causality and interpersonal attitudes. *Journal of Personality and Social Psychology*, 1969, *11*, 133–147.

Treisman, A. Verbal cues, language, and meaning in attention. *American Journal of Psychology*, 1964, *77*, 206–214. (a)

Treisman, A. The effect of irrelevant material on the efficiency of selective listening. *American Journal of Psychology*, 1964, *77*, 533–546. (b)

Valins, S., & Nisbett, R. Attribution processes in the development and treatment of emotional disorders. In E. E. Jones, D. Kanouse, H. H. Kelley, R. Nisbett, S. Valins, & B. Weiner (Eds.), *Attribution*, Morristown, N.J.: General Learning Press, 1972.

Weiner, B., Frieze, I., Kukla, A., Reed, L., Rest, S., & Rosenbaum, R. M. Perceiving the causes of success and failure. In E. E. Jones, D. E. Kanouse, R. E. Nisbett, S. Valins, & B. Weiner (Eds.), *Attribution*. Morristown, N.J.: General Learning Press, 1972.

Wicker, A. W. Attitudes versus actions: The relationship of verbal and overt behavioral responses to attitude objects. *Journal of Social Issues*, 1969, *25*, 41–78.

Wortman, C. Some determinants of perceived control. *Journal of Personality and Social Psychology*, 1975, *31*, 282–294.

Wortman, C., Costanzo, P., & Witt, T. The effects of anticipated performance on the attributions of causality to self & others. *Journal of Personality & Social Psychology*, 1973, *27*, 372–381.

3 Affective Consequences of Causal Ascriptions

Bernard Weiner
Dan Russell
David Lerman
University of California at Los Angeles

In previous publications (Weiner, 1972, 1974), a model of achievement-related behavior has been offered that evolved from an analysis of perceived causality (see Table 3.1). This paper briefly examines each of the associations specified within that model and indicates our confidence in the proposed theoretical linkages. A designated relation between locus of causality and the intensity of affective experiences is then analyzed in detail, for a flaw in the prior reasoning has become evident. To illustrate the shortcomings of the prior conception and to provide support for a new way of thinking about the relation between causal ascriptions and emotions, a recently completed investigation is reported. Finally, the significance of this new theoretical approach for theories of emotion and for conceptions of achievement strivings is discussed.

AN ATTRIBUTIONAL MODEL OF ACHIEVEMENT STRIVINGS

The model shown in Table 3.1 can be best understood by beginning with the list of the perceived causes of success and failure. As indicated in Table 3.1, in achievement-related contexts, four causes — ability, effort, task difficulty, and luck — are used to interpret the outcome of an achievement-related event. That is, in attempting to explain a prior success or failure, individuals estimate their own or a performer's level of ability, the amount of effort that was expended, the difficulty of the task, and/or the magnitude and direction of experienced luck. There surely are other perceived causes of success and failure such as fatigue, mood, illness, and the biases of others, as well as causes that are unique to specific situations. In a cross-cultural study it was even reported that patience (Greece and Japan)

59

TABLE 3.1
The Current Attributional Model of Achievement Motivation

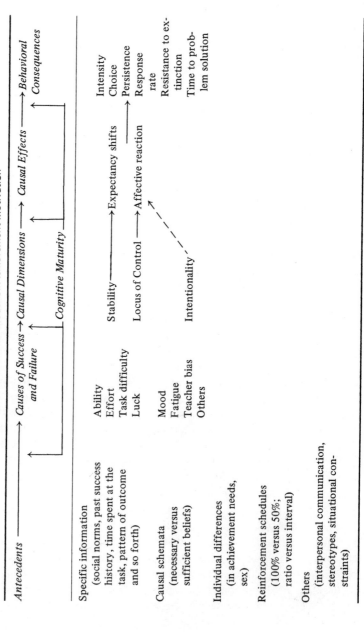

Antecedents ⟶	*Causes of Success and Failure* ⟶	*Causal Dimensions* ⟶	*Causal Effects* ⟶	*Behavioral Consequences*
		Cognitive Maturity		
Specific information (social norms, past success history, time spent at the task, pattern of outcome and so forth)	Ability Effort Task difficulty Luck	Stability ⟶ Expectancy shifts		Intensity Choice Persistence Response rate
Causal schemata (necessary versus sufficient beliefs)	Mood Fatigue Teacher bias Others	Locus of Control ⟶ Affective reaction		Resistance to extinction Time to problem solution
Individual differences (in achievement needs, sex)		Intentionality		
Reinforcement schedules (100% versus 50%; ratio versus interval)				
Others (interpersonal communication, stereotypes, situational constraints)				

and tact and unity (India) are perceived as causes of success (Triandis, 1972). But in our culture and within the confines of academic, occupational, and athletic accomplishment, it clearly has been documented that ability and effort are typically perceived as the dominant causes of success and failure, and task difficulty and luck are perceived among the other causes of achievement outcomes (Elig & Frieze, 1975; Frieze, 1976). We therefore feel somewhat justified in having focused attention on these particular causes, although it is evident that other interesting causal interpretations have been relatively neglected and that in any particular situation other causes instead of (in addition to) ability, effort, task difficulty, and luck may be most dominant. Researchers in this area must consider the situation that is being examined and use appropriate causes, rather than indiscriminately using the four that were originally posited. For example, it is not reasonable to expect luck to be perceived as the main cause of occupational success for a brain surgeon!

The causes of success and failure have been subsumed within a three-dimensional taxonomy (see Weiner, 1974). One dimension is the internal—external description of causes often associated with the field of locus of control (Rotter, 1966). Ability, effort, mood, and patience, for example, are properties internal to the person; whereas task difficulty, luck, and teacher bias are external or environmental causes. A second dimension of causality characterizes causes on a stable (invariant) versus unstable (variant) continuum. Ability, the difficulty of a task, and patience are likely to be perceived as relatively fixed, whereas luck, effort, and mood are more unstable. Luck implies random variability, effort may be augmented or decreased from one episode to the next, and mood typically is conceived as a temporary state. A third dimension of causality, first proposed by Heider (1958) and then incorporated within the achievement scheme by Rosenbaum (1972), is labeled "intentionality." Some causes, such as effort or the bias of a teacher or supervisor, are likely to be perceived as intentional; whereas, for example, ability, mood, or the difficulty of a task are unintentional, uncontrollable causes. Although the intentionality dimension of causality has not been subject to a great deal of research within Weiner's attributional framework, an application of this concept in the area of helping behavior has proven valuable (Barnes & Ickes, in press; Ickes & Kidd, 1976). And, of course, intention has long been a key term in the psychological analysis of moral judgment (see Maselli & Altrocchi, 1969).

The more recent introduction of intentionality into the model created some difficulties, inasmuch as the dimensions of causality no longer appeared to be orthogonal. Intentionality implies volitional control, which implies instability. Unstable causes need not be perceived as controllable (e.g., fatigue and mood), but causes subject to volitional control are likely to be perceived as unstable. Furthermore, effort is the single internal cause that appears to be under volitional control, although it could be contended that traits such as patience are perceived as controllable, particularly given other- rather than self-perception.

The possibility that patience may be perceived as intentional (controllable) or unintentional (not controllable) warns us that the placement of a cause within a dimension is not invariant over time or between people. For example, luck may be perceived as an internal, stable factor (he is "lucky") or as an external, unstable cause (a "chance" event). Even effort is not always considered volitional ("I just can't seem to study for this course"). Experimenters often can influence the perceived dimensional placement of a cause. For example, Valle and Frieze (1976) depicted task difficulty as an unstable cause by specifying that salespeople would be changing their sales territory, which was the operational anchor for level of task difficulty. Inasmuch as attribution theorists are concerned with phenomenal causality, the variability of the perception of causes within the dimensions is a fact that must be taken into account, rather than a theoretical shortcoming. Furthermore, the prediction of behavior must be based upon the subjective meaning of the causes to the individual.

Other significant dimensions of causality that identify the general properties of causes are likely to emerge with further analysis. But at the present time it appears that the main dimensions of causality in achievement-related contexts may have been identified. These dimensions are "second-order" concepts (Schütz, 1967, p. 59). That is, they are concepts used by attribution theorists to organize the causal concepts of the layman. As such, the dimensions are not strictly part of a "naive" psychology, or an ethnoscience. Rather, they are elements of the trained scientist's language. Perhaps the difficulty caused by the intent dimension of causality is that it is both part of the naive causal language as well as the trained scientist's system.

It is unlikely that empirical techniques such as factor analysis or multidimensional scaling will produce the same dimensions as those generated by the logical analyses that have already been completed. Discrepancies between the findings of the empirical techniques and the logical analyses do not invalidate the latter (or the former). Rather, such differences shed interesting light upon the discrepancies between the organizational schemes of the layman and those imposed by the scientist. For example, a "naive" person might not spontaneously recognize that luck, mood, and effort are similar because they are unstable, and thus a stability dimension of causality may not emerge in a multidimensional scaling procedure. But a scientist could find it useful to subsume these terms within the same causal category because they relate in an identical manner to a particular class of behaviors.

Continuing with an analysis of Table 3.1, the left-hand column of that table lists some of the antecedents or the determinants of causal judgments. We will not take up space in the present context to discuss these associations (see Weiner, 1974, 1976). However, it has become apparent that the linkages of specific informational cues and causal schemata to causal inferences are quite strong, whereas the presumed causal biases ascribed to individual differences in gender or in achievement needs remain somewhat tenuous.

Consequences

We now turn to the consequences of causal attributions, which are the main concern of this chapter. The principles of causal inference summarily outlined above have been integrated within an expectancy-value conception of motivation (see Atkinson, 1964; Weiner, 1972, 1974). Expectancy-value theorists maintain that the intensity of aroused motivation is determined jointly by the expectation that a response will lead to a goal and by the attractiveness of that goal object. Causal ascriptions for success and failure influence both these factors. Concerning expectancy of success, it has been found that ascription of an outcome to a stable factor such as ability increases expectancy of success after a success, and decreases expectancy of success after a failure, more than does an ascription to an unstable cause, such as luck. Stated somewhat differently, if one anticipates that conditions will remain unchanged, then the prior outcome at a task will be foreseen again with an increased degree of certainty. But if conditions are perceived as changeable, then there is some doubt that the prior success or failure will be repeated. The well-documented "gambler's fallacy" is one instance of this general principle. The relations between causal attributions and expectancy of success have been confirmed in so many studies that Weiner, Nierenberg, and Goldstein (1976) concluded: "We now consider this relationship to be proven. It is unfortunate from our point of view that psychologists continue to discuss locus of control in relation to expectancy of success and continue to confound the internal aspects of perceived control with the volitional and stable dimensions of causality [p. 65]."

In addition to influencing expectancy of success, it has been postulated that causal attributions in part determine the affective consequences of success and failure, and hence the "attractiveness" of the goal. Weiner (1976), for example, contended that:

> Pride and shame, as well as interpersonal evaluation, are absolutely max-
> imized when achievement outcomes are ascribed internally and are min-
> imized when success and failure are attributed to external causes. Thus,
> success attributed to high ability or hard work produces more pride and
> external praise than success that is perceived as due to the ease of the task
> or good luck. In a similar manner, failure perceived as caused by low ability
> or lack of effort results in greater shame and external punishment than
> failure that is attributed to the excessive difficulty of the task or bad
> luck. In sum, locus of causality influences the affective or emotional
> consequences of achievement outcomes [p. 183].

The evidence concerning the proposed relations between locus of control and affect is discussed in the next section of this paper.

Finally, numerous investigations have demonstrated that both expectancy and affect influence a variety of behaviors. It is believed that because attributions modify expectancy and affect, they also play an important role in determining the speed of performance, choice, persistence, and other indices of motivated behavior (see Weiner, 1974; Weiner & Sierad, 1975).

ATTRIBUTION AND AFFECTIVE INTENSITY

Now that the entire attributional model of achievement strivings has been presented, it is fitting to return to the central topic of this paper — the relation between attribution and affect. Recall it was proposed that internal attributions magnify affective reactions, whereas emotional responses are minimized given external causal ascriptions. Figure 3.1 summarizes the hypothesized relations between attributions and emotions.

Figure 3.1, which first appeared in Weiner, Frieze, Kukla, Reed, Rest, and Rosenbaum (1971), was accepted as valid by many workers in the attributional area. This approval appears to have been due to three related reasons. First of all, Fig. 3.1 is intuitively compelling. One clearly does not experience pride in success when, for example, receiving an "A" from a teacher who gives only that grade or when defeating a tennis player who loses all his or her other matches. In these instances the causes of success are external to the actor. On the other hand, an "A" from a teacher who gives few high grades or a victory over a highly rated tennis player following a great deal of practice generate great positive affect. In these instances, the causes of success are likely to be perceived as some personal characteristic(s), such as high ability and/or great effort expenditure. The current movement in industry stressing that the worker should be given personal credit for the product is another example that intuitively is guided by a perceived attribution—affect linkage. In sum, as captured in Fig. 3.1, internal attributions magnify emotional experience, whereas external ascriptions minimize affective reactions.

Secondly, the analysis depicted in Fig. 3.1 is in accord with the theory of achievement motivation formulated by Atkinson (1964). Atkinson post-

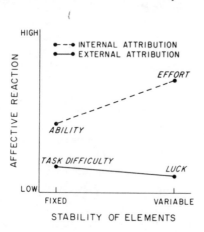

FIG. 3.1.   Hypothesized relationships between causal attributions for success and failure and the magnitude of affective reactions.

ulated that the incentive values of success (pride) and failure (shame) are inversely related to the probability of success and failure at a task. More specifically, one experiences greatest pride when succeeding at a difficult task and greatest shame following failure at an easy task. Inasmuch as success at a difficult task and failure at an easy task produce internal attributions, or feelings of self-responsibility (Kun & Weiner, 1973; Weiner & Kukla, 1970), Atkinson's ideas fit perfectly within the attributional conception.

The final reason for the acceptance of Fig. 3.1 is that data are scattered throughout the psychological literature supporting the attribution—intensity of affect linkage. For example, Feather (1967) found that self-reports of attraction for success and repulsion for failure are greater given skill-related (internal attribution) than chance-related (external attribution) tasks; Lanzetta and Hannah (1969) observed that greater punishment is administered for failure at an easy task than for failure at a difficult task, presumably because only failure at an easy task might be ascribed to a lack of effort; equity theorists such as Leventhal and Michaels (1971) report that inputs (effort) and assigned rewards covary; and Storms and McCaul (1976) as well as many others contend that internal ascriptions for aberrant behaviors magnify anxiety and feelings of inadequacy.

In sum, the relations illustrated in Fig. 3.1 were, for the most part, believed to be valid. The only controversy generated by the figure was the contention that effort attributions have greater affective consequences than ability attributions. This presumption has been tested in the context of interpersonal evaluation, for interpersonal evaluation and intrapersonal affective experience were believed to be governed by the same principles. In the particular reference experiment under consideration, subjects typically were asked to pretend that they were teachers and were to provide evaluative "feedback" to their pupils (see Eswara, 1972; Kaplan & Swant, 1973; Rest, Nierenberg, Weiner, & Heckhausen, 1973; Weiner & Kukla, 1970; and Weiner & Peter, 1973). The pupils were characterized in terms of effort, ability, and their performance on an exam. The data from these investigations conclusively demonstrated that effort is of greater importance than ability in determining rewards and punishments. That is, high effort was rewarded more than high ability, and low effort was punished more than low ability.

There appear to be two reasons for the discrepancy between ability and effort as determinants of reward and punishment. First, effort attributions elicit strong moral feelings — trying to attain a socially valued goal is something that one "ought" to do. Secondly, rewarding and punishing effort is instrumental to changing behavior, inasmuch as effort is believed to be subject to volitional control. On the other hand, ability is perceived as nonvolitional and relatively stable and thus should be insensitive to external control attempts.

Everyday observations also seem to support the expressed beliefs concerning the relative affective (evaluative) significance of ability versus effort attributions. For example, the handicapped person who struggles to complete a race and the retarded child who persists to finish an objectively easy task elicit great social approval. On the other hand, the gifted athlete who refuses to practice and the bright dropout generate great social disapproval, as though it were "immoral" not to utilize one's capacities. Thus, effort rather than ability appears to be the main determinant of evaluation.

Other investigators, however, correctly opposed the position of Weiner and his colleagues regarding the relative importance of ability versus effort attributions (see Covington & Beery, 1976; Nicholls, 1976). These individuals posed the following question: "If effort is so rewarded and ascriptions to this factor generate so much pride, then why do students often hide their efforts and refuse to admit that they studied hard?" To solve the dilemma raised by this question, Covington and Beery (1976) proposed that pupils strive to maintain a self-concept of high ability. Failure ascribed to a lack of effort and success in spite of low effort promote the view that one has ability (see Kun & Weiner, 1973). On the other hand, Covington and Beery also accept the findings reported by Weiner and others that teachers reward and punish primarily on the basis of effort expenditure. Thus, a conflict of attributional values and goals is established between pupils and teachers that could retard academic learning and school satisfaction. This analysis strongly implies that self- versus other-perception and affect versus evaluation must be distinguished when examining the consequences of causal attributions. In addition, Nicholls (1976) argued that if a task is related to long-term goals, then ability is valued more than effort as a determinant of success. Conversely, he also suggests that if performance at a task has no long-term consequences, then effort is more esteemed than ability as a cause of success.

In sum, at times ability attributions and at other times effort attributions magnify affective reactions of actors and observers (evaluators). What is now needed is a specification of the conditions under which each of these relations hold and a clearer distinction between reward from others and personal emotional reactions. But note that even among psychologists involved in the controversy regarding the relative importance of ability versus effort ascriptions, there was no doubt expressed concerning the main implication of Fig. 3.1 − internality maximizes affect and externality minimizes emotional reactivity.

THE PROBLEM

An indication that the conception shown in Fig. 3.1 was inadequate was briefly alluded to by Weiner (1975, 1976), although on those occasions the significance of the shortcoming was not fully recognized. Weiner (1975) stated: "Of course, failure ascribed to external factors [such as teacher bias] may produce emotional reactions, such as anger and frustration. But external attributions minimize *achievement-related affects* of pride and shame [p. 61]."

Thus, it was recognized that *some* affects are augmented by external, as opposed to internal, attributions. However, these emotional reactions were dismissed as unimportant, because they were considered to be unrelated to achievement concerns. Achievement motivation was uniquely linked with one particular affective dimension, labeled *pride–shame*. Yet there is neither experimental nor anecdotal evidence supporting the belief that the affect one experiences in a "pure" achievement setting is pride or shame. It is equally reasonable to propose that following success or failure one experiences feelings of competence (incompetence), safety (fear), contentment (agitation), or gratitude (vindictiveness). In addition, all of these affects, including pride and shame, might be experienced in contexts unrelated to achievement [e.g., pride is likely to be felt after winning an election, which is defined by Veroff (1957) and others as a power concern; one may feel competent in a social setting that generates affiliative concerns; and so on]. In sum, there is no clear reason to accept the position voiced by McClelland, Atkinson, Clark, and Lowell (1953) that pride and shame are the dominant (only?) emotions associated with achievement concerns. Indeed, it does not even appear that these affects are singularly associated with achievement-related sources of motivation. Therefore, the statement quoted from Weiner (1975) suggesting that anger and frustration may be intensified by external attributions cannot be dismissed on the grounds that these affects fall outside the domain of achievement strivings.

If the foregoing paragraph contains some truth, then Fig. 3.1 is inaccurate, for internal attributions do not magnify achievement-related emotional reactions – they augment *some* emotional reactions. On the other hand, external ascriptions intensify other affective reactions. Furthermore, if attributions mediate between achievement-related outcomes and emotional expression, then causal ascriptions might not only influence the intensity of emotional feelings but also the quality of affective life. For example, success ascribed to ability versus luck versus the help of a teacher could uniquely generate the following affective reactions:

Attribution	Emotion
Ability	Competence
Luck	Surprise
Teacher Help	Gratitude

The research study that follows is a first attempt to examine the association between attributions for success and failure and the intensity as well as the quality of emotional experiences.

A RESEARCH INVESTIGATION

General Procedure

Our general procedure was first to compile dictionary lists of potential affective reactions to success and failure in an academic context. Those lists are given in Table 3.2. The lists are formidable but by no means exhaustive (we tried to

TABLE 3.2

Selected Affective Reactions for Success and Failure

Success-Related Affects

affable^u	cheerful	ebullient^u	glad	lighthearted	safe
amazed	comfortable	ecstatic	good	merry	sanguine^u
amused	competent	effervescent^u	grateful	mirthful^u	satisfied
animated^u	composed	elated	gratified	modest	secure
appreciative	conceit	enjoyment	happy	optimistic	sedate^t
astonished	confident	enthralled^u	hopeful	overjoyed	surprised
awe	considerate	enthusiastic	humble^t	overwhelmed	thankful
benevolent	contented	excited	indifferent^t	placid^u	thoughtful
blitheful^u	delighted	exhilaration	jaunty^u	pleasant	thrilled
blissful	delightful	exquisite^u	jolly	pleased	titillated^u
boisterous	delirious	exultant^u	jovial	proud	unctuous^u
calm	dignified	frenzied^t	jubilant	rapturous^u	uproarious
carefree	dumbfounded^t	gay	kind	relaxed	wonderful
charmed	eager	genial^u	kindhearted	relieved	wonderment
					zestful

Failure-Related Affects

afraid	convulsive[u]	enraged	hurt	prostrate[u]	surprised
aggravated	crestfallen[u]	exasperated	hysterical[t]	rabid[u]	spiteful
aggressive	crushed	ferocious	inadequate	regretful	tense
agitated	cynical[u]	flustered	incompetent	repugnance[u]	terrified
agony	deflated	forlorn[u]	indifferent[t]	resentful	thoughtful
aimless	defected	frantic	indignant[u]	resigned	tormented
alarmed	demoralized	frenzied	inflamed	restless	tremulous[u]
amazed	depressed	fretful	infuriated	revengeful	troubled
angry	despair	frightened	intimidated	sad	uncheerful
animosity[u]	desperate	frustrated	irritated	scared	uncomfortable
annoyed	despondent[u]	fuming	insecure	scornful	unctuous[u]
anxious	diffident[u]	furious	jittery	sedate[t]	uneasy
apathetic	discontent	gloomy	joyless	self-pity	unexcited
appalled	disdain[u]	glum	languish[u]	shaken	unhappy
apprehensive	disgruntled[u]	grim	lousy	shocked	unnerved
aroused	disgust	grouchy	mad	solemn	unrelieved
astonished	disheartened	guilty	miserable	sore	unsatisfied
bewildered	dismayed	hateful	nauseated	sorrowful	upset
bitter	displeasure	heartsick	nervous	sorry	vexed[u]
blue	dissatisfied	heavyhearted	offended	startled	vicious
calm[t]	disquieted[u]	helpless	overwhelmed	stern[t]	vindictive
choler[u]	distressed	hopeless	panic	stolid[u]	woeful
collected[t]	disturbed	horrified	passive[t]	strained	wonderment
composed[t]	dumbfounded	hostile	pensive[u]	stunned	worried
concern	embarrassed	humble	perturbed	sullen	wretched

[u] = unknown.

[t] = trivial.

exclude words with identical connotations), and we are striving to include other pertinent emotional reactions in achievement-related contexts. One interesting fact that we noted in our search was the preponderance of affective labels associated with failure (see also Plutchik, 1962).

Our next step was to identify the causal attributions for achievement performance. Fortunately, a relatively complete list already had been compiled by Frieze and her colleagues (Elig & Frieze, 1975; Frieze, 1976). We selected 10 ascriptions for success and 11 for failure that accounted for the vast majority of free-response ascriptions in these prior investigations (the attributions for success and failure are respectively given in Tables 3.3 and 3.6).

We then proceeded with the viewpoint of a "naive, skeptical phenomenologist [Davitz, 1969, p. 5]." We merely gave a cause for success or failure within a brief story format, randomly listed the affective reactions that we had found, and asked subjects to report the affective intensity that would be experienced in this situation. Responses were made on a rating scale anchored at the extremes (not at all–extremely). A typical story was:

> Francis studied intensely for a test he took. It was very important for Francis to record a high score on this exam. Francis received an extremely high score on the test. Francis felt that he received this high score because he studied so intensely. How do you think Francis felt upon receiving this score?

This procedure is fraught with danger, even as a starting point. First of all, we assumed that individuals would project their own emotional experiences or those observed in others upon the characters in the stories. Second, we assumed that our labels reflect the "real" experiences of the subjects. Finally, we assumed that repression and suppression, memory distortion, response sets, experimenter demands, and individual differences in affective labeling as well as in the subjective meaning of these labels would not render our results meaningless (see Davitz, 1969). In sum, there obviously are many limitations of this initial study; we hope that other methodologies will evolve that permit better "royal roads" (or even dirty alleys) to inner experience. Our best defense of the methodology that we employed is the systematic, significant, and heuristic findings that will be reported.

This methodology also was dictated in part by pilot work revealing that individuals apparently do not have a rich language available to describe their affective experiences. Operant verbal procedures therefore were believed to be inadequate. Of course, it could be argued that the existence of a language or a label is a necessary condition for a particular emotional experience and that our procedure was therefore faulty. But it seemed to us that the opportunity for expression had to be made available, and for this reason a vocabulary of emotions was provided for the subjects. Unfortunately, this reactive and respondent procedure undoubtedly encourages subjects to report affects that are not experienced or are experienced in a manner not truly captured by the labels. Future studies relying upon introspective reports will have to determine how to steer a proper course between the Scylla (insufficient language) and Charybdis (reactive measure) of studies of emotion. At this point in time it does not seem likely that behaviorally oriented methodologies are adequate for the difficult problem posed in this investigation, although the possible value of such procedures should not be ruled out.

The subjects for the study were 90 male and female students at the University of California, Los Angeles, who were paid for their participation. Each subject was given one attribution for a successful outcome followed by the randomized list of positive affects, and one attribution for a failed outcome followed by the list of negative affects. The subjects were tested in groups not exceeding 10. For scoring purposes the affective rating scale was divided into 9 equal intervals. It also was possible for subjects to indicate that they did not know the meaning of a given word.

RESULTS AND DISCUSSION

Success

Of the 85 affective labels, 16 were eliminated from the final analysis because at least 10% of the subjects acknowledged that they were unfamiliar with the terms. These words are identified with the superscript u (unknown) in Table 3.2. In addition, five affects were eliminated from the final analysis because they were weakly elicited (mean intensity rating less than 2 over all causal attributions; $\bar{X} < 5$ within each ascription). These affects are identified with the superscript t (trivial) in Table 3.2.

Table 3.3 lists the 10 affects reported as being most intensely experienced for each of the 10 attributions for success. Perhaps the most evident fact in Table 3.3 is the overlap between the lists: *pleased* and *happy* appear on all 10 of the attributional lists, *satisfied* on 8, *good* and *contented* on 7, and so on for *proud, cheerful, delighted, glad,* etc. These might be called "outcome-dependent" affects, for they represent a general positive reaction to success, regardless of the "why" of success. Thus, in experiments asking subjects to give a global, self-report affective rating following success, differences in affective reports as a function of the attribution might be minimal (see, for example, Ruble, Parsons, & Ross, 1976).

Table 3.3 indicates that the most intensely experienced affects are indiscriminately reported for each attribution. We then identified those affects that are *discriminately* reported as a function of the causal attribution for success (see Table 3.4). The reader should note that many of these affects also are reported as intensely experienced. An affect is considered to be discriminating if its mean intensity rating within a given attribution significantly ($p < .01$) exceeds its

TABLE 3.3
Success: Top 10 Intensity Ratings by Attribution;
Listed Are the Affects and the Mean Intensity Ratings.

Ability	$\bar{X}$	Unstable Eff.[a]	$\bar{X}$	Stable Eff.	$\bar{X}$	Task Diff.[b]	$\bar{X}$	Mood	$\bar{X}$
pleased	8.8	good	8.9	satisfied	8.3	pleased	8.1	cheerful	8.6
satisfied	8.7	happy	8.9	good	7.9	contented	8.0	good	8.6
confident	8.7	satisfied	8.8	pleased	7.9	happy	7.7	delighted	8.4
competent	8.6	delighted	8.7	secure	7.9	secure	7.7	happy	8.4
happy	8.4	gratified	8.7	comfortable	7.8	satisfied	7.7	pleasant	8.3
good	8.1	pleased	8.5	competent	7.8	optimistic	7.6	pleased	8.2
proud	8.1	competent	8.5	contented	7.7	safe	7.6	glad	8.1
contented	8.0	relieved	8.4	proud	7.7	comfortable	7.4	enthusiastic	8.1
cheerful	7.7	elated	8.4	happy	7.6	hopeful	7.4	gay	8.0
thrilled	7.7	cheerful	8.4	pleasant	7.6	affable	7.3	jovial	7.9

Personality	$\bar{X}$	Other's Eff.	$\bar{X}$	Other's Mot.[c] and Pers.[d]	$\bar{X}$	Luck	$\bar{X}$	Int. Mot.[e]	$\bar{X}$
pleased	8.8	pleased	8.6	good	8.6	happy	8.8	pleased	8.7
happy	8.8	happy	8.4	happy	8.5	thankful	8.8	happy	8.7
contented	8.8	appreciative	8.4	delighted	8.4	delighted	8.5	satisfied	8.6
proud	8.8	satisfied	8.4	satisfied	8.4	pleased	8.5	proud	8.4
satisfied	8.6	proud	8.2	cheerful	8.3	relieved	8.5	confident	8.4
glad	8.4	good	8.2	pleased	8.1	glad	8.4	contented	8.3
confident	8.4	secure	8.0	glad	8.1	cheerful	8.4	competent	8.2
competent	8.4	enthusiastic	8.0	appreciative	8.1	excited	8.3	optimistic	8.2
thrilled	8.3	delighted	7.9	gratified	8.0	enthusiastic	8.1	good	8.2
gratified	8.1	contented	7.9	contented	8.0	thrilled	8.1	secure	8.0

[a]Eff. = effort. [c]Mot. = motivation. [e]Int. Mot. = intrinsic motivation.
[b]Diff. = difficulty. [d]Pers. = personality.

72

TABLE 3.4
Discriminating Affects Reported as a Function of
the Causal Attribution for Success

Ability	t	$\bar{X}$	Personality	t	$\bar{X}$
confident	6.7	(8.7)	conceit	9.2	(6.9)
competent	5.6	(8.6)	ecstatic	6.4	(8.1)
pleased	4.4	(8.8)	proud	6.3	(8.1)
satisfied	3.6	(8.7)	dignified	6.0	(6.8)
proud	2.9	(8.1)	contented	4.8	(8.8)
Unstable Eff.[a]			*Other's Eff.*		
uproarious	8.4	(7.0)	appreciative	7.2	(8.4)
delirious	6.8	(5.1)	composed	5.6	(7.4)
delighted	5.6	(8.7)	relaxed	4.4	(7.6)
good	5.3	(8.9)	grateful	4.0	(6.9)
competent	5.0	(8.5)	proud	3.2	(8.2)
Stable Eff.			*Other's Mot.* *and Pers.[c]*		
calm	2.8	(5.4)	grateful	7.6	(7.8)
relaxed	2.8	(7.2)	modest	7.2	(6.4)
			appreciative	6.0	(8.1)
			thoughtful	5.6	(6.6)
			charmed	4.8	(6.2)
Task Diff.[b]			*Luck*		
hopeful	4.0	(7.4)	surprised	14.0	(7.0)
composed	3.2	(6.8)	astonished	12.0	(6.6)
safe	2.8	(7.6)	wonderment	11.6	(6.6)
amused	2.4	(5.0)	thankful	10.0	(8.8)
dignified	2.0	(5.8)	awe	8.0	(5.1)
Mood			*Int. Mot.[d]*		
pleasant	5.9	(8.3)	benevolent	6.8	(6.6)
cheerful	5.8	(8.6)	wonderment	6.8	(5.4)
gay	5.8	(8.0)	amused	6.4	(6.0)
jovial	5.0	(7.9)	awe	6.4	(4.7)
zestful	4.4	(7.6)	delirious	6.0	(4.9)

[a]Eff. = effort.

[b]Diff = difficulty.

[c]Other's motivation and personality.

[d]Intrinsic motivation.

mean intensity rating across the remaining nine combined attributions. Table 3.4 lists the highest five discriminating affects for each attribution in order of *t*-value magnitude, with the mean intensity rating following the discriminative *t* value. Thus, for each affect a relative (*t*-value) and an absolute (mean intensity) index is provided.

The data in Table 3.4 suggest that a number of attributions give rise to a rather unique quality of emotional experience, although it is evident that the interpretation of any given attribution is not completely unambiguous. We therefore offer the following interpretations with some trepidation.

Six of the attributions (unstable effort, stable effort, personality, other's effort, other's motivation and personality, and luck) generate distinct types of emotions, whereas two others (ability and task difficulty) permit a comparative analysis. Of the two remaining attributions, intrinsic motivation yields ambiguous data, and the affective reactions to mood seem confounded, because they merely could be a continuation of the specified cause, rather than a consequence of the ascription of success to mood.

Unstable effort.   The highest affective discriminators given success attributed to immediate and intense effort expenditure are *uproarious, delirious,* and *delighted.* In addition, other discriminating emotions for unstable effort ascriptions are *ecstatic, elated, exhilarated, jubilant, overjoyed, thrilled,* and *zestful.* (These emotions are not among the top five discriminators and do not appear in Table 3.4.) It therefore appears that success perceived as due to intense effort generates heightened excitement and a state of activation, along with augmented positive affect.

Stable effort.   Stable effort ascriptions are discriminatively linked with only two emotions: calmness and relaxation. These reactions, accompanied by the positive affects listed in Table 3.3, remind one of the phrase "basking in glory," which has been used to characterize the reactions to success of individuals high in achievement needs (Atkinson, 1964).

The distinction between the affective consequences of temporary versus stable effort had not been previously taken into consideration in the achievement literature and was an unexpected revelation to us. In earlier work (e.g., Rest et al., 1973), it was anticipated that these ascriptions might generate disparate expectancies of success because they differ in perceived stability. However, the causes were not distinguished in terms of emotional consequences.

Personality.   Three of the discriminating affects given an ascription of success to personality are *conceit, proud,* and *dignified.* This implies that an ascription to personality enhances self-evaluation. It is of interest to note that *conceit* did not appear on any other attributional list as a discriminating affect.

Other's effort; other's motivation and personality. Given both these ascriptions, *grateful* and *appreciative* are dominant discriminative emotional consequences following success. It therefore might be argued that these two causal ascriptions should be combined. However, they do somewhat differ in the other affects that they elicit.

Ascription of success to others implies that one is aware of individuals who are available to help and even to fulfill one's duties. This is suggested by the affects of *composed* and *relaxed,* which describe reactions given success because of other's efforts. In addition, *carefree* is a significant emotional reaction (although not among the top five discriminators), given only causal ascriptions to other's efforts and other's motivation and personality. It could be contended that psychologists have overemphasized the negative consequences of perceived external causality, perhaps because the external cause in so many research studies is a barrier in the environment that impedes (controls) goal attainment. Yet it also is evident that the external environment may be facilitative rather than inhibitory, and ascriptions of success to a concerned other or a cooperating group may be quite positively valued. Indeed, given a Marxist orientation, this could be regarded as the most positive causal ascription. The consequences of passing perceived control from the individual to a group deserve more attention from psychologists.

Luck. The three main discriminative reactions when success is ascribed to luck are *surprised, astonished,* and *wonderment,* all with t values exceeding 10. Not far behind in the discriminating list are *awe* and *amazed.* Perhaps the clearest finding in our data is the unique emotional reaction given a luck ascription for success.

Ability versus task difficulty. Ability and task difficulty comprise two of the three components within Heider's (1958) naive analysis of action and jointly determine whether one "can" attain a goal. Weiner (1974) previously maintained that both these causal ascriptions are stable. Hence, following success one should anticipate continued success given an ascription to either of these two causes. However, it also was argued that because the ascriptions differ in locus of causality, there would be greater positive affect given the ability (internal), than the task difficulty (external), ascription.

The present data both support and disconfirm the above contentions. The main affective responses given a success ascribed to ability are *confident* and *competent;* the emotional labels most highly associated with success ascribed to task difficulty (ease) are *hopeful, composed,* and *safe.* Both these sets of reactions imply the anticipation of future success, which is consistent with the assumed relation between causal stability and expectancy. But the quality of affective reactions to the causal ascriptions is quite disparate.

TABLE 3.5
Attributions and Dominant Discriminating Affects for Success

Attribution	Affect
Unstable effort	Activation, augmentation
Stable effort	Relaxation
Own personality	Self enhancement
Other's effort and personality	Gratitude
Luck	Surprise
Ability versus task difficulty (ease)	Competence versus safety

Summary and Implications

The dominating discriminative affects outlined in the prior paragraphs are summarized in Table 3.5. Ability and task difficulty are linked in the table, because their interpretation is clearest given a mutual comparison. It is evident that disparate attributions for success give rise to distinct emotional feelings.

The four dimensions of affect previously identified by Davitz (1969) — namely, competence, activation, hedonic tone (relaxation), and relatedness — are represented in Table 3.5 and result from contrasting causal ascriptions. The linkages are ability ascriptions (competence), unstable effort (activation), stable effort (relaxation), and other's effort or other's motivation and personality (relatedness). Davitz (1969) also suggested that competence and relatedness involve the social environment, whereas activation and relaxation are relatively independent of social processes. These notions are sensible given the attributional framework, inasmuch as ability judgments often require social comparison data and, of course, help from others is an interpersonal action. On the other hand, the antecedent cues for, and the consequences of, perceived effort expenditure appear to be less dependent upon the social environment.[1]

Given these findings, it is now possible to reconstruct Fig. 3.1 so that it is consistent with our current manner of thinking. The abscissa of Fig. 3.1, which reads "affective reaction," should be modified to include the specific emotion being considered. Furthermore, the absolute intensity and the discriminatory power of the affects must be distinguished. At this point in time it seems more heuristic to be concerned with the discriminating affects.

---

[1]A factor analysis of the emotional ratings also was performed but is not reported in detail here because of space limitations and uncertainty both in the interpretation of the factors and their reliability. For success, five factors each accounted for more than 4% of the rating variance. In order of interpretive certainty, the factors, percentage of variance, and affects with the highest loadings were: luck (11%; *surprised, astonished, wonderment*); positive interpersonal feelings (6%; *kind, considerate, benevolent*); positive expectancy (6%; *confident, competent, optimistic*); arousal (9%; *uproarious* and *delirious* as loading positively and *solemn* and *relaxed* as loading negatively); and a general positive affect (37%; *delighted, happy, overjoyed*).

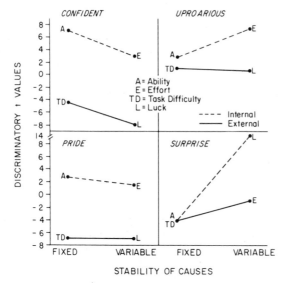

FIG. 3.2. Selected affective reactions to success as a function of the four main causal attributions. Discriminative *t* values are plotted.

In Fig. 3.2, discriminatory values for four illustrative affects (*confident, pride, uproarious,* and *surprise*) are plotted in relation to the four main ascriptions for success shown in Table 3.1. For ease of comparison, Figs. 3.1 and 3.2 take the same form, although it would be more accurate to construct Fig. 3.2 as a bar graph. It can be observed in Fig. 3.2 that *confident* is augmented given internal attributions, is greatest given an ability attribution, and is least given a causal ascription of success to luck; *pride* also is augmented given internal attributions, although there is little distinction within the two internal or the two external causes; an *uproarious* feeling is only evident given a causal ascription to effort; and *surprise* is experienced only given an attribution to luck. It is thus quite apparent that the patterns of results are not adequately captured by Fig. 3.1, although that figure approximates the data for some of the affects, such as *pride*.

Failure

Table 3.2 also lists the 150 affects associated with failure in achievement-related contexts. Using the same criteria as those imposed for success, 21 of the words were eliminated from the final analysis because they were not comprehended, and 8 were discarded because of insufficient intensity ratings.

Table 3.6 lists the 10 affects reported as most intensely experienced given disparate attributions for failure. Here again a large overlap in the lists can be observed, or "outcome-dependent" affects that are experienced regardless of the

TABLE 3.6
Failure: Top 10 Intensity Ratings by Attribution;
Listed Are the Affects and the Mean Intensity Ratings

Ability	$\overline{X}$	Unstable Eff.[a]	$\overline{X}$	Stable Eff.	$\overline{X}$	Task Diff.[b]	$\overline{X}$
concerned	8.1	sorry	8.6	troubled	8.3	displeasure	8.7
uncheerful	8.1	lousy	8.5	displeasure	8.2	uncheerful	8.6
dissatisfied	8.0	unhappy	8.3	unsatisfied	8.1	unhappy	8.4
upset	7.7	regretful	8.2	uncheerful	8.0	sad	8.2
discontent	7.7	troubled	8.0	depressed	8.0	upset	8.1
anxious	7.6	displeasure	7.9	unhappy	7.9	troubled	8.1
disheartened	7.6	anxious	7.9	upset	7.9	discontent	7.9
tense	7.5	miserable	7.9	disheartened	7.7	disheartened	7.9
inadequate	7.5	disturbed	7.9	disturbed	7.6	disturbed	7.6
joyless	7.5	ashamed	7.8	dissatisfied	7.6	glum	7.7

Mood		Personality		Other's Eff.		Other's Mot.[c] and Pers.[d]	
uncheerful	8.4	displeasure	6.9	uncheerful	8.2	disturbed	8.0
unhappy	8.4	unsatisfied	6.9	discontent	8.0	concerned	7.7
disgust	8.2	upset	6.7	dissatisfied	7.9	dissatisfied	7.7
disturbed	8.2	unhappy	6.7	bitter	7.9	upset	7.6
lousy	8.1	joyless	6.6	miserable	7.9	dismayed	7.6
miserable	8.1	resigned	6.4	unhappy	7.8	aggravated	7.6
concerned	8.1	grim	6.4	unsatisfied	7.8	disturbed	7.6
upset	8.1	lousy	6.4	concerned	7.7	frustrated	7.5
displeasure	8.0	frustrated	6.4	frustrated	7.7	discontent	7.5
unsatisfied	8.0	uncheerful	6.2	irritated	7.6	unhappy	7.4

Luck		Int. Mot.[e]		Fatigue–Ill.[f]			
frustrated	9.0	unhappy	7.9	dissatisfied	8.2		
concerned	8.6	sad	7.9	unhappy	8.2		
dissatisfied	8.4	depressed	7.7	sorry	8.1		
irritated	8.2	uncheerful	7.7	lousy	8.1		
perturbed	8.2	displeasure	7.7	sad	8.0		
unsatisfied	8.2	blue	7.6	sullen	8.0		
shaken	8.1	upset	7.5	displeasure	7.9		
displeasure	8.1	dissatisfied	7.4	shaken	7.9		
unhappy	8.1	uncomfortable	7.2	miserable	7.7		
dejected	8.0	uneasy	7.1	uncheerful	7.7		

[a]Eff. = effort.
[b]Diff. = difficult.
[c]Mot. = motivation.
[d]Pers. = personality.
[e] = Intrinsic motivation.
[f]Ill. = Illness.

reason for failure (*uncheerful* and *displeasure* are on 6, *upset* on 5, and so on.)[2] Although the overlap apparently was greater for success than for failure, statistical analyses not reported here revealed similar relationships between intensity of affect and discriminatory value for both success and failure.

Table 3.7 presents the list of the top five discriminating affects within each of the causal ascriptions. Eight of the attributions (ability, unstable effort, stable effort, personality, other's effort, other's motivation and personality, luck, and intrinsic motivation) generate a somewhat meaningful pattern. Of the three remaining attributions, fatigue—illness and task difficulty yield ambiguous data, and mood was again eliminated from possible interpretation because of the antecedent-consequence confound mentioned previously.

Ability. The most dominant affective reactions to failure given a lack of ability are *incompetent* and *inadequate.* As was true in the interpretation of success given an ascription to high ability, the internal, stable cause of ability suggests that the person should expect the specified outcome (now failure) also to occur in the future. The affects of *incompetent* and *inadequate* also define a particular self-concept as well as an associated emotional state.

Unstable effort; stable effort. Ascription of failure to a lack of effort generates reactions of guilt and shame. The indication of a relation between achievement and moral evaluation also was previously documented by Weiner and Peter (1973). However, unstable versus stable effort ascriptions also are associated with unique affects. Unstable effort seems to be linked with fear (*scared, frightened,* and *afraid* are among the top 10 discriminators, perhaps because of the knowledge that the performer will be held responsible for the actions). On the other hand, stable effort seems more closely related to depression (*hopeless, depressed,* and *disheartened* are among the top 10 discriminators). These reactions could be mediated by the stability of the cause and the perceived likelihood of future failures.

Personality; intrinsic motivation. *Resigned* and *apathetic* appear as affective reactions given these two causal ascriptions. In addition, other discriminating affects suggest depression, giving up, and a low expectancy of future success. The general relation between depression and causal ascriptions is discussed in greater detail later in this chapter.

---

[2]Some investigators (e.g., Nisbett, Borgida, Crandall, & Reed, 1976) have reported that individuals often do not alter their reactions to stressful events when receiving social norm information concerning the reactions of others. The investigators therefore conclude that social norms are not used or processed by the actor. It may be, however, that normative information alters attributions but not affective reactions in the situation under study. For example, knowing that other first-year faculty members also are unhappy may change attributions from personal to situational but would not be likely to change feelings of unhappiness, which in this context could be outcome dependent and attributionally independent.

TABLE 3.7
Discriminating Affects Reported as a Function of
the Causal Attribution for Failure

Ability	t	$\bar{X}$	Personality	t	$\bar{X}$
incompetent	10.1	(7.5)	resigned	9.6	(6.4)
inadequate	8.0	(7.5)	apathetic	7.6	(5.5)
aimless	7.6	(6.1)	incompetent	4.8	(6.3)
panic	7.2	(7.1)	aimless	4.4	(5.4)
humble	6.8	(5.8)	solemn	2.1	(5.3)
Unstable Eff.[a]			*Other's Eff.*		
ashamed	9.2	(7.8)	ferocious	9.2	(5.9)
scared	8.8	(7.5)	revengeful	9.0	(6.0)
sorry	8.4	(8.6)	aggressive	8.6	(7.0)
panic	8.0	(7.4)	furious	8.6	(7.2)
guilty	7.2	(7.9)	bitter	6.4	(7.9)
Stable Eff.			*Other's Mot.[b]* *and Pers.*		
humble	9.6	(6.4)	revengeful	10.0	(6.3)
guilty	7.4	(7.0)	surprised	8.4	(6.0)
troubled	5.7	(8.3)	vicious	7.6	(5.9)
hopeless	5.6	(6.9)	wonderment	6.1	(5.1)
ashamed	4.0	(6.4)	fuming	6.8	(6.9)
Task Diff.[c]			*Luck*		
stunned	8.0	(7.1)	astonished	12.6	(7.1)
unexcited	7.7	(5.9)	overwhelmed	11.8	(6.6)
dumbfounded	6.9	(6.2)	surprised	10.8	(6.6)
thoughtful	6.8	(6.3)	stunned	10.0	(7.6)
displeasure	6.5	(8.7)	horrified	9.3	(7.0)
Mood			*Int. Mot.[d]*		
disgust	6.7	(8.3)	sad	4.7	(7.9)
horrified	6.3	(6.3)	resigned	3.9	(5.3)
bewildered	6.0	(6.0)	helpless	3.4	(6.8)
frenzied	5.8	(6.0)	apathetic	3.2	(4.4)
tormented	5.0	(6.8)	blue	3.2	(7.6)
Fatigue–Ill.[e]					
offended	7.0	(6.1)			
unnerved	6.5	(7.0)			
sorry	6.3	(8.1)			
shaken	6.3	(7.9)			
sullen	5.9	(8.0)			

[a]Eff. = effort. [c]Diff. = difficulty.

[b]Mot. = motivation; Pers. = personality. [d]Int. mot. = Intrinsic motivation.

[e]Ill. = illness.

TABLE 3.8
Attributions and Dominant Discriminating Affects for Failure

Attribution	Affect
Ability	Incompetence
Unstable effort; stable effort	Guilt
Personality; intrinsic motivation	Resignation
Other's efforts; other's motivation and personality	Aggression
Luck	Surprise

Other's effort; other's motivation and personality. The manifest feeling when failure is ascribed to others is aggressive retaliation. Recall that success ascribed to others generates the opposite reactions of gratitude and appreciation.

Luck. Again luck ascriptions engender feelings of surprise. However, in opposition to the reactions given success, in this case surprise is accompanied by negative affects (*stunned, horrified*). Luck is the only causal ascription in which success and failure yield the same discriminating affects (although we did not provide positive emotions as reactions to failure or negative affects as possible reactions given success).

Summary

The most discriminating affects given diverse causal ascriptions for failure are summarized in Table 3.8. It is again evident that disparate causal attributions yield disparate emotional reactions.[3] In Fig. 3.3, the discriminating t values of four affects (*despair, shame, hostility,* and *surprise*) are plotted in relation to the four main ascriptions for failure outlined in Table 3.1. These particular affects were selected, because they permit some comparisons with the success data and because they illustrate some unexpected findings. Concerning *despair*, this affect is experienced given all attributions except effort, which is under volitional control and implies that failure might be averted in the future; *shame*, on the other hand, is salient only given a causal ascription to a lack of effort; *hostility* is most enhanced given ability and luck ascriptions; and *surprise* is primarily experienced when attributions are to external factors, particularly luck. The data for *shame* resemble the hypothesized relations depicted in Fig.

---

[3]A factor analysis of the emotional ratings also was performed for failure. Five factors emerged that each accounted for more than 4% of the rating variance. In order of interpretive certainty, the factors, percentage of variance, and affects with the highest loadings were: luck (5%; *surprised, astonished, amazed*); fear (4%; *afraid, scared, panic*); and three additional factors that cannot clearly be interpreted but appear to relate to the intensity of affect (34%, 10%, and 6%; *crushed, unhappy, miserable; sad, gloomy, heavyhearted; annoyed, unsatisfied, discontent*).

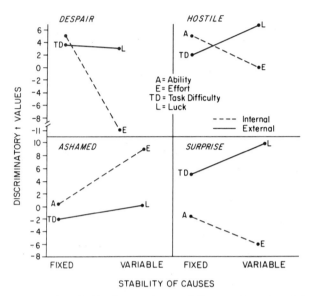

FIG. 3.3   Selected affective reactions to failure as a function of the four
main causal attributions. Discriminative *t* values are plotted.

3.1, and in conjunction with the pride affective ratings, these data indicate that
Fig. 3.1 reasonably predicts the *pride-shame* affective reports. But the data patterns for the other affects reveal quite different stories.

A Conceptual Problem

The data presented in the prior pages indicate that Table 3.1, which presents the
attributional model of motivation guiding the work of Weiner and his colleagues,
is in need of alteration. In Table 3.1, affective reactions are directly linked to the
locus of control dimension of causality. However, it appears that affects often
(but not necessarily always) are directly tied to the causes, without locus of
control serving a mediating role. This creates some theoretical difficulties, for
it is quite evident that the internal—external dimension of causality influences a
wide array of thoughts and actions. How, then, locus of control should be conceptualized within this attributional approach to motivation remains a problem
for the future.

EMOTIONS

The field of emotions is immense, embracing the insights of many important
historical figures, including Darwin (the innate bases of emotional expression),
Freud (affective transformations), and James (the temporal relation of bodily
processes and emotion). There is no comprehensive theory that can account for

even a small portion of the data in this field. The ideas and data presented in this paper also are microscopic, and no attempt will be made to offer a more general theory of emotions. Nonetheless, the present approach is quite relevant to issues of general importance in the contemporary study of emotion.

Schachter's Two-Factor Theory

The notion that a general state of internal arousal determines affective expression plays a prominent role in the psychological study of emotions, with Cannon (1927) perhaps the first influential holder of this belief. Schachter (1964) and Schachter and Singer (1962), following in the tradition of Cannon, also assume that arousal is a necessary antecedent of emotional expression. However, they do not accept arousal as a sufficient explanatory construct, for given the same level of arousal, different emotional reactions are possible. High arousal, for example, might be accompanied by feelings of intense love *or* intense hate. Arousal theory as first formulated by Cannon did not provide the conceptual tools to distinguish sufficiently between the varieties of emotional experience.

Schachter (1964) supplemented arousal theory so it would be able to differentiate between emotions. He proposed that emotions are a function of two factors: (1) level of arousal; and (2) cognitions about the arousing situation that provide the "steering" or directionality for emotional experience. The two-factor theory and the study by Schachter and Singer (1962) have proven astonishingly popular and have greatly contributed to the experimental study of emotions. Yet this work appears to have fundamental methodological and theoretical shortcomings (for methodological issues see, for example, Plutchik & Ax, 1967). Here we will consider only the theoretical speculations concerning the arousal—emotion sequence and the place of arousal in a theory of emotions.

Arousal. Schachter (1964) contends that the perception of an external event or, for example, the ingestion of a particular drug may produce arousal. The social situation then provides the cues to interpret (ascribe the cause for) the arousal and thus label its origin. The labeling, in turn, guides the appropriate emotional experience. But work by investigators such as Lazarus (1966, 1968) strongly suggests that arousal itself is a product of cognitive factors. More specifically, Lazarus reports that how one interprets a stressful situation influences the amount of fear (arousal) that is generated. In a similar manner, in this paper we report data that whether success is followed by activation or calmness depends in part on the perception of the cause of success.

It presently is not known whether arousal precedes, accompanies, or follows an emotional experience or if all of the above may be true given disparate situations. The empirical findings reported by Schachter and Singer may indeed be valid, yet there may be other instances in which arousal is not a necessary condition for emotional experience. One implication of the preliminary data presented here is that cognitions are *necessary and sufficient* causes of emotion. Feeling competent because of perceived high ability or feeling gratitude because of an ascription to the help of others does not appear to require a prior state of arousal that the individual must seek to interpret. On the contrary, the degree of arousal seems to be determined by the prior causal cognition. A change in the conception of emotions so that it follows a historical sequence of *drive → drive and cognition → cognition only* is reminiscent of other historical progressions in the field of motivation, such as the *drive → drive and incentive → incentive only* temporal movement in the explanation of motivated behavior (Bolles, 1975). Many motivational theorists have come to believe that the drive concept is a "way station" in psychological explanation (see Atkinson, 1964; Bolles, 1975; Weiner, 1972). We suggest this may be the case in the study of emotions.

The position that cognitions are sufficient determinants of affect is consistent with the theoretical ideas of Valins (1966). Valins has suggested that arousal is merely another source of information used to elicit appropriate emotional reactions. But there is evidence questioning the conclusion of his investigations (Goldstein, Fink, & Mettee, 1972), and the issues remain unsettled.

The Physiology of Emotions

There is a belief that disparate emotional states are associated with distinctive internal conditions. For example, it has been shown that anger and aggression produce different hormonal secretions than fear, anxiety, and depression (Ax, 1953; Schildkraut & Kety, 1967). Lazarus (1968) states:

> It seems to me quite unlikely that the physiological states connected with anger, fear, depression, grief, sexual excitement, guilt, disgust, shame, or what have you, can logically be expected to be all the same, granted that these states have in common some core of excitement or mobilization which we call arousal. If we take on the problem of positively toned emotions, the issue gets even stickier. It seems shortsighted to focus exclusively on the excitement which these reactions have in common. Each of these, in addition, involves particular kinds of organized behavioral activity, or their absence, and, as such, there ought logically to be considerable physiological and psychological differentiation between them [p. 206].

The quote from Lazarus indicates that the possibility of a general arousal state characterizing all emotions is not denied. Rather, it is suggested that emotions are also linked with unique internal patterns, just as intelligence often is conceived as a general system with distinctive components.

To date, the physiological patterns associated with the vast array of human emotions have yet to be identified. It may indeed be that states such as rage and sexual excitement, which also are manifested in infrahuman species, will be characterized by particular and distinctive internal states. But it seems highly questionable that emotional distinctions apparently unique to humans, such as competence versus safety or even pride versus shame, will be differentiated on the basis of physiological or hormonal conditions. The psychology of emotions should proceed independent of the physiology of emotions, and an isomorphism between these two levels of analysis should not be anticipated, although at times circumscribed bridges might be uncovered.

Relations to Clinical Psychology

The attributional position advocated here is that emotions primarily are determined by cognitive processes. The notion that "you feel the way you think" (Ellis, 1974, p. 312) is not new to the field of psychology. For example, many individuals in the field of behavioral change have contended that if thoughts can be altered, then feelings and action also will undergo modification (see Beck, 1974; Ellis, 1974; Kelly, 1955). Rational—emotive therapy as described by Ellis is particularly consistent with the ideas offered in this paper.

The attributional conception of emotions also is relevant to the etiology of depression, particularly as characterized by Seligman (1975). Seligman contends that depression follows from the perception that actions and outcomes are independent. That is, one has "learned helplessness." Our data, however, suggest that disparate affective consequences are generated as a function of *why* actions and outcomes are perceived as independent. For example, lack of perceived response instrumentality because of interference from others versus low ability versus an overly difficult task produce quite disparate emotional experiences. In our data, the depression-related labels of hopeless, helpless, and depressed, and related affects such as resigned and aimless, most appear when there is an internal, stable attribution (ability, stable effort, personality, and intrinsic motivation) for failure. We anticipate that causal attributions in part influence the various manifestations of depressive disorders and the course of recovery through their differential influence upon affective reactions and goal expectancies. In accordance with this position, Klein, Fencil-Morse, and Seligman (1976) and Tennen and Eller (1977) found that only subjects making internal ascriptions for failure exhibited learned helplessness. Of course, this does not preclude a relation between particular types of depression and particular types of external ascriptions.

ACHIEVEMENT MOTIVATION AND EMOTION

In the minds of many psychologists the fields of motivation and emotion are united — they have the same linguistic root, psychology courses and texts carry a joint motivation—emotion identification, and there is a psychological journal with the title *Motivation and Emotion*. But this linkage is more apparent than real. The systematic study of psychogenic needs such as achievement, affiliation, and power has advanced with only lip service attention paid to the affects, such as the phrase "pride in accomplishment" that is associated with achievement strivings. There are investigations in the field of emotions and investigations in the field of motivation, but few analyses of their interrelationships (see, for example, Tomkins, 1962, 1963, who asserts that affects provide the motor for action). Yet even our tenuous attempt presented in this paper demonstrates the important role that an understanding of affect might play in the explanation of human motivation.

Individual Differences in Achievement Needs

In our data neither stable nor unstable effort attributions, which are believed characteristic ascriptions for success among individuals high in achievement needs (Weiner, 1974), are discriminably linked to feelings of pride. But it has been contended that it is the anticipation of pride that gives rise to achievement-oriented behaviors (Atkinson, 1964). Could it possibly be, then, that other affects play a more important role than pride in the realm of achievement strivings?

Given the tentative findings reported here, one might contend, for example, that persons high in achievement needs anticipate a quiet contentment and relaxation after success. This idea reminds one of the so-called arousal jag postulated by Berlyne (1960), who suggests that you hit yourself on the head because it feels so good when you stop! This could imply that the highly achievement-oriented person works so hard because it feels so good when he or she is finished. Of course, it is not our intent to bring the cognitive (attributional) analysis of emotions and motivation within a drive-reduction framework. Nevertheless, the notion of the anticipation of relaxation does call into mind the premise of drive theory. The general message of this line of thought is that a more detailed examination of affects in achievement-related contexts could provide a better foundation for the understanding of achievement strivings and individual differences in achievement needs.

The Assessment of Achievement Needs

The Thematic Apperception Test (TAT) is the most frequently used instrument to infer the magnitude of achievement needs. The TAT motivational scoring system contains a number of scoring categories, including one labeled "nurturant press." Nurturant press is inferred in a TAT protocol if the actor perceives that another person is aiding in the attainment of an achievement-related goal. But attributions for success to others produce feelings of thankfulness and gratitude, not feelings of pride. Hence, given the conception of achievement strivings adhered to by Atkinson (1964) and McClelland et al. (1953), it seems inconsistent to include this scoring category when inferring the presence of achievement needs unless the anticipation of gratitude is accepted as one of the springs of achievement strivings.

Expectancy and Affect

Atkinson (1964) has postulated that the affective consequence associated with success (pride) is inversely related to the probability of success at a task. Thus, expectancy and value are not independent. On the other hand, in the attributional model followed by Weiner and his colleagues, it has been assumed that expectancy and affect are independent, inasmuch as they are influenced by orthogonal dimensions of causality. However, this position cannot be entirely correct, for it has been conclusively determined that one experiences more pride for success at a difficult task than for success at an easy task, inasmuch as these events elicit disparate degrees of self-attribution. In addition, pride is not the only affect that is influenced by the expectancy of success. For example, success at a task with a low probability of success generates attributions to luck (Feather, 1969), which has been shown here to be linked with the experience of surprise. The variety of emotions that are influenced by the probability of success will have to be examined more closely.

SUMMARY

In previous work it was contended that affective magnitude is influenced by ascriptions of success and failure to internal versus external causes. However, this formulation failed to recognize the diverse consequences that causal attributions have for emotional reactions. The investigation reported here demonstrates that there are qualitative differences in feelings as a function of causal ascriptions for success and failure. It is therefore suggested that future work focus upon the cognitive (attributional) determinants of emotion, guided by the principle that cognitions are necessary and sufficient causes of emotional experience. Furthermore, it is contended that the conception of achievement motivation be broadened to encompass other affects in addition to pride and shame: the incentive values of achievement success can range, for example, from competence and relaxation to safety and activation.

ACKNOWLEDGMENTS

The writing of this chapter was supported by Grant MH25687-03 to the senior author from the National Institute of Mental Health.

REFERENCES

Atkinson, J. W. *An introduction to motivation*. Princeton, N.J.: Van Nostrand, 1964.
Ax, A. F. The physiological differentiation of fear and anger in humans. *Psychosomatic Medicine*, 1953, *15*, 433-442.

Barnes, R. D., & Ickes, W. Effects of the perceived intentionality and stability of another's dependency on helping behavior: A field experiment. *Journal of Personality and Social Psychology*, in press.

Beck, A. T. Cognition, affect, and psychopathology. In H. London & R. E. Nisbett (Eds.), *Thought and feeling*. Chicago, Ill.: Aldine, 1974.

Berlyne, D. E. *Conflict, arousal, and curiosity*. New York: McGraw–Hill, 1960.

Bolles, R. C. *Theory of motivation* (2nd ed.). New York: Harper & Row, 1975.

Cannon, W. B. The James–Lange theory of emotions: A critical examination and an alternative theory. *American Journal of Psychology*, 1927, *39*, 106–124.

Covington, M. V., & Beery, R. G. *Self-worth and school learning*. New York: Holt, Rinehart & Winston, 1976.

Davitz, J. R. *The language of emotion*. New York: Academic Press, 1969.

Elig, T. W., & Frieze, I. H. A multi-dimensional scheme for coding and interpreting perceived causality for success and failure events: The Coding Scheme of Perceived Causality (CSPC). *SAS: Catalog of Selected Documents in Psychology*, 1975, *5*, 313.

Ellis, A. Rational-emotive therapy, In A. Burton (Ed.), *Operational theories of personality*. New York: Brunner/Mazel, 1974.

Eswara, H. S. Administration of reward and punishment in relation to ability, effort, and performance. *Journal of Social Psychology*, 1972, *87*, 139–140.

Feather, N. T. Valence of outcome and expectation of success in relation to task difficulty and perceived locus of control. *Journal of Personality and Social Psychology*, 1967, *7*, 552–561.

Feather, N. T. Attribution of responsibility and valence of success and failure in relation to initial confidence and task performance. *Journal of Personality and Social Psychology*, 1969, *13*, 129–144.

Frieze, I. H. Causal attributions and information seeking to explain success and failure. *Journal of Research in Personality*, 1976, *10*, 293–305.

Goldstein, D., Fink, D., & Mettee, D. R. Cognition of arousal and actual arousal as determinants of emotion. *Journal of Personality and Social Psychology*, 1972, *21*, 41–51.

Heider, F. *The psychology of interpersonal relations*. New York: Wiley, 1958.

Ickes, W. J., & Kidd, R. F. An attributional analysis of helping behavior. In J. H. Harvey, W. J. Ickes, and R. F. Kidd (Eds.), *New directions in attribution research* (Vol. 1). Hillsdale, N.J.: Lawrence Erlbaum Associates, 1976.

Kaplan, R. M., & Swant, S. G. Reward characteristics of appraisal of achievement behavior. *Representative Research in Social Psychology*, 1973, *4*, 11–17.

Kelly, G. A. *The psychology of personal constructs*. New York: W. W. Norton, 1955.

Klein, D. C., Fencil-Morse, E., & Seligman, M. E. P. Learned helplessness, depression, and the attribution of failure. *Journal of Personality and Social Psychology*, 1976, *33*, 508–516.

Kun, A., & Weiner, B. Necessary versus sufficient causal schemata for success and failure. *Journal of Research in Personality*, 1973, *7*, 197–207.

Lanzetta, J. T., & Hannah, T. E. Reinforcing behavior of "naive" trainers. *Journal of Personality and Social Psychology*, 1969, *11*, 245–252.

Lazarus, R. S. *Psychological stress and the coping process*. New York: McGraw-Hill, 1966.

Lazarus, R. S. Emotions and adaptation: Conceptual and empirical relations. In W. J. Arnold (Ed.), *Nebraska Symposium on Motivation*. Lincoln, Neb.: University of Nebraska Press, 1968.

Leventhal, G. S., & Michaels, J. W. Locus of cause and equity motivation as determinants of reward allocation. *Journal of Personality and Social Psychology*, 1971, *17*, 229–235.

Maselli, M. D., & Altrocchi, J. Attribution of intent. *Psychological Bulletin*, 1969, *71*, 445–454.

McClelland, D. C., Atkinson, J. W., Clark, R. A., & Lowell, E. L. *The achievement motive.* New York: Appleton-Century-Crofts, 1953.

Nicholls, J. G. Effort is virtuous, but it's better to have ability: Evaluative responses to perceptions of effort and ability. *Journal of Research in Personality,* 1976, *10,* 306–315.

Nisbett, R. E., Borgida, E., Crandall, R., & Reed, H. Popular inducation: Information is not necessarily informative. In J. S. Carroll & J. W. Payne (Eds.), *Cognition and social behavior.* Hillsdale, N. J.: Lawrence Erlbaum Associates, 1976.

Plutchik, R. *The emotions: Facts, theories and a new model.* New York: Random House, 1962.

Plutchik, R., & Ax, A. F. A critique of *Determinants of emotional state* by Schachter and Singer (1962). *Psychophysiology,* 1967, *4,* 79–82.

Rest, S., Nierenberg, R., Weiner, B., & Heckhausen, H. Further evidence concerning the effects of perceptions of effort and ability on achievement evaluation. *Journal of Personality and Social Psychology,* 1973, *28,* 187–191.

Rosenbaum, R. M. A dimensional analysis of the perceived causes of success and failure. Unpublished doctoral dissertation, University of California, Los Angeles, 1972.

Rotter, J. B. Generalized expectancies for internal versus external control of reinforcement. *Psychological Monographs,* 1966, *80,* 1–28.

Ruble, D. N., Parsons, J. E., & Ross, J. Self-evaluative responses of children in achievement settings. *Child Development,* 1976, *47,* 990–997.

Schachter, S. The interaction of cognitive and physiological determinants of emotional state. In L. Berkowitz (Ed.), *Advances in experimental social psychology* (Vol. 1). New York: Academic Press, 1964.

Schachter, S., & Singer, J. E. Cognitive, social, and physiological determinants of emotional state. *Psychological Review,* 1962, *69,* 379–399.

Schildkraut, J. J., & Kety, S. S. Biogenic amines and emotion. *Science,* 1967, *156,* 21–30.

Schütz, A. *Collected papers. I. The problem of social reality.* The Hague: Martinus Nijhoff, 1967.

Seligman, M. E. P. *Helplessness.* San Francisco: Freeman, 1975.

Storms, M. D., & McCaul, K. D. Attribution processes and emotional exacerbation of dysfunctional behavior. In J. H. Harvey, W. J. Ickes, & R. F. Kidd (Eds.), *New directions in attribution research* (Vol. 1). Hillsdale, N.J.: Lawrence Erlbaum Associates, 1976.

Tennen, H., & Eller, S. J. Attributional components of learned helplessness and facilitation. *Journal of Personality and Social Psychology,* 1977, *33,* 265–271.

Tomkins, S. S. *Affect, imagery, consciousness* (Vols. 1 and 2). New York: Springer, 1962, 1963.

Triandis, H. *The analysis of subjective culture.* New York: Wiley–Interscience, 1972.

Valins, S. Cognitive effects of false heart-rate feedback. *Journal of Personality and Social Psychology,* 1966, *4,* 400–408.

Valle, V. A., & Frieze, I. H. Stability of causal attributions as a mediator in changing expectations for success. *Journal of Personality and Social Psychology,* 1976, *33,* 579–587.

Veroff, J. Development and validation of a projective measure of power motivation. *Journal of Abnormal and Social Psychology,* 1957, *54,* 1–8.

Weiner, B. *Theories of motivation: From mechanism to cognition.* Chicago: Rand–McNally, 1972.

Weiner, B. (Ed.). *Achievement motivation and attribution theory.* Morristown, N.J.: General Learning Press, 1974.

Weiner, B. An attributional interpretation of expectancy-value theory. In B. Weiner (Ed.), *Cognitive views of human motivation.* New York: Academic Press, 1975.

Weiner, B. An attributional model for educational psychology. In L. Shulman (Ed.), *Review of research in education* (Vol. 4). Itasca, Ill.: Peacock, 1976.

Weiner, B., Frieze, I., Kukla, A., Reed, L., Rest, S., & Rosenbaum, R. M. *Perceiving the causes of success and failure.* Morristown, N.J.: General Learning Press, 1971.

Weiner, B., & Kukla, A. An attributional analysis of achievement motivation. *Journal of Personality and Social Psychology,* 1970, *15,* 1–20.

Weiner, B., Nierenberg, R., & Goldstein, M. Social learning (locus of control) versus attributional (causal stability) interpretations of expectancy of success. *Journal of Personality,* 1976, *44,* 52–68.

Weiner, B., & Peter, N. A cognitive–developmental analysis of achievement and moral judgments. *Developmental Psychology,* 1973, *9,* 290–309.

Weiner, B., & Sierad, J. Misattribution for failure and enhancement of achievement strivings. *Journal of Personality and Social Psychology,* 1975, *31,* 415–421.

4 Attributional Egotism

Melvin L. Snyder
Dartmouth College

Walter G. Stephan
University of Texas at Austin

David Rosenfield
Southern Methodist University

INTRODUCTION

Kim and Chris are playing Scrabble, a game that involves drawing lettered tiles and making words from them. The score for each word is based partly on the point value of the letters used. When the game ends, Kim has 174 points; Chris has 192 and wins. Kim says that the only reason Chris won is because Chris was lucky enough to draw more of the high-point letters. Chris says that is nonsense; the smartest player won.

Kim and Chris may be exhibiting what we have called egotism (Snyder, Stephan, & Rosenfield, 1976). We defined egotism as the tendency to take credit for good outcomes and deny blame for bad ones. When the interests of people are opposed as in a competitive relationship, egotism also may include refusing to recognize others for their accomplishments and finding them at fault when they perform poorly. We consider egotism to be a motivational phenomenon in which attributions about good and bad outcomes serve to protect or enhance one's self-esteem.

Do people really account for events in an egotistic fashion? To most readers the answer may be all too obvious. Of course they do. The matter, however, is not that simple. Egotism does not always occur, and we believe it is important to try to understand why. Toward this end, we have examined factors that contribute to egotistic motivation and factors that may inhibit egotism.

We also consider whether some other motive besides a desire to protect or enhance self-esteem might account for egotism such as cognitive dissonance or the desire to feel in control of one's environment. Some have suggested that the evidence for egotism can be understood by using an assortment of principles that are purely cognitive in nature and that a motivational explanation is not required (e.g., Miller & Ross, 1975). We will address these issues and review the evidence. But first we will explain why we chose the term *egotism.*

Why the Term Egotism?

We chose the term *egotism,* because other possible terms do not quite hit the mark. For example, egotism is related to the concept of defensive attribution discussed by Shaver (1970). However, that concept includes other phenomena such as concern for physical safety and belief in a just world, which involve somewhat different motivations. In addition, it focuses on attributions to others. Finally, the *defensiveness* label is used by Shaver with respect to undesirable outcomes, whereas we are also concerned with the possibility that people take more credit for good outcomes than they deserve or than others will give them.

Jones (1908) introduced the psychoanalytic concept of rationalization and argued that feelings influence judgments and beliefs. Egotism is a kind of *rationalization,* but that term suffers from being far too broad. Moreover, like the defensiveness label, it is used primarily in reference to undesirable outcomes.

Egotism can also be distinguished from both of the meanings of *egocentricity* found in Heider (1958). In one case, he uses the term to refer to the impact of needs or wishes on attribution (p. 118), and thus — like rationalization — that term subsumes egotism and encompasses many other phenomena. And again, like rationalization, it lacks precision. Heider (1958, p. 158) also uses the term to refer to the individual's presumption that his or her reactions are attributable to the nature of the world rather than to personal idiosyncrasies. If one believes that one's own response is the natural reaction, then others who are confronted with the same reality should respond in the same way — unless *they* are somehow deviant. Consequently, one may believe that others share one's opinions more than they really do.

Having said what egotism is not, we now say more about what it is.

More About the Nature of Egotism

At the risk of belaboring the obvious, we will now try to be as explicit as we can about the nature of egotism and what we believe motivates it. This discussion will provide a conceptual framework that we will use to present the evidence and to examine theoretical and methodological issues. To begin with, we assume that people are motivated to maintain the best possible opinion of themselves (cf. Adler, 1956; Festinger, 1954; Heider, 1944). Heider (1944) believed that the attributions for success and failure that we have labeled egotism serve to protect and enhance self-esteem. What does the occurrence of a good or bad outcome have to do with self-esteem? Asking the indulgence of those for whom the connection is self-evident, we will use two works by Heider (1944, 1958) to explicate the relationship.

In the search to understand the causes of events, people seem disposed to explain them in terms of the personal characteristics of those associated with the events. Heider (1944, 1958) suggests some reasons for this. First, explaining events in terms of persons permits a simpler organization of the world than does an extended and differentiated causal analysis. It is easier to view the person as the cause than to track down and keep in mind numerous other factors. Another reason is that behavior tends to engulf the field, by which Heider (1958, p. 54) meant that the person is more salient as a cause than is the environment (see, for example, Snyder & Frankel, 1976). Third, attribution to persons makes irreversible events appear to be reversible, even if only symbolically. For example, Heider suggests that revenge is a symbolic reversing of events. A need for justice may promote attribution to persons in another way. According to Heider (1958, p. 235), for justice to reign, our perception of a person's worth must be consistent with the value of the outcomes he or she receives. In order to preserve a belief that the world is just, we may tend to perceive those who receive good outcomes as good persons and those who receive bad outcomes as bad. Lerner and his colleagues have gathered considerable support for this idea (cf. Lerner, Miller, & Holmes, 1976).

Thus, success and failure may be taken as signs of an individual's merits beyond what a rational analysis would call for. Outcomes, therefore, may influence the esteem in which the person is held by others and that he feels for himself. The impact of the outcome on self-esteem may be greater to the extent that the outcome is attributed to the person. The desire to deny responsibility for bad outcomes, may really be the desire to protect self-esteem from the negative implication of producing bad effects. And taking credit for good acts can be regarded as an attempt to enhance self-esteem. Thus, egotism may be a way to enhance or preserve self-esteem.

We hypothesize that threat to self-esteem depends on two necessary factors. One is that the outcome must be attributed to the person. The other is that the attribution made must be relevant to the person's self-esteem. If either factor is absent, there is no threat. We will call these factors the outcome/attribution factor and the attribution/self-esteem factor. If both are present to some degree, the threat to self-esteem depends on the strength of each factor. The threat is greater, the stronger the attribution of the outcome to the person and the stronger the connection between that attribution and the roots of the person's self-esteem.

For example, if someone plays a game and loses, threat to self-esteem depends both on attributing the failure to oneself and on believing one's failure says something negative about an important aspect of self. If one believes the game is a matter of chance and the outcome is independent of what one does, the first factor is absent and there should be no threat. Even if one believes the game depends on skill and that the blame for the loss rests squarely on one's shoulders, there is no threat if the game is perceived as irrelevant to important aspects of self. Someone may believe that losing at bridge is one's own fault but may not regard poor ability at card games as important. Skill at card games is irrelevant to this person's self-esteem. Motivation for egotism should be nil if either factor is absent.

A number of variables that enhance egotism can be understood as influencing one or the other factor. To illustrate and to anticipate the presentation of the evidence for egotism: A personal attribution for failure on a test has a stronger connection to self-esteem when the test is reputed to be a valid measure of intelligence and adjustment than when the validity of the test is not yet established (Miller, 1976). This manipulation can be understood as operating on the attribution/self-esteem factor. So does a manipulation of whether a test measures the masculine or feminine personality, with the strength of the connection depending on the subject's sex (Rosenfield & Stephan, in press).

Making self salient promotes attribution to self (Duval & Hensley, 1976; Storms, 1973). Thus, making a person more aware of self may enhance egotism by strengthening the outcome/attribution factor — the attribution of the outcome to self (Federoff & Harvey, 1976). That factor can also be bolstered by informing the person that he or she has failed on other tasks as well as the one for which attributions are made (Stevens & Jones, 1976). In Kelley's (1967) terms, failure is not distinctive to a single task and is more likely to be attributed to the person. Egotism in response to failure consists of negating the first of the two factors — denying the attribution of failure to *self.* But we should not be surprised to see negating of the second one as well — that is, a denial of the implication of one's failure for one's general self-worth.

Let us now consider egotism and success. The general tendency to attribute outcomes to persons — whether because of a sense of justice, because the world is simpler that way, or because behavior engulfs the field — have favorable implications for self-esteem in the case of success. Self-esteem is enhanced to the extent that each factor is strong; that is, success is attributed to self, and the attribution contributes to self-esteem. Egotism involves strengthening the first factor. It might be more likely to occur when the second factor is strong. If the second factor is absent — if attributing success to self says nothing important about self — then there is little motivation for self-attribution of success.

In short, our conceptual framework states that the relation of outcomes to self-esteem depends on two factors. A negative outcome threatens self-esteem if the outcome is attributed to self and if, in turn, the self-attribution has implications for self-esteem. Egotism consists of weakening the first factor, the attribution of the outcome to self.[1] A positive outcome provides an opportunity to enhance self-esteem. The same two factors are involved. Egotism is manifested in strengthening the first link — the attribution of the outcome to self — but it only makes sense to do so if the second link is there and the attribution will enhance self-esteem.

Thus far, the discussion has focused on variables that influence motivation to make egotistical attributions. But as is the case with any other motive, whether the individual acts on egotistic desires will be influenced by: (1) his or her perception of the probability of successfully doing so (cf. Jones & Wortman, 1973); and (2) the strength of conflicting motives. We believe that many of the factors affecting the manifestation of egotism can be organized into those two standard categories. The following is an attempt to do so.

Probability That Egotism Will Work

Whether egotism occurs should depend on the probability that it will successfully defend or enhance self-esteem. Chances for success are diminished if the egotistic attribution lacks plausibility or if it may be contradicted by others' explanations for outcomes or by one's own subsequent behavior.

Plausibility of acceptable causes.   As Heider (1958) put it, rationalizations must be plausible as well as personally acceptable. Snyder, Stephan, and Rosenfield (1976) suggested that there must be some ambiguity about the roles played by various potential causes in producing outcomes. In the Scrabble game with which we began the chapter, Kim attributed losing to bad luck in drawing letter tiles. The role of luck is ambiguous. Thus, this excuse was plausible as well as being more acceptable to Kim as an explanation than lack of skill. If Kim took a long time to make words, then low effort was not a plausible excuse.

---

[1]To some, it may seem illogical to require as a necessary condition for egotistic motivation that the negative outcome be attributed to self and then say that egotism consists of denying exactly that. The paradox is a familiar one. For example, in the case of cognitive dissonance, a necessary condition is the perception of dissonant elements, but that perception may motivate a process of dissonance reduction and ultimately negate itself (Festinger, 1957).

Even though we see no logical problem in hypothesizing that the attribution is made and then denied, and even though we formulate egotism in that way, we must acknowledge that there is no evidence for this sequence. (But see the discussion of possible sequential effects in the section on cognitive dissonance.) Perhaps the critical element is not that the individual makes a personal attribution but rather that he or she perceives it as eminently makeable or apt to be made by others.

Luck would not have been so plausible if Kim played and lost many games with Chris. Luck would presumably even out over the long haul. Similarly, when the task is the same for two people and one fails while the other succeeds, task difficulty as an explanation for failure is a poor excuse. If an attribution lacks plausibility, there is little chance that it will successfully satisfy the egotistical motive, and it is unlikely to be made.

Likelihood of being contradicted by others.   Even when one can make a plausible excuse or boast, one may show restraint if one anticipates the possibility of being contradicted. Bradley (in press), Gunn (1976), Stephan (1975), and Wortman (1970) discuss failures to find egotism that may be the result of subjects' expecting another to evaluate their performance. However, the data supporting this suggestion are rather inconsistent. In one study there was no evidence for egotism (Jones, Rock, Shaver, Goethals, & Ward, 1968). In another, subjects showed a counterdefensive pattern (Feather & Simon, 1971). In others, it depends on which conditions are compared (Beckman, 1973; Ross, Bierbrauer, & Polly, 1974). In still others, egotism occurred despite the presence of an observer whom subjects might easily have suspected would evaluate them (Harvey, Arkin, Gleason, & Johnston, 1974; Harvey, Harris, & Barnes, 1975).

None of these studies manipulated expectations about being evaluated. Wells (1977) has done so. He gave subjects false feedback about how helpful they had been relative to others. Compared to observers, subjects who were told they had been helpful attributed their behavior more to personal characteristics, whereas those told they had not been helpful emphasized the role of the situation. But when both observers and subjects anticipated discussing with the other person their interpretation of the level of helping, they did not differ in the attributions they made. Although not the only explanation, it may well be that anticipation of discussing attributions tempers egotism because of fear of being contradicted by others (cf. Cialdini, Levy, Herman, & Evenbeck, 1973; Hass, 1975; Hass & Mann, 1976). It is not clear whether the effect obtained extends to privately made assessments as well as publicly displayed ones.

Likelihood of being contradicted by one's subsequent performance.   One's egotistic attributions may be contradicted, not only by someone else, but also by one's own subsequent behavior (cf. Wortman, 1970). If one succeeds, it is dangerous to boast if there is a risk of imminent failure in the same endeavor. Likewise, one who is failing may be reluctant to write off outcomes as due to luck or as irrelevant to important self-concerns if success still seems possible. To blame bad luck for early failure may make it harder to take credit for ultimate success. When no further action can be taken and all the outcomes are determined, egotism may be more likely. Relevant here is a study by Wortman, Costanzo, and Witt (1973). They gave subjects an alleged test of social perceptiveness. Subjects received either positive or negative feedback, and some expected additional testing of that trait; the rest did not. Regardless of feedback, if subjects thought there would be further testing they attributed less social perceptiveness to themselves.

Conflicting Motives

Even if one is motivated to be egotistical and even if the probability of success is high, one may not be egotistical if a conflicting motive takes precedence.

Desire to be accurate. When motives influence attributions, unless they cancel out some other bias, the result is likely to be an inaccurate picture of self or of the task. One attributes more ability to self than is warranted. One exaggerates or underplays the role that luck may have had in determining outcomes. Egotism may be dampened if the desire to be accurate is high. Jones and Gerard (1967) argued that there is a basic antinomy between being flexible and open to new information versus being self-protective and closed-minded. Which side of the antinomy one is on depends on the utility of new information for deciding a course of action. The person is open to new information before action is taken, when the information may guide the decision. When the decision is completed, however, they suggest that the person will become defensive and self-protective. Information inconsistent with the person's choice will be rejected. Indeed, at that point, according to Jones and Gerard (1967), contradictory information "can only arouse subjective strain or displeasure [p. 220]."

One implication of the desire to be accurate is that egotism may not occur prior to decisions in which it might frustrate this need. Expecting to continue the activity (e.g., the test, task, or game) for which egotistic attributions might be made may place the person in a predecisional phase in which accurate information is useful. This suggests an additional reason why, in the study by Wortman et al. (1973), subjects who anticipated further testing of social perceptiveness claimed less of that trait. Besides not wanting to get egg on their face if on the subsequent test they fail, they may also have desired an accurate assessment of their ability. Such an assessment may be useful, e.g., it might inform them how hard they will need to try in order to succeed on the forthcoming tests.

However, all forms of egotism were not absent in the Wortman et al. (1973) study among subjects who anticipated further testing. Although they claimed less social perceptiveness than those who thought the testing was over, they also rated their test questions as more difficult. In addition they considered their own questions to be harder than another subject's. Regardless of whether one succeeds or fails, emphasizing difficulty level puts self in a better light. However, expecting further test questions inhibited the more direct form of egotism, a claim of ability.

An experiment on response to evaluators after one fails is also relevant to the present discussion (Jones & Ratner, 1967). The issue in such studies is whether one responds more favorably to evaluators who say nice things about oneself or to those who say negative things, consistent with one's failure. In the Jones and Ratner study, all subjects "failed" a test. Subsequently, they were asked to select a difficulty level for a task to which the ability tested was relevant. Some subjects received feedback from others on the ability and responded to it before deciding on the level of task difficulty. Others decided, then got the feedback, and responded to it. Subjects received positive feedback from one person, negative from another. Subjects who had yet to make a decision were more favorable toward the responses of the negative evaluator. This may indicate appreciation of accurate information. If, however, the decision had already been made, subjects reacted more favorably to the responses of the positive evaluator. Although by no means a study of attributions, the data show that the resolution of the conflict between need for accuracy and self-enhancement can turn on which side of the decision point — and thus the basic antinomy — the person is on.

Self-presentational motives.   Bradley (in press), Gunn (1976), Stephan (1975), and Wortman (1970) each suggest that egotism may fail to occur because the person wishes to appear modest or fears being perceived as egotistical or defensive. Motivation to be modest, they suggest, may result from the subject expecting to be evaluated.

Another self-presentational motive that can conflict with egotism is the desire to portray oneself as a cooperative subject. There may be some reluctance to tell the experimenter that one really did not try very hard and that is why one failed, or to say one succeeded without really trying. The temptation to do so is there because with performance constant, the less effort exerted, the more ability is implied (Heider, 1958). However, it may be undiplomatic to make such a claim to the experimenter.

We have noted a number of potential constraints on egotism. There are conditions that may lower the probability of successfully defending self-esteem such as the plausibility of egotistic attributions and the likelihood of being contradicted by others or by one's own performance. Other motives may often conflict with egotism. These include the desire to be accurate, to appear modest, to avoid looking defensive, and to appear cooperative. We are reluctant to affirm that these factors will always inhibit egotism. These factors variously imply anticipation of the consequences of one's behavior, social sensitivity, tact, and intelligence. To take them into account may require that people be more efficient, more skilled, and — if it is not a contradiction in terms — more rational in making egotistical attributions than they really are. In the heat of the moment after victory or defeat, under the influence of alcohol or other drugs, because of distraction, and so forth, one may ignore factors which should temper egotism.

EVIDENCE FOR EGOTISM

In 1972, we were inspired to collaborate in order to demonstrate egotism, because though our intuitions affirmed its occurrence, studies purporting to show it had an alternative explanation. In several studies the attributions of successful people were compared with those who were unsuccessful. Stronger self-attributions by successful subjects might indicate egotism, but it also might simply illustrate a tendency to make stronger attributions for intentional acts than unintentional ones (Heider, 1958, p. 112; cf. Miller & Ross, 1975). If good outcomes are generally intended and bad ones are not, then taking credit for good outcomes and denying blame for bad ones may simply reflect the influence of intention on attribution, rather than egotism. Ross et al. (1974) point this out with respect to success and failure. One solution to this problem is to compare the attributions of actors and observers, both of whom are likely to believe success is intended (Snyder et al., 1976; Stephan, Rosenfield, & Stephan, 1976).

This solution, however, is inadequate if the study includes only a negative outcome. In that case, egotism predicts stronger personal attributions by an observer than by the actor. But Jones and Nisbett (1971) have proposed that even for neutral acts observers will make stronger attributions to the actor than the actor will. When the outcome is positive, however, Jones and Nisbett's proposal cannot account for the egotistic pattern. Now, the actor makes a stronger personal attribution than the observer. Thus, if we compare actors and observers as a way around the confound of outcome with intention, it is important that we vary the desirability of outcomes.

We had doubts that just any good or bad outcome would provoke egotism, though at the time we had not explicitly stated the two-factor formula connecting outcome to self-esteem by way of attribution. The procedure we used, however, is consistent with that formulation. We were talking in terms of concern for self-esteem or ego involvement. Based on intuition, personal experience, and social comparison theory (Festinger, 1954), we thought that competition was likely to bring out egotism. So we had people compete.

Competition can be understood as promoting the first factor — attribution of the outcome to the person. Festinger (1954) pointed out that competition is the primary way by which people evaluate their abilities. In this regard, one purpose of competition is to make ability attributions. Typically, people compete with the same rules applying to all. As long as the structure of the game is not biased toward one player or the other, outcomes can be readily understood as reflecting ability. Still, in many forms of competition, luck and effort also can contribute to outcomes; so that the loser has a handy excuse.

How much the loser desires to make an excuse may depend on the other factor in our formula — the connection between the attribution made and self-esteem. To strengthen that connection, we told subjects — who were male undergraduates at the University of Texas at Austin — that they would be playing a game that involved abilities important in making business decisions.

We (Snyder et al., 1976) used a matrix game with a general format and rules like the Prisoner's Dilemma game (e.g., Lave, 1965). For those unfamiliar with such games, we will provide a brief description. On each trial one player chose a row of the matrix; the other chose a column. The selection was made blind to the other player's choice. The cell at the intersection of the chosen row and the chosen column indicated the points each player received. The matrices we used had three rows and three columns. Point values were assigned to each of the nine cells of the matrix so that, across the cells, there was a strong negative correlation between the outcome of the row player and the column player. In other words, when one player won points, the other lost points. Unlike the Prisoner's Dilemma, this game provided no opportunity for gaining points by cooperation.

There were three trials on each of three matrices for a total of nine. Those randomly assigned to the winner condition won all but trials 2 and 6. Losers lost all but those two trials. When the game was over and subjects made attributions, each served as both actor (accounting for his own outcome) and observer (accounting for his opponent's outcome). The dependent measures were based on Heider (1958) and Weiner, Frieze, Kukla, Reed, Rest, and Rosenbaum (1971). Each player assessed the contribution of skill, effort, task difficulty, and luck to his own score and to his opponent's.

In accounting for the loser's low score, the loser (in the role of actor) blamed bad luck more and lack of skill less than did his winning opponent (in the role of observer of the loss). In accounting for the winner's high score, the winner (as actor) emphasized the role of skill and downplayed good luck, compared to what the loser (as observer) had to say about it. The evidence clearly supports egotism.

The results cannot be understood in terms of stronger attributions for intended outcomes than unintended ones. Each of the critical comparisons was between the attributions of actors and observers, all of whom we presume believe that the actor intended to win rather than to lose.

Nor can the results be understood as part of a general tendency for an observer to make stronger personal attributions than the actor. When the actor won, the actor made stronger personal attributions than the observer. Jones and Nisbett (1971) were correct in their speculation that the general tendency of which they spoke would be reversed when the actor performs a praiseworthy act.

In sum, Snyder et al. (1976) found evidence for egotism when male subjects competed in a matrix game. The next study to be presented examines whether females also display egotism.

Doubts have been raised about whether females make egotistic attributions (e.g., Deaux, 1976). To explore this issue, Stephan, Rosenfield, and Stephan (1976) used the procedure of Snyder et al. (1976) and had subjects of both sexes compete in all combinations of subject sex and opponent sex. Females did show egotism, but only if the opponent was also a female. Why was egotism limited to female opponents? The social comparison tendencies ordinarily induced by competition may be muted if the opponent is dissimilar (Festinger, 1954). Indeed, Zanna, Goethals, and Hill (1975) have shown that subjects prefer to make social comparisons within their own sex. Thus, female subjects may have regarded the outcome of a game with a male as implying little about their ability. The predilection to attribute outcomes to ability may have been weaker than when the opponent was also female. We are speculating that one of the two factors we posit as necessary for egotistical motivation was weak — the tendency to attribute outcomes to self.

It is also possible that the nature of the game contributed to the pattern of results. Competitive mathematical games are perceived as masculine tasks (Broverman, Vogel, Broverman, Clarkson, & Rosenkrantz, 1972). If female subjects do perceive the game as masculine, then attributions made for outcomes may have had few implications for important aspects of self. That is to say that the other factor postulated as essential to egotistical motivation is weak — the connection between the attribution made and the roots of self-esteem. For this factor to account for results, it is necessary to presume that the sex of the opponent influenced the perception of the nature of the task. Playing against a male opponent may have made females perceive the task as more masculine than playing a female opponent, because competition with a male may make salient other masculine aspects of the situation.

Unlike females, male subjects were egotistic regardless of the sex of their opponents. Why do they show egotism with a female opponent when social comparison tendencies should be weak? Perhaps the prospect of being bested by a female threatens male egos. Males may perceive females as weak opponents, and to lose to a weak adversary may reflect more unfavorably on self than to lose to a strong one.

To sum up, though there are ambiguities in explaining aspects of the data, Stephan et al.'s results are important for at least three reasons:

1. They demonstrate egotism among females.
2. They replicate Snyder et al. (1976).
3. They suggest the importance of both social comparison and the self-esteem relevance of the task for the manifestation of egotism.

In terms of the two-factor formula for egotistical motivation, we would argue that social comparison may enhance the tendency to make attributions from outcomes and that task relevance might strengthen the connection of the attribution to the roots of the person's self-esteem.

Although these results are suggestive, the experiment did not directly manipulate either factor, as its primary purpose was to compare egotism in males and females. We will now examine studies that bear more closely on the two-factor formula.

The Connection Between Attributions and Self-Esteem: Evidence for the Attribution/Self-Esteem Factor

We consider three studies that have manipulated the relevance of personal attributions for self-esteem. The first is by Rosenfield and Stephan (in press).

Rosenfield and Stephan (in press).   We used subjects of both sexes and manipulated whether the task was alleged to be a test of the masculine or feminine personality. We hypothesized that egotism would be greater when the test was said to assess the personality characteristics of one's own gender, because males are more concerned about being masculine and females about being feminine.

The task was a design-matching test in a multiple-choice format. The same task was described in two different ways. The masculine task description emphasized geometric shapes, mathematical abstraction, and the positive correlation with intelligence. It was also said to predict success in business and finance. In contrast, the feminine task description emphasized delicate designs, subtle cues, and correlations with sensitivity to others' feelings and sympathy for their wants and needs.

Subjects made a total of 15 design matches. They were told the average number correct was 8. There was no direct one-on-one competition, though subjects were given feedback on how they did compared to others. After the task was explained to the subject and described as either masculine or feminine, and before actually beginning the task, subjects were asked some questions. A check on the manipulation of subjects' perception of the nature of the task indicated the manipulation had the expected effect. We were curious about the relation of performance expectancy to attributions, so there were two questions on performance expectancy. One question asked how many of the 15 items they expected to get correct; the other asked what percentage of the people working on this task they thought would outscore them. There were also two questions designed to assess ego involvement. One asked, "How important is it for you to do well on this task?" The other inquired, "How much do you value the characteristics that are involved in performing this task?" These two questions are pertinent to the second factor in the formula for egotistic motivation — the connection between attributions and self-esteem.

Subjects in the success condition got 13 of the 15 items correct, missing 1 in the first block of 5 and 1 in the third block of 5. They were told they scored in the top 10% of the people who worked on the task. Failure subjects got 4 matches correct: 1 in the first block, 2 in the second, and 1 in the third. They were informed that they were in the bottom 20% of subjects who worked on the task.

The dependent measures used the same general format as in Snyder et al. (1976) and Stephan et al. (1976). Egotism was defined in terms of relative emphasis on internal factors of skill and effort by those who succeeded and emphasis on the external factors of task difficulty and luck for those who failed. For each subject, attributions to the two internal causes were added together, and from that sum the attributions to the two external causes were subtracted. As predicted, there was a significant triple-order interaction between sex of subject, success/failure, and sex typing of the task. When the task was described as masculine, males attributed success more internally and failure more externally than females. When the same task was described as feminine, females attributed success more internally and failure more externally than males.

In addition to the index of internal minus external attributions, the same triple-order interaction was significant for the separate measures of attribution to skill and attribution to the task. These results were consistent with the pattern for the index. For skill attributions, females tended to be less egotistic than males when the task was described as masculine but more so when the task was said to be feminine. Attributions to the task were as follows: Subjects who succeeded on a task relevant to their gender — as opposed to the other gender — tended to say they were helped less by the ease of the task. This seems to be a subtle, if slightly illogical, way of claiming credit for success.

Although these triple-order interactions are significant, it is difficult to determine just where egotism is occurring. None of the simple effects are statistically significant. However, some of the simple interactions are, and the pattern of responses is just as predicted.

To assess the effects of expectancy and ego involvement on attributions, the two measures of each, taken before the subject began the task, were transformed to z scores and combined to give a single expectancy and a single ego involvement index. For both males and females, ego involvement and expectancy of success were greater when the task was relevant to the subject's own sex. Moreover, these indices correlated with the attributional index of internal causes minus external ones, albeit ego involvement much more strongly than expectancy.

One way to test the extent of the contributions of ego involvement and expectancy to the obtained results is to perform an analysis of covariance with the indices of those two factors as covariates. To the extent those two factors are responsible for the results, eliminating them statistically with a covariance procedure should weaken the results. It did. The critical three-way interaction between subject sex, task description, and outcome was no longer significant.

Further investigation revealed that across cells, 48% of the variance is accounted for by the two covariates, 6% uniquely by expectancy, 38% uniquely by ego involvement, and 4% was accounted for by expectancy and ego involvement jointly. It would appear that ego involvement and to a lesser extent, expectancy, account for the effects of task description on the egotism displayed by each sex. We have more to say about expectancy later on. This internal analysis supports our interpretation of the effects of task description as largely due to its influence on the strength of the second factor in our formula — the connection between attributions for outcomes and self-esteem.

Nicholls (1975).   Nicholls (1975) included a variable in his study that pertains to the attribution/self-esteem factor. He gave fourth graders an angle-matching task and manipulated the value of attaining a high score. In one case the task was said to measure future intelligence. In the other, it simply measured ability to make accurate size judgments at a distance. On the presumption that perceived intelligence has a large impact on self-esteem, when the test is alleged to predict intelligence, threat to self-esteem and, therefore, egotism should be enhanced.

The test description manipulation interacted with success and failure feedback to determine ability attributions. When the task measured intelligence rather than size perception, successful subjects tended to make stronger ability attributions, and failure subjects made weaker ones. As in Rosenfield and Stephan's (in press) study, attributions were more egotistical, the stronger their connection to self-esteem. In one case, the connection is stronger when the task is pertinent to the subject's gender. In the other case, the connection is stronger when the task measures intelligence. Differences between studies in procedures, manipulations, and subjects give us some confidence in the external validity of the findings.

Both studies, however, share a feature that does permit a class of alternative interpretations. In both, the manipulation of ego involvement — the attribution/self-esteem factor — occurred before the subject began the task. And in both, feedback was given as the task proceeded. Thus, in both it is possible that the manipulations had an effect on effort or some other aspect of performance. The next study eliminates this possibility.

Miller (1976).   Miller (1976) eliminated differences in effort or any other feature of performance as an explanation for the impact of ego involvement on egotism. Miller did so by manipulating both ego involvement and outcome (success versus failure) after subjects had completed the task. The task was a multiple-choice test of social perceptiveness. Subjects answered questions based on case studies. Then they were told either that the test was well validated as a measure of "intelligence, personal and marital happiness, and job satisfaction [p. 902] " or that the test was new and its validity "was still very much in doubt [p. 902]." Subjects thereafter received false success or failure feedback, indicating either that they had answered 24 of the 30 questions right and scored at the 80th percentile or that they had answered only 8 right and scored at the 20th percentile.

Subjects who failed thought performance reflected ability to a lesser degree if validity was described as high rather than low. But those who succeeded judged performance to reflect ability more if the experimenter said the test's validity had been established. A similar interaction occurred on attributions to luck. Test validity made no difference for successful subjects, whereas among failing subjects those in the high-validity condition attributed outcomes more to luck. Failure subjects also claimed to exert less effort if they were in the high-validity condition. This is remarkable in that ego involvement and feedback were manipulated after the subject had finished the test. Miller's study thereby eliminated any confounding of those variables with performance. Thus, differencies in performance cannot account for the effect of the two variables on attributions.

Taking together Miller's results with those of Nicholls (1975) and Rosenfield and Stephan (in press) we have evidence that egotistic motivation is enhanced by strengthening the attribution/self-esteem factor.

The Connection Between Outcomes and Attributions: Evidence for the Outcome/Attribution Factor

Having cited support for the attribution/self-esteem factor, we now will consider two studies that provide support for the role of the other factor we hypothesized to be necessary for the arousal of motivation for egotism — the connection between outcome and attribution. In the case of failure, the threat to self-esteem will be greater to the extent that the subject perceives the outcome as attributable to self.

Stevens and Jones (1976). Stevens and Jones (1976) manipulated the distinctiveness (Kelley, 1967) of failure. In some cases, subjects failed only on a single task yet succeeded on others, so that failure was distinctive to a particular task. Other times subjects failed at all tasks; now failure was not distinctive to just one of them. In Kelley's (1967) scheme, low distinctiveness implies that the attribution should be made not to the task but to the person, as the person produces the same outcome across several tasks. However, subjects for whom failure was not distinctive to a single task were unwilling to make stronger attributions to ability than those who failed on only one task. Instead, they made a stronger attribution to luck. This is exactly the opposite of what subjects should say when distinctiveness is low. Such attributions seem to be motivated by the desire to make as weak as possible a self-attribution when the pattern of outcomes points the blame at self.

Federoff and Harvey (1976). Federoff and Harvey's (1976) study also can be understood as manipulating the outcome/attribution factor. They varied objective self-awareness (Duval & Wicklund, 1972; Wicklund, 1975). Subjects administered therapy to a (fictitious) phobic patient. Subjects in the self-aware condition were told the session was being videotaped and a camera was focused on them. This is one of several procedures used to make subjects self-conscious (cf., Wicklund, 1975). Others include actually playing back a videotape of the subject, forcing the subject to look in a mirror, playing back an audio tape of the subject's voice, and so forth. There is evidence that such procedures do enhance the tendency to make attributions to self (Duval & Hensley, 1976; Storms, 1973). Thus, we might expect that when the therapy fails to ameliorate the patient's phobia, subjects who are self-aware will feel more of a threat to self-esteem. This appears to be the case.

Self-awareness strengthened attribution to self when the therapy was successful, but it weakened attribution to self when it was unsuccessful. Self-awareness seems to mobilize subjects to deny responsibility for the failure of therapy. Other explanations for these results are possible, and we will consider a somewhat different interpretation below, but they can be understood as support for the hypothesized role of the outcome/attribution factor — stronger motivation for egotism, the more failure appears to be attributable to self.

There are some data that reflect indirectly on the outcome/attribution factor, and furthermore, they point to an alternative explanation for learned-helplessness effects in terms of egotism. The essential idea of learned helplessness is that experience with lack of control leads one to give up, even in situations where control may be possible, because one learns that outcomes are independent of what one does (Seligman, 1975). Most studies of helplessness in humans confound lack of control with failure at a task and control with success at the task. If people tend to give up on a task following a failure experience, it may result from a desire to avoid a further blow to self-esteem rather than from feeling helpless. Not to try and to fail provides a ready-made excuse for failure, whereas to try and still fail makes the failure much more attributable to lack of ability (Heider, 1958).

Frankel (1977) tested the idea that a decline in performance after experiencing no control (i.e., failure) may result from the threat of further failure. He reasoned that a way to eliminate that threat is to describe the task as very difficult. Now failure is attributable to the task instead of self, and trying poses no threat to self-esteem. He found, just as he predicted, that describing a task as very difficult significantly improved the performance of subjects who were "helpless" (had just failed). In contrast, describing the same task as very difficult to success subjects nonsignificantly impaired their performance.

These data are exciting because they suggest that much of the human learned-helplessness literature can be understood in terms of egotism and failure threat rather than simply expectations of no control. We must admit, though, that the data are only indirect support for the outcome/attribution factor, because performance is measured rather than attributions.

The two studies that reflect more directly on the outcome/attribution factor find that people downplay their own personal contribution to failure more in the presence of a variable that ordinarily enhances attribution to the person — self-awareness in Federoff and Harvey (1976), low distinctiveness (Kelly, 1967) in Stevens and Jones (1976). There is also support for the attribution/self-esteem factor. At this point, there is no evidence that clearly bears on whether both are necessary. All that can be said for now is that, at least under some conditions, strengthening one of them increases egotistic attributions.

Some Observations on the Egotism Literature

The sophisticated reader realizes that the collection of studies cited so far as evidence for egotism is only the tip of the iceberg. Are we leaving other studies submerged because they leave us cold? No, we find many of them provocative, if not definitive. Results differ from one study to the next, and the phenomenon of egotism falls somewhere between being elusive and pervasive. We believe that much of the inconsistency in studies of egotism can be understood in terms of the variables discussed in the beginning of this chapter: the motivation for egotism as a function of the outcome/attribution factor and the attribution/self-esteem factor, the probability that egotism will successfully protect or enhance self-esteem, and the presence of conflicting motives (cf. Bradley, in press; Gunn, 1976; Stephan, 1975; Wortman, 1970). Another variable worthy of consideration is how interpersonal relationships may influence egotism.

Interpersonal relationships. Does the individual in a group of peers working toward a common end perceive self as less responsible for failure than they are? The results are mixed, but there appears to be a meaningful pattern. When subjects communicate with one another, there is no egotism (Caine & Schlenker, 1976; Schlenker & Miller, 1976, in press; Schlenker, Soraci, & McCarthy, in press). However, when subjects do not have the opportunity to communicate, but rather their separate responses are combined to determine their common outcome, egotism occurs in four studies (Schlenker, 1975; Schlenker & Miller, 1976; Stephan, Presser, Kennedy, & Aronson, in press; Wolosin, Sherman, & Till, 1973, Experiment 2) though not in a fifth (Forsyth & Schlenker, 1976).

Communication might weaken individually oriented egotism in a number of ways. It might increase diffusion of responsibility (cf. Latané & Darley, 1970) and thus have weakened the outcome/attribution factor. Or, it might have increased liking for other members of the group, perhaps by making them more familiar (Zajonc, 1968). Consequently, subjects may have made attributions that were egotistic on behalf of the group as well as themselves. A study by Regan, Straus, and Fazio (1974) supports this idea. They manipulated liking for another and had subjects watch that person succeed or fail. Liking for the target person produced attributions that were favorable to him, in other words, attributions that were egotistic from his perspective. Also consistent with this train of thought is a line of research on attribution and ethnocentrism. Attributions are more favorable toward members of the in-group (Mann & Taylor, 1974; Stephan, 1977; Taylor & Jaggi, 1974). It may be because members of the in-group are liked more than out-group members.

When is the individual the wrong unit of analysis?   We are reminded that a decision to consider the individual the unit of analysis is somewhat arbitrary. Individuals may identify with others or sympathize with their interests. Sometimes the individual may be too small a unit. On the other hand, when we argue for consideration of the attribution/self-esteem factor, we are in a sense saying the individual is too large a unit. Self-esteem is based only on certain aspects of self. Attributions about other aspects may not threaten self-esteem.

There is evidence that the unit of analysis that people use is influenced by egotistic motivation. Cialdini, Borden, Thorne, Walker, Freeman, and Sloan (1976) found signs of stronger identification with one's school when the football team won than when it lost. Indices of identification were the wearing of apparel with the school name and the use of *we* in describing the game's outcome. This effect seems to be enhanced by a threat to self-esteem, as it was significantly stronger when subjects were first asked to describe a "nonvictory" and then a victory than when the order of questions was reversed. These data can be understood as an attempt to modulate attributions of responsibility made on a primitive basis – mere association of person and event (Heider, 1958).

A resolution of the unit issue might also help clarify results in studies of cooperation among unequals. In many studies, the subject is asked to teach a student or to treat a patient. (e.g. Beckman, 1973; Harvey et al., 1975). More often than not, egotism occurs, but we might be in a better position to predict when it will if we knew whether subjects identified with, or felt any affection for, the student or patient.

THEORETICAL PERSPECTIVES ON EGOTISM

There are at least three orientations that provide perspectives on the idea of egotism: control motivation, cognitive dissonance theory/balance theory, and objective self-awareness theory.

Control Motivation

Wortman (1976) has considered the proposition that the desire to feel in control of one's environment may influence attributions. In the case of success, egotism and this motivation pull in the same direction — toward self-attribution. To take credit implies one could produce the good outcome again. They also both agree that failure may be accounted for by low effort. If so, failure reflects little on ability, which satisfies egotistic desires, and the implication is that greater effort would bring success, which gratifies the need to feel in control (cf. Dweck, 1975). However, to ascribe failure to chance or task difficulty is not only to excuse oneself but also to say that what happened was beyond one's control. Control and egotistic motivation, therefore, part company on the roles of chance and task difficulty in accounting for failure.

Circumstances may determine whether it is more important to feel in control or to feel blameless. In the case of a severe negative outcome the person may be less concerned with preserving self-esteem than with maintaining a sense of control over that particular outcome. This may explain self-blame on the part of parents of leukemia victims (Chodoff, Friedman, & Hamburg, 1964). A patient may blame self with the naive, unspoken hope that the outcome can be reversed, that the act can be undone, that restitution can be made.

When the particular negative outcome is not especially important in and of itself, there is a case to be made that control motivation, as well as egotism, predicts attributions to luck. But control motivation is satisfied only if one convinces oneself that an unusual run of bad luck was the cause of the negative outcome and that with the average luck one can ordinarily expect, success is within one's grasp. That sort of luck attribution is consistent with control motivation, as opposed to the conclusion that outcomes are determined generally by chance events.

Cognitive Dissonance Theory and Balance Theory

There are parallels between the factors contributing to egotistic motivation and factors contributing to cognitive dissonance. We will discuss these parallel factors and then ask whether they are sufficiently strong to consider egotism a special case of dissonance reduction.

Festinger (1957) said that two elements are dissonant if "one does not, or would not be expected to, follow from the other [p. 15]." Intuitively, failure seems dissonant and success consonant. At least, it seems safe to say that failure is typically more dissonant than success. But can we specify the element with which failure is dissonant and success is consonant?

Heider's (1958) balance theory can help us do so, as it also concerns feelings and beliefs that follow psychologically from one another. If a person has a unit relationship with something, meaning that something belongs to or is otherwise associated with him or her, then one tends to like that something and vice versa. That is, liking something tends to bring about owning it or associating with it. By definition, people like positive outcomes and dislike negative ones. Therefore, one will tend to perceive positive outcomes as belonging to self but not negative ones. This corresponds to the definition of egotism we offered at the beginning of this chapter as taking credit for good outcomes and denying blame for bad ones.

However, Heider (1958, p. 210) noted this follows from balance theory only if one presupposes a positive attitude toward self. If balance theory is to account for egotism, then people who dislike themselves should fail to be egotistic. There is some evidence for this. People with low self-esteem presumably like themselves less than do highs. Lows display less egotism than highs (Feather, 1969; Fitch, 1970). This finding holds with statistical controls for expectations (Bussanich, 1976; Rosenfield & Stephan, 1976). The answer we derive from balance theory to the question of with what is failure dissonant and with what is success consonant is liking for self. One can understand this as applying to self the sense of justice we spoke of early in the chapter. In a just world, failure implies that one deserves to fail and is therefore unworthy.

Having used balance theory to elaborate on how failure might be dissonant, we can pursue the parallels between egotistical motivation and cognitive dissonance. As we do this, we will presume the person likes him- or herself. One of the factors we hypothesized to be necessary to arouse egotistic motivation for a negative outcome is the attribution of that outcome to the person. With a presumption of liking for the self, a negative outcome is a dissonance-arousing one. Wicklund and Brehm (1976) reviewed the dissonance literature and noted that the major development in the theory has been the recognition of the role of the individual's feeling responsible for the dissonance-arousing event. The person must perceive a "causal-link [p. 70]" between self and that event. This statement differs in no important way from the description of the outcome/attribution factor.

Is there also a parallel for the attribution/self-esteem factor? There is. To restate the factor: The attribution made must be relevant to self-esteem. Festinger (1957) said that elements must be relevant to one another in order to be either dissonant or consonant. Of course if the attribution made for failure is relevant to self-esteem, it is very likely to be dissonant with self-esteem, assuming one likes him- or herself. We conclude that failure is a source of dissonance if: (1) failure is attributed to self; (2) the attribution is inconsistent with liking self; and (3) the person likes self.

Why is egotism not ordinarily discussed in terms of cognitive dissonance? The answer may be that it is only in recent years that responsibility has supplanted volition (Brehm & Cohen, 1962) as a necessary condition for dissonance (Wicklund & Brehm, 1976). When choice was the central concept, it did not seem to apply to failure. One does not choose to fail. One's choices may, of course, lead to failure, and consequently, one may feel responsible.

We find it of interest that both of the factors in the formula for egotistic motivation have parallel concepts in cognitive disssonance theory. Indeed the parallels are so close that we are tempted to regard egotism as a special case of cognitive dissonance. At a procedural level studies of egotism differ from the typical dissonance paradigms. We have defined egotism in the face of failure as denial of responsibility, a mode of dissonance reduction that is explicitly precluded in most dissonance research so that subjects will use another mode such as attitude change. Despite this procedural difference, one can understand motivation for egotism as a specific manifestation of cognitive dissonance.

Where does this lead us? Can dissonance theory suggest new and interesting predictions for egotism? There are many possibilities worth exploring. We will discuss sequential effects, because not only do they follow from dissonance theory results but also because we find them intuitively appealing.

Festinger (1964) suggested that for a period of time before dissonance reduction, the person may focus on the dissonant elements. The person experiences the tension state of dissonance as he or she selects among ways of reducing it. The implication for egotism is that it may take time. It may show up more clearly after a short delay.

There is also evidence that dissonance reduction may sometimes only be temporary (Walster, 1964). Wicklund and Brehm (1976) suggested that attitudes may revert as the dissonance-arousing cognitions lose salience. Perhaps the same thing happens in the case of egotism. When one loses a game of bridge, cards are a matter of luck. If one wins on the next occasion, the previous loss may be forgotten, and now skill is the critical factor. Egotism may be abandoned when emotions are cooler and the outcome at issue looms less large and other matters capture one's attention.

In sum, thinking about egotism in terms of dissonance theory suggests an investigation of sequential effects with egotism taking time to develop and then after a while dissipating unless the dissonance-arousing cognitions remain salient (Wicklund & Brehm, 1976).

Can dissonance theory account for egotism with respect to success? The answer is yes, if one argues that taking credit for success is consonant with liking self. A different position is possible. Aronson (1969) has argued that the aspect of self with which failure is dissonant is positive expectancies about one's performance. This version of dissonance theory predicts that success that exceeds expectations will be rejected. It cannot account for egotism when the person performs better than expected.

It is entirely possible that failure is dissonant with both liking for self and with high expectations for performance. There is no a priori way to decide which causes more dissonance. What is clear is that attempts to relate egotism to expected performance have not been stunningly successful. In Stephan et al.'s (1976) study, which had male and female subjects compete with subjects of both sexes, subjects were asked how many moves out of nine they expected to win. Expectancies were unrelated to attributions, and an analysis of covariance using expectancy as a covariate had no impact on the significance levels of the findings. We reported above that Rosenfield and Stephan (in press) did find, in an internal analysis, that expectancy played a role in subject's attributions but a much weaker one than ego involvement. An account of egotism using an expectancy version of dissonance theory and, for that matter, any other way to account for egotism in terms of expectancy (see Brewer, 1977; Miller & Ross, 1975) is troubled by these data.

Objective Self-Awareness Theory

We said above that self-awareness may increase egotistic motivation in the case of failure. This may happen because making self salient increases the tendency to attribute outcomes to self. Self-esteem is threatened, and the person denies responsibility. Self-awareness theory suggests a different mechanism (cf. Wicklund, 1975). Self-awareness promotes discrepancy reduction in general and reduction of cognitive dissonance in particular (Wicklund & Duval, 1971, Experiment II; Insko, Worchel, Songer & Arnold, 1973). Thus, the finding that self-awareness enhances egotism (Federoff & Harvey, 1976) is consistent with the dissonance explanation for egotism.

Self-awareness theory suggests a new element with which failure is discrepant. In addition to high expectations for performance and liking for self, failure is discrepant from the ideal self. Thus, according to self-awareness theory, someone who did not like him- or herself, but still held success as an ideal, might be more egotistic when self-aware.

How does self-awareness facilitate discrepancy reduction? It leads the person to focus on the discrepancy, which then looms larger and creates more negative affect. This negative affect motivates discrepancy reduction. When one fails, self-awareness may make the failure seem worse and the person feel worse unless a plausible excuse is found.

Most experiments manipulate self-awareness by informing the subject he or she is being videotaped, confronting the subject with a mirror image, and so forth. The relevance of self-awareness theory to egotism may be questioned on grounds that video cameras, mirrors, and other self-awareness props are not often in evidence. But self-awareness theory is relevant for several reasons. First, the presence of others may enhance self-awareness (Duval & Wicklund, 1972) and thus also egotistic motivation. Secondly, Jones and Gerard (1967) argued that awareness of the self is provoked by potential and actual inconsistency. Failure may produce self-awareness even in the absence of props or audiences.

Finally, the tendency to attribute outcomes to self when self is salient (Duval & Hensley, 1976; Storms, 1973) suggests a mechanism for how egotism works. Surely the environment influences the focus of attention, but just as surely attention is at least partially under voluntary control, including attention to self (Shaver, 1976). Perhaps egotism occurs because success leads one to pay more attention to self, whereas failure, after initially provoking self-awareness, leads to avoidance of self-focus (cf. Gibbons & Wicklund, 1976).

SUMMARY

We defined egotism as taking credit for good outcomes and denying blame for bad ones in order to enhance or preserve self-esteem. We distinguished it from defensive attribution, rationalization, and egocentricity. Two factors were hypothesized as necessary to arouse egotistic motivation when one fails. There must be some tendency to attribute the outcome to self. And that attribution must be relevant to self-esteem. There is some evidence for both these factors. When one succeeds, we posited that one will be motivated to attribute the outcome to self to the extent that doing so is relevant to self-esteem and thus enhances it.

Whether people actually make egotistic attributions when motivated to do so depends on the probability of successfully defending or enhancing self-esteem and the presence of conflicting motives. Another complicating factor is the person's identification with, and affection for, others.

We considered egotism from other theoretical perspectives. We noted that egotism motivation and the motive to perceive self as in control make many of the same predictions. They differ in that control motivation suggests people will not attribute negative outcomes to high task difficulty or bad luck.

Dissonance theory, we noted, has factors that closely parallel the two we suggested as necessary for egotistic motivation. Other parallels are worth exploring. It may be useful to consider egotism as a form of dissonance reduction.

Self-awareness theory points out the role that salience of self may play in motivating the reduction of negative discrepancies. Egotism can be conceived of as discrepancy reduction.

ACKNOWLEDGMENTS

In addition to thanking the editors, John Harvey, William Ickes, and Robert Kidd, we would like to express our appreciation to Amanda Merrill, William N. Morris, and Robert A. Wicklund for valuable comments and suggestions.

REFERENCES

Adler, A. *The individual psychology of Alfred Adler* (H. L. Ansbacher & R. R. Ansbacher, Eds.). New York: Harper & Row, 1956.

Aronson, E. The theory of cognitive dissonance: A current perspective. In L. Berkowitz (Ed.), *Advances in experimental social psychology* (Vol. 4). New York: Academic Press, 1969.

Beckman, L. Teachers' and observers' perceptions of causality for a child's performance. *Journal of Educational Psychology*, 1973, *65*, 198–204.

Bradley, G. W. Self-serving biases in the attribution process: A re-examination of the fact or fiction question. *Journal of Personality and Social Psychology*, in press.

Brehm, J. W., & Cohen, A. R. *Explorations in cognitive dissonance*. New York: Wiley, 1962.

Brewer, M. B. An information-processing approach to attribution of responsibility. *Journal of Experimental Social Psychology*, 1977, *13*, 58–69.

Broverman, J. K., Vogel, S. R., Broverman, D., Clarkson, F. E., & Rosenkrantz, P. S. Sex-role stereotypes: Current appraisal. *Journal of Social Issues*, 1972, *28*, 59–78.

Bussanich, J. *The influence of locus of control, self-esteem and success-failure feedback on attribution of responsibility*. Unpublished doctoral dissertation, University of Texas at Austin, 1976.

Caine, B. T., & Schlenker, B. R. *Role position and group performance as determinants of egocentric perceptions in cooperative groups*. Unpublished manuscript, University of Florida, 1976.

Chodoff, P., Friedman, S., & Hamburg, D. Stress defenses and coping behavior: Observations in parents of children with malignant disease. *American Journal of Psychiatry*, 1964, *120*, 743–749.

Cialdini, R. B., Borden, R. J., Thorne, A., Walker, M. R., Freeman, S., & Sloan, L. R. Basking in reflected glory: Three (football) field studies. *Journal of Personality and Social Psychology*, 1976, *34*, 366–375.

Cialdini, R B., Levy, A., Herman, P., & Evenbeck, S. Attitudinal politics: The strategy of moderation. *Journal of Personality and Social Psychology*, 1973, *25*, 100–108.

Deaux, K. Sex: A perspective on the attribution process. In J. H. Harvey, W. J. Ickes, & R. F. Kidd (Eds.), *New directions in attribution research* (Vol. 1). Hillsdale, N.J.: Lawrence Erlbaum Associates, 1976.

Duval, S., & Hensley, V. Extensions of objective self-awareness theory: The focus of attention-causal attribution hypothesis. In J. H. Harvey, W. J. Ickes, & R. F. Kidd (Eds.), *New directions in attribution research* (Vol. 1). Hillsdale, N.J.: Lawrence Erlbaum Associates, 1976.

Duval, S., & Wicklund, R. A. *A theory of objective self awareness*. New York: Academic Press, 1972.

Dweck, C. S. The role of expectations and attributions in the alleviation of learned help-lessness. *Journal of Personality and Social Psychology*, 1975, *31*, 674–685.

Feather, N. T. Attribution of responsibility and valence of success and failure in relation to initial confidence and task performance. *Journal of Personality and Social Psychology*, 1969, *13*, 129–144.

Feather, N. T., & Simon, J. G. Attribution of responsibility and valence of outcome in relation to initial confidence and success and failure of self and other. *Journal of Personality and Social Psychology*, 1971, *18*, 173–188.

Federoff, N. A., & Harvey, J. H. Focus of attention, self-esteem, and the attribution of causality. *Journal of Research in Personality*, 1976, *10*, 336–345.

Festinger, L. A theory of social comparison processes. *Human Relations*, 1954, *7*, 117–140.

Festinger, L. *A theory of cognitive dissonance*. Stanford, Calif.: Stanford University Press, 1957.

Festinger, L. *Conflict, decision and dissonance*. Stanford, Calif.: Stanford University Press, 1964.

Fitch, G. The effects of self-esteem, perceived performance, and choice on causal attributions. *Journal of Personality and Social Psychology*, 1970, *16*, 311–315.

Forsyth, D. R., & Schlenker, B. R. *Attributing the causes of group performance: Effects of performance quality, task importance, and future testing*. Unpublished manuscript, University of Florida, 1976.

Frankel, A. Unpublished doctoral dissertation, Dartmouth College, 1977.

Gibbons, F. X., & Wicklund, R. A. Selective exposure to self. *Journal of Research in Personality*, 1976, *10*, 98–106.

Gunn, S. P. *Motivational bias and self-attribution.* Unpublished manuscript, Florida State University, 1976.

Harvey, J. H., Arkin, R. M., Gleason, J. M., & Johnston, S. Effect of expected and observed outcome of an action on the differential causal attributions of actor and observer. *Journal of Personality,* 1974, *42,* 62–77.

Harvey, J. H., Harris, B., & Barnes, R. D. Actor–Observer differences in the perceptions of responsibility and freedom. *Journal of Personality and Social Psychology,* 1975, *32,* 22–28.

Hass, R. G. Persuasion or moderation? Two experiments on anticipatory belief change. *Journal of Personality and Social Psychology,* 1975, *31,* 1155–1162.

Hass, R. G., & Mann, R. W. Anticipatory belief change: Persuasion or impression management? *Journal of Personality and Social Psychology,* 1976, *34,* 105–111.

Heider, F. Social perception and phenomenal causality. *Psychological Review,* 1944, *51,* 358–374.

Heider, F. *The psychology of interpersonal relations.* New York: Wiley, 1958.

Insko, C. A., Worchel, S., Songer, E., & Arnold, S. E. Effort, objective self-awareness, choice, and dissonance. *Journal of Personality and Social Psychology,* 1973, *28,* 262–269.

Jones, E. Rationalization in everyday life. *Journal of Abnormal Psychology,* 1908, *3,* 161–169.

Jones, E. E., & Gerard, H. B. *Foundations of social psychology.* New York: Wiley, 1967.

Jones, E. E., & Nisbett, R. E. The actor and the observer: Divergent perceptions of the causes of behavior. In E. E. Jones, D. Kanouse, H. H. Kelley, R. E. Nisbett, S. Valins, & B. Weiner (Eds.), *Attribution: Perceiving the causes of behavior.* Morristown, N.J.: General Learning Press, 1971.

Jones, E. E., Rock, L., Shaver, K. G., Goethals, G. R., & Ward, L. W. Pattern of performance and ability attribution: An unexpected primacy effect. *Journal of Personality and Social Psychology,* 1968, *10,* 317–340.

Jones, E. E., & Wortman, C. *Ingratiation: An attributional approach.* Morristown, N.J.: General Learning Press, 1973.

Jones, S. C., & Ratner, C. Commitment to self-appraisal and interpersonal evaluations. *Journal of Personality and Social Psychology,* 1967, *6,* 442–447.

Kelley, H. H. Attribution theory in social psychology. In D. Levine (Ed.), *Nebraska Symposium on Motivation,* 1967. Lincoln, Neb.: University of Nebraska Press, 1967.

Latané, B., & Darley, J. M. *The unresponsive bystander: Why doesn't he help?* New York: Appleton–Century–Crofts, 1970.

Lave, L. B. Factors affecting co-operation in the prisoner's dilemma. *Behavioral Science,* 1965, *10,* 26–38.

Lerner, M. J., Miller, D. T., & Holmes, J. G. Deserving and the emergence of forms of justice. In L. Berkowitz & E. Walster (Eds.), *Advances in experimental social psychology* (Vol. 9). New York: Academic Press, 1976.

Mann, J. F., & Taylor, D. M. Attribution of causality: Role of ethnicity and social class. *Journal of Social Psychology,* 1974, *94,* 3–13.

Miller, D. T. Ego involvement and attributions for success and failure. *Journal of Personality and Social Psychology,* 1976, *34,* 901–906.

Miller, D. T., & Ross, M. Self-serving biases in the attribution of causality: Fact or fiction? *Psychological Bulletin,* 1975, *82,* 213–225.

Nicholls, J. G. Causal attributions and other achievement-related cognitions: Effects of task outcome, attainment value, and sex. *Journal of Personality and Social Psychology,* 1975, *31,* 379–389.

Regan, D. T., Straus, E., & Fazio, R. Liking and the attribution process. *Journal of Experimental Social Psychology,* 1974, *10,* 385–397.

Rosenfield, D., & Stephan, W. G. *The relationship between self-esteem, achievement motivation and internal–external locus of control on attributions of causality.* Unpublished manuscript, 1976. (Available from David Rosenfield, Psychology Department, Southern Methodist University, Dallas, Texas 75275.)

Rosenfield, D., & Stephan, W. G. Sex differences in attributions for sex-typed tasks. *Journal of Personality,* in press.

Ross, L., Bierbrauer, G., & Polly, S. Attribution of educational outcomes by professional and nonprofessional instructors. *Journal of Personality and Social Psychology,* 1974, *29,* 609–618.

Schlenker, B. R. Group member's attributions of responsibility for prior group performance. *Representative Research in Social Psychology,* 1975, *6,* 96–108.

Schlenker, B. R., & Miller, R. S. *Group dissent and group performance as determinants of egocentric perceptions.* Unpublished manuscript, University of Florida, 1976.

Schlenker, B. R., & Miller, R. S. Group cohesiveness as a determinant of egocentric perceptions in cooperative groups. *Human Relations,* in press.

Schlenker, B. R., Soraci, S., & McCarthy, B. Self-esteem and group performance as determinants of egocentric perceptions in cooperative groups. *Human Relations,* in press.

Seligman, M. E. P. *Helplessness.* San Francisco: Freeman, 1975.

Shaver, K. G. Defensive attribution: Effects of severity and relevance on the responsibility assigned for an accident. *Journal of Personality and Social Psychology,* 1970, *14,* 101–113.

Shaver, P. *Self-awareness theory: Problems and prospects.* Paper presented at the meeting of the New England Social Psychological Association, Hanover, New Hampshire, October 9, 1976.

Snyder, M. L., & Frankel, A. Observer bias: A stringent test of behavior engulfing the field. *Journal of Personality and Social Psychology,* 1976, *34,* 857–864.

Snyder, M. L., Stephan, W. G., & Rosenfield, D. Egotism and attribution. *Journal of Personality and Social Psychology,* 1976, *33,* 435–441.

Stephan, C., Presser, N. R., Kennedy, J. F., & Aronson, E. Attributions to success and failure in cooperative, competitive and interdependent interaction. *European Journal of Social Psychology,* in press.

Stephan, W. G. Actor vs. observer: Attributions to behavior with positive and negative outcomes and empathy for the other role. *Journal of Experimental Social Psychology,* 1975, *11,* 205–214.

Stephan, W. G. Stereotyping: The role of ingroup–outgroup differences in causal attributions for behavior. *Journal of Social Psychology,* 1977, *101,* 255–266.

Stephan, W. G., Rosenfield, D., & Stephan, C. Egotism in males and females. *Journal of Personality and Social Psychology,* 1976, *34,* 1161–1167.

Stevens, L., & Jones, E. E. Defensive attribution and the Kelley cube. *Journal of Personality and Social Psychology,* 1976, *34,* 809–820.

Storms, M. D. Videotape and the attribution process: Reversing actors' and observers' points of view. *Journal of Personality and Social Psychology,* 1973, *27,* 165–175.

Taylor, D. M., & Jaggi, V. Ethnocentrism and causal attribution in a south Indian context. *Journal of Cross-Cultural Psychology,* 1974, *5,* 162–171.

Walster, E. The temporal sequence of post-decision processes. In L. Festinger (Ed.), *Conflict, decision and dissonance.* Stanford, Calif.: Stanford University Press, 1964.

Weiner, B., Frieze, I., Kukla, A., Reed, L., Rest, S., & Rosenbaum, R. M. Perceiving the causes of success and failure. In E. E. Jones, D. Kanouse, H. H. Kelley, R. E. Nisbett, S. Valins, & B. Weiner (Eds.), *Attribution: Perceiving the causes of behavior.* Morristown, N.J.: General Learning Press, 1971.

Wells, G. *Anticipated discussion of interpretation eliminates actor–observer differences in the attribution of causality.* Unpublished manuscript, Ohio State University, 1977.

Wicklund, R. A. Objective self-awareness. In L. Berkowitz (Ed.), *Advances in experimental social psychology* (Vol. 8). New York: Academic Press, 1975.

Wicklund, R. A., & Brehm, J. W. *Perspectives on cognitive dissonance.* Hillsdale, N.J.: Lawrence Erlbaum Associates, 1976.

Wicklund, R. A., & Duval, S. Opinion change and performance facilitation as a result of objective self-awareness. *Journal of Experimental Social Psychology,* 1971, *7,* 319–342.

Wolosin, R. J., Sherman, S. J., & Till, A. Effects of cooperation and competition on responsibility attribution after success and failure. *Journal of Experimental Social Psychology,* 1973, *9,* 220–235.

Wortman, C. B. *A theory of defensive attribution.* Unpublished manuscript, Duke University, 1970.

Wortman, C. B. Causal attributions and personal control. In J. H. Harvey, W. J. Ickes, & R. F. Kidd (Eds.), *New directions in attribution research* (Vol. 1). Hillsdale, N.J.: Lawrence Erlbaum, Associates, 1976.

Wortman, C. B., Costanzo, P. R., & Witt, T. R. Effect of anticipated performance on the attributions of causality to self and others. *Journal of Personality and Social Psychology,* 1973, *27,* 372–381.

Zajonc, R. B. Attributional effects of mere exposure. *Journal of Personality and Social Psychology Monograph Supplement,* 1968, *9,* 1–27.

Zanna, M. P., Goethals, G. R., & Hill, J. F. Evaluating a sex-related ability: Social comparison with similar others and standard setters. *Journal of Experimental Social Psychology,* 1975, *11,* 86–93.

5 Attributional Styles

William Ickes
Mary Anne Layden
University of Wisconsin

As a way of introducing the ideas to be discussed in this chapter, we invite the reader to consider the following letter to "Dear Abby." The letter appeared in a local newspaper under the headline, *She can't see beyond her nose:*

Dear Abby:

I am a 34-year-old woman who has divorced three husbands. (Not my fault. I always picked losers.)

My problem is my nose. I had plastic surgery on it when I was 18, and the doctor botched the job, so at 21 I had it reshaped and then it was worse. I think it makes me look stuck up and keeps me from making friends.

I went to a well-known local plastic surgeon, and I offered to pay him in full in advance but he refused to take me as a patient! He said he didn't think any plastic surgeon could please me because I had "emotional and social problems" I should face up to instead of blaming everything on my nose. Then he insulted me further by suggesting that I use my money to see a psychiatrist!

Abby, there is nothing wrong with my mind. It's my nose! Will you please recommend a good plastic surgeon? I can afford to go anywhere.

Determined in Hartford

(published in The Capital Times, Madison, Wisconsin, September 29, 1976.)

119

This letter is a rather remarkable statement, for in a few brief paragraphs it manages to convey the essence of one person's attributions about some of the major life outcomes that she has experienced. It offers some penetrating insights into the way the writer construes the causes, and therefore the meaning, of her frustrations. And there is reason to assume that the letter was meant to be taken seriously — that the writer *believed* the things she wrote. But given this assumption — that the writer's beliefs about her life were sincerely expressed — the letter strikes us as so tragicomic that we hardly know whether to laugh or cry.

Obviously, the woman who wrote this letter had problems. But to what cause(s) should these problems be ascribed? If we accept her own attributions at face value, the causes of her problems would seem to be external and almost entirely beyond her control. Her life, as she describes it, is a succession of doomed encounters with husbands who are "losers" and doctors who are either incompetent or insensitive and insulting. Her current worries seem to focus mostly on her belief that fate, with an assist from medical malpractice, has left her with a nose that stigmatizes and alienates her socially. If these attributions are correct, we might assume that with a genuinely good nose job and a little luck at avoiding the "losers" in the future, there is nothing to keep this person from enjoying a happy and rewarding life.

Or is there? As readers who are not emotionally involved with the events she describes, we may have our doubts. Indeed, *our* attributions about the validity of *her* attributions may be influenced, as Kelley (1967) has proposed, by the consistency with which she invokes external causes to account for her problems, by the apparent lack of consensus that these attributions are correct, and by the absence of any contrasting, internal attributions that would highlight the distinctiveness of the particular events to which external causes are ascribed. According to Kelley's analysis, this pattern of data should lead us to infer that her external attributions are not entirely valid (that is, entity-based), but instead derive from a predisposition she has to externalize the blame for her negative outcomes. Once having made this inference, we may be tempted to conclude that she is "determined" not so much to get a good nose job as to avoid confronting the possibility that she may be partly responsible for some of the problems she has experienced.

In striking contrast to the apparent attributional style of our letter writer, a much different style is suggested by Beck's (1967) description of the phenomenology of a clinically depressed patient:

A patient reported the following sequence of events occurring within a period of half an hour: His wife was upset because the children were slow in getting dressed. He thought, "I'm a poor father because the children are not better disciplined." He noticed a leaky faucet and thought that this showed he was a poor husband. While driving to work, he thought, "I must be a poor driver or other cars would not be passing me." As he arrived at work, he noticed some other personnel had already arrived. He thought,

"I can't be very dedicated or I would have come earlier." When he noticed folders and papers piled up on his desk, he concluded, "I'm a poor organizer because I have so much work to do [p. 235]."

It is clear that the attributions this man uses to explain his negative outcomes are quite different from those of the woman in our first example. He seems as determined to internalize the blame for his negative outcomes as she is to externalize the blame for hers. If we apply Kelley's (1967) analysis to the data this patient has provided, we find that his attributions of self-blame are consistent, nondistinctive, and (if we may speculatively go beyond these data a bit) probably nonconsensual. Again we are likely to suspect that the attributions in question are not entirely reality-based but derive in this case from a general predisposition the attributor has to internalize the blame for his negative outcomes.

It is interesting that fairly strict applications of the covariance analysis that Kelley has proposed as a rational model of the attribution process have led us to conclude that individuals may have characteristic styles of attribution that are, perhaps, not so rational. We should note, however, that these examples represent extreme cases. Moreover, as data we present in this chapter shows, they are also unrepresentative of the commonly reported sex differences in attribution, because it is men who typically externalize their negative outcomes and women who internalize them. At best, these examples are far from conclusive as data and are probably no more convincing than any evidence of an anecdotal type.

From an empirical standpoint, then, on what basis can we decide whether or not such styles of attribution actually exist? The purpose of this chapter is to describe how we came to ask ourselves this question and how, subsequently, we attempted to answer it. We will begin by considering some of the existing research that led us to think seriously about the possible relationship between personality and attributional style.

AN EMPIRICAL PUZZLE

As we were reading the research in an area where a number of different literatures converged, we became aware of an interesting empirical puzzle. First, we found evidence in a number of studies that a major individual difference variable — self-esteem — was related to differences in performance as well as to differences in attribution. Then, we found evidence in a second series of studies that sex — another important individual difference variable — was also related to performance and attribution in a manner quite similar to that of self-esteem. A third set of studies demonstrated a link between self-esteem and sex-role identification, whereas studies in the first two sets revealed some intriguing and fairly robust relationships between attribution and performance.

The apparent convergence of these literatures looked very promising at first glance. However, a more careful look convinced us that the puzzle was still incomplete and that some of its most important pieces were quite obviously missing. For example, studies that investigated the interrelation of self-esteem, sex, and attribution were not to be found, and neither were studies that attempted to assess the independent effects of each of these variables on performance when the two remaining variables were held constant. The pieces of the puzzle that *were* available to us are described in the following sections.

Self-Esteem Differences

The results of a few scattered studies have revealed differences in actual performance, rated performance, and expectancy of future performance for groups differing in self-esteem. In general, these data suggest that the performance of low self-esteem subjects is adversely affected by failure feedback, whereas the performance of high self-esteem subjects is not (Cruz Perez, 1973). Moreover, in their ratings of their actual and future-expected performance, low and high self-esteem subjects again evidence a differential sensitivity to failure feedback: both kinds of ratings are depressed for low self-esteem subjects but not for high self-esteem subjects. Following success feedback, however, the converse is true: both actual and expected performance ratings increase for high self-esteem subjects but not for low self-esteem subjects (Ryckman & Rodda, 1972; Shrauger & Rosenberg, 1970). The impression left by these results is that low self-esteem subjects are likely to accept failure feedback as informative about their future performance but are likely to discount success feedback; thus, only the effects of the failure feedback are reflected in their actual performance. Conversely, high self-esteem subjects appear to accept success feedback but discount failure feedback; thus, information about a failure does not change their expectations about future performance nor affect their actual performance.

So much for the way self-esteem relates to performance. How does it relate to attribution? The data that bear on this question indicate that attribution as well as performance differences have been found for groups differing in self-esteem. Such differences, however, have not always been described in attributional terms. For example, in a study published in 1956, Solley and Stagner recorded the spontaneous remarks of subjects working on insolvable anagrams. They found that high self-valuing subjects made more comments such as, "Is this a word?", "Is this English?"; whereas low self-valuing subjects made more comments such as, "I must be stupid." Although the authors discussed these results in terms of the subjects' continuing or ceasing to attend to the task, it is clear that these findings could also be interpreted in attributional terms. To be specific, high self-esteem subjects were apparently attributing their failure to external causes, whereas low self-esteem subjects were attributing their failure to some internal deficiency.

Gilmore and Minton (1974) did not measure the self-esteem of their subjects, but they did divide the subjects into groups who were either high or low in initial confidence about their performance on the task. After performing, subjects were asked to attribute their success or failure either to ability (an internal factor) or to luck (an external factor). Highly confident subjects attributed success more internally and failure more externally than did subjects low in confidence. Similarly, Fitch (1970) found evidence that high self-esteem subjects attributed failure more to external causes than did low self-esteem subjects. He also found that high self-esteem subjects attributed success more to internal causes than did low self-esteem subjects, although this result was not significant.

The study that is clearly the best in this area was conducted by Maracek and Mettee (1972). These authors were the first to examine the effect of attributions on the performance of subjects who differ in self-esteem. They found that low self-esteem subjects would avoid continued success on a task when it was attributed to skill, but not when it was attributed to luck. When the attribution to luck was given by the experimenter, low self-esteem subjects showed substantially improved performance after success feedback. These results not only suggest that attributions may mediate the "meaning" of the feedback; they also suggest that such "meanings" must be congruent with the person's self-concept before the feedback will have any impact on subsequent performance (cf. Heider, 1958, p. 172). Presumably, "being lucky" is not inconsistent with the self-concept of low self-esteem subjects, but "being skillful" is.

Taken in sum, the research in the area of self-esteem and attribution indicates a fairly consistent pattern: high self-esteem subjects appear to internalize their success outcomes and externalize their failure outcomes more than low self-esteem subjects do. It should be noted, however, that there are at least two problems with these studies that make this conclusion somewhat tentative. First, it is probably inappropriate to make statements about *general* tendencies that characterize the attributional styles of low and high self-esteem subjects on the basis of data collected in such highly constrained situations. In the research just reviewed, subjects gave (or were given) attributions only for their performance on a specific laboratory task. To increase our confidence that these specific differences provide evidence of general differences in attributional style, it would be desirable to measure subjects' attributions about positive and negative outcomes across a range of achievement and nonachievement situations. Second, although the Maracek and Mettee (1972) study makes an important first step toward clarifying the effect of attribution on performance, the attributions employed were not ones that had been generated by the subject; instead, they were given to the subject by the experimenter. This raises the important question of whether attributions given to the subject function in the same manner as those that the subject produces himself.

Sex Differences

A number of studies provide evidence of sex differences in performance and attribution. Although the results of these studies do not perfectly parallel those obtained for self-esteem, the overall correspondence is quite striking. In general, the pattern of data suggests that the actual performance, expected performance, and causal attributions of males tend to resemble those of high self-esteem subjects, whereas the responses of females on the same measures resemble those of low self-esteem subjects.

Some studies indicate that masculinity is perceived to be associated with successful performance and that males have a higher expectancy for success than females. For example, Feather and Simon (1975) asked subjects to rate the femininity of males and females who succeeded or failed at masculine or feminine professions. They found a significant main effect for outcome: those who failed were seen as more feminine than those who succeeded, regardless of the sex of the target or the sex-typing of the profession involved. A number of related studies (Crandall, 1969; Deaux & Emswiller, 1974; Deaux & Farris, 1974; Deaux, White, & Farris, 1975; Feather, 1969; Montanelli & Hill, 1969) have shown that males have a higher expectancy of success than do females over a wide range of skill and achievement tasks.

The males' higher expectancy for success may lead them to greater persistence on skill tasks, as suggested by the research of Battle (1965) and Crandall (1969). More recently, Deaux, White, and Farris (1975) found that males more often choose skill games and persist at them longer, whereas females prefer games of chance. Unlike low self-esteem subjects, but very much like high self-esteem subjects, males persist longer than females at skill games even when they are failing.

Deaux and Emswiller (1974) examined the different attributions made for successful and unsuccessful performance by males and females on masculine or feminine tasks. Similar to results found for high self-esteem subjects, they found a bias to see good performance by males as due to skill (a factor that is internal and stable), no matter what the task. Deaux and Farris (1974) had previously reported that when a given performance outcome is expected (as success apparently is with males), internal–stable ratings tend to be higher (see Deaux, 1976, for an excellent review of the work in this area).

Correlations of Sex-Role Identification
And Self-Esteem

The studies cited above indicate that task-relevant competence and skill are typically seen as masculine attributes. Since the perception that one is competent is a primary component of self-esteem (Coopersmith, 1967; Wylie, 1961), it might be expected that a person's self-esteem would also vary according to his or her sex-role identification. Data collected by Wetter (1975), the present authors (unpublished data, 1975), and Spence, Helmreich, and Stapp (1975) tend to confirm this expectation.

Wetter (1975), reporting data for large samples drawn from both high school and college populations, found positive correlations between self-esteem and self-ratings of masculinity (rs ranged from .32 to .40), but negative or nonsignificant correlations between self-esteem and self-ratings of femininity (rs ranged from −.11 to .06). Similarly, the present authors found a strong correlation ($r = .49$) between self-esteem and masculinity as defined by the Bem Sex Role Inventory, but only a weak correlation between femininity and self-esteem ($r = .13$). And though Spence et al. (1975) found that *both* masculine *and* feminine ratings were positively correlated with self-esteem, the correlation was considerably higher for the masculine than for the feminine self-ratings, despite the fact that the measure of "self-esteem" employed was really a measure of social competence. Because a measure of social competence would emphasize such feminine-identified characteristics as sociability and expressiveness (Bem, 1974; Broverman, Vogel, Broverman, Clarkson, & Rosenkrantz, 1972; Ickes & Barnes, 1977, in press; Spence et al., 1975), it is hardly surprising that such a measure would correlate positively with rated femininity.

In summary, it appears that self-esteem is correlated with a masculine sex-role identification but is uncorrelated with a feminine sex-role identification. This difference appears to be evident to the degree that self-esteem is measured as a global construct, rather than being limited to the domain of perceived social competence.

What About Locus of Control?

The literature we have reviewed suggests that sex and self-esteem may be related to dispositional biases in attributing one's positive and/or negative outcomes to an internal or external locus of causality. Given the relatively small number of studies we have cited, however, the reader may be wondering why we have chosen to consider only some of the studies from the voluminous literature on locus of control. Since the degree to which a person believes in an internal versus an external locus of control is assumed by Rotter (1966) and others (see Phares, 1976) to be a relatively stable individual difference variable, and since the locus-of-control literature has yielded some differences for sex (Feather, 1967, 1968; Ryckman & Sherman, 1973) and self-esteem (Fish & Karabenick, 1971; Platt, Eisenman, & Darbes, 1970; Ryckman & Sherman, 1973), our failure to include this literature in our review may appear to be a glaring oversight.

We think that there may be some insight in our oversight, so we will take a moment to clarify our position on this matter. Put simply, there are a number of good reasons to believe that the variable of internal/external locus of causality to which we have previously referred is not synonymous with the variable of internal/external locus of control. The theoretical and operational definitions of the two concepts differ in several important respects, making the results obtained in one area not clearly applicable to the other. The following are just a few of the ways in which the two constructs differ.

First, locus-of-control scales often confound locus of control with locus of causality by using: (1) items that imply causality but no control of an event; (2) items that imply control but no causality of an event; or (3) items that imply both (Rotter, 1966; Mischel, Zeiss, & Zeiss, 1974; Crandall, Katkovsky, & Crandall, 1965). Second, the confusion between control and causality is particularly evident in the research that deals with negative events. It is not clear whether internal control of negative outcomes means that the subjects *caused* the negative event (Mischel et al., 1974), or whether it means that a negative outcome can be escaped or avoided and therefore "controlled" (see Glass & Singer, 1972; Wortman, Panciera, Shusterman, & Hibscher, 1976). Third, Rotter's (1966) scale, the one most widely used, suffers from additional problems to the ones mentioned above. The scale collapses over positive and negative outcomes, thus reflecting an implicit assumption that subjects would perceive a similar degree of internal versus external control for both. Moreover, some items are written in the first person whereas others are written in the third person, reflecting a second implicit assumption that whatever subjects see as the locus of control for other people's outcomes will also be seen as the locus of control of their own. The studies we have already cited in this chapter tend to invalidate the first of these assumptions, and findings relevant to actor/observer differences in attribution (for a review, see Monson & Snyder, 1977) tend to invalidate the second.

For these reasons, among others, it is nearly impossible to meaningfully compare and integrate the findings of the locus-of-control literature with the findings relevant to perceived locus of causality. And that's why the locus-of-control literature has not been reviewed here!

PERSONAL CHARACTERISTICS, ATTRIBUTION, AND PERFORMANCE: MORE PIECES OF THE PUZZLE

Taken collectively, like the pieces of a puzzle that is being completed, the various findings we have just reviewed suggested to us that there were some interesting similarities between the individual difference variables of sex and self-esteem. In addition to self-esteem being correlated with a masculine sex-role identification, sex and self-esteem appeared to resemble each other closely in their relationship to attributional preference and performance. The general picture implied by these data was certainly an intriguing one; but although its essential outlines seemed to be apparent, there were still some major gaps in the design. Because of our desire to clarify and add greater structural detail to this emerging pattern, we conducted the series of studies described below. In the first two studies we investigated the interrelation of self-esteem, sex and attributional preference. In the third study we attempted to determine whether changes in attributional preferences are associated with corresponding changes in self-esteem. In the fourth and final study of this series we assessed the independent effects of self-esteem, sex, and attributional style on behavioral reactions to experimentally induced "failure" in a standard performance situation.[1]

Studies 1 and 2:
Self-Esteem, Sex, and Attributional Style

Study 1. Our first study was designed to examine the specific differences in attributional style exhibited by groups differing in sex and level of self-esteem. The subjects in this study were 268 male and 339 female undergraduates who were tested in large groups, in hour-long sessions. During each session the subjects were asked to complete a number of pencil-and-paper measures; these included a self-esteem inventory designed for use with a college population (Morse & Gergen, 1970) and a measure of attributional style developed by the authors. The two measures were separated by a number of other scales so that subjects would not be likely to see them as being related.

The attributional style measure consisted of 12 items. Each item contained a brief description of the outcome of a hypothetical event (for example, "You got an 'A' on a class project") followed by four possible causes of the outcome ("you worked hard to prepare this project," "the project was relatively easy," "the project is one in which you have considerable skill," or "you were lucky and happened to do a project that corresponded to the professor's interest"). The subjects' task was to read each item, imagine that they had experienced the outcome described, and then choose what they would consider to be the single most probable cause of the outcome. The items were carefully written to minimize potential sex bias in the subjects' responses. Additionally, they were designed to represent a wide range of conceptually distinct situations, requiring subjects to make attributions not only about physical and intellectual performance but also about interpersonal relationships, moral behaviors, and moods. Within each type of situation, the outcomes of the events were counterbalanced so that half were positive and half were negative.

The four possible causes provided for each outcome corresponded (in a random order) to the four types of causes in the two-dimensional taxonomy proposed by Weiner and his associates (Weiner, Frieze, Kukla, Reed, Rest, & Rosenbaum, 1971). Thus, for each outcome described, an example of a cause in each of the following (causal locus x stability) categories was offered as an alternative: internal–stable, internal–unstable, external–stable, external–unstable. Because

---

[1]Collectively, the series of studies reported here were conducted by the second author under the first author's supervision as the basis of her master's thesis.

of a number of practical considerations, we did not attempt to incorporate "intentionality," a third causal dimension proposed by Rosenbaum (1972), into the response alternatives.

The attribution measure was scored by summing the number of times the subject chose each of the four types of causes for the two separate categories of positive and negative outcomes. This response format proved to be problematic because of the nonindependence of the four scores within the positive- and negative-outcome categories. The nonindependence of these scores made it necessary to analyze the data "in pieces," by means of a series of multivariate analyses in which only three of the four scores within each outcome category were analyzed. As it was also desirable to compensate for possible statistical bias resulting from extremely unequal cell frequencies (due to the fact that all subjects whose self-esteem scores fell within a standard deviation of the mean were defined as "moderates"), the alpha value required for significance was set at a much more conservative level (.001) than is commonly used.

The results of the study indicated a general tendency for subjects to ascribe positive outcomes more to internal than to external causes, but to ascribe negative outcomes more to external than to internal causes ($p < .001$; see Table 5.1). However, this two-way interaction of causal locus and valence of outcome was further qualified by the subjects' level of self-esteem ($p < .001$). For positive outcomes, high self-esteem subjects strongly preferred to ascribe internal rather than external causes. This preference, which was somewhat attenuated for moderate self-esteem subjects, was even more attenuated for low self-esteem subjects. For negative outcomes, however, the opposite pattern of attributional preferences was found. The high self-esteem subjects strongly preferred to ascribe external rather than internal causes — a tendency that was again relatively weaker but in the same direction for subjects with moderate self-esteem. For low self-esteem subjects this tendency was not merely further attenuated but was actually *reversed* such that internal causes were preferred to external causes in accounting for negative outcomes. This reversal suggests that it is in their response to nega-

TABLE 5.1
Mean Number of Internal (I) and External (E) Attributions for Items
with Positive (+) and Negative (−) Outcomes (Study 1)

| | Types of Attributions | | | |
	+I	+E	−I	−E
High self-esteem Ss	3.96	2.04	2.20	3.80
Moderate self-esteem Ss	3.65	2.35	2.65	3.35
Low self-esteem Ss	3.37	2.63	3.28	2.72
Overall means	3.65	2.35	2.67	3.33

tive outcomes that low self-esteem subjects may differ most in their attributional style from subjects with high self-esteem.

A further insight into these divergent styles of attribution can be obtained by comparing the percentage of subjects in the various self-esteem groups who: (1) favored internal attributions for positive outcomes but external attributions for negative outcomes (a +I−E pattern); or (2) favored external attributions for positive outcomes but internal attributions for negative ones (a +E−I pattern). Of the nine possible patterns of responses, including those in which internal and external causes are preferred equally (+E−E, +I−I, +I−E, +E−I, +=−=, +=−I, +=−E, +I−=, +E−=), fully 44% of the 107 high self-esteem subjects exhibited the +I−E pattern, in full contrast to only 27% of the 395 moderate self-esteem subjects, and only 18% of the 105 subjects with low self-esteem. Conversely, although 11% of the low self-esteem subjects used the +E−I pattern, this pattern was used by only 4% of moderate self-esteem subjects, and by only *one* of the 107 subjects who were high in self-esteem.[2] In fact, when the data are collapsed across the categories in which negative outcomes are "internalized," they reveal that only 14% of high self-esteem subjects tended to internalize negative outcomes, whereas 22% of the moderates and 45% of the lows did so. These differences, which are highly significant by chi-square analyses ($ps < .001$), indicate a strong association between self-esteem and attributional style.

In addition to demonstrating systematic preferences regarding the causal locus of positive and negative outcomes, subjects also demonstrated systematic preferences regarding the stability of the attributed causes. The data indicated that they strongly preferred to ascribe the outcomes of the hypothetical situations to unstable rather than to stable causes. The mean number of unstable and stable causes chosen were 7.66 and 4.34, respectively ($p < .001$). This main effect for stability of causes was qualified, however, by significant two-way interactions with causal locus ($p < .001$) and with valence of outcome ($p < .001$). The means for the first of these interactions revealed that for outcomes perceived as internally caused, subjects had a strong preference for ascribing unstable ($\bar{X}$ = 4.70) rather than stable ($\bar{X}$ = 1.64) causes; however, for outcomes perceived as externally caused, subjects showed no preference for stable ($\bar{X}$ = 2.70) over unstable ($\bar{X}$ = 2.96) causes. The means for the second interaction revealed a somewhat similar pattern: for negative outcomes, unstable causes ($\bar{X}$ = 4.16) were strongly preferred over stable causes ($\bar{X}$ = 1.84), but for positive outcomes the preference for unstable ($\bar{X}$ = 3.51) over stable ($\bar{X}$ = 2.49) causes was con-

---

[2]Although the percentages of subjects exhibiting the +E−I style were fairly low across all self-esteem levels, the overall difference among these levels was significant by chi-square analysis. Because we consider the +E−I style to be a relatively pathological one, we would not expect its incidence to be high in a "normal" population.

siderably attenuated. Thus, although the outcomes of the events described were generally attributed to unstable causes, this attributional tendency was especially pronounced for outcomes that were negative and/or perceived as internally caused.

Study 2. Before we attempt to address the question of how all of these results might be interpreted, it will be useful first to examine the findings obtained in a second study for purposes of comparison. The second study was designed to replicate our initial results with an extensively revised version of the attributional style measure. To minimize the chances that some of the initial results might be due to the influence of a particular item or items rather than to more general attributional preferences, items on the first version for which response variance was low (one or two of the causes were chosen almost exclusively by subjects) were either discarded or rewritten. In addition, new items were added to expand the revised measure from 12 to 24 descriptions of hypothetical event-outcomes for which attributions were to be made. Care was again taken in the writing or revising of the items to ensure that potential sex biases were not "built into" the items' content. Finally, to avoid the nonindependence problem inherent in the first version, the response format of the second was changed so that instead of choosing the most probable of the four alternative causes, the subjects rated the probability of each of the causes on a five-point scale ranging from "not at all probable" (1) to "extremely probable" (5). In all other respects, the study was conducted in the same manner as the first, with 238 male and 268 female undergraduates serving as subjects.

In general, the results of the second study were quite consistent with those of the first. The data indicated a significant tendency to rate positive outcomes as more internally caused ($\overline{X}$ = 31.45) than externally caused ($\overline{X}$ = 28.60), whereas negative outcomes were rated as more externally caused ($\overline{X}$ = 25.69) than internally caused ($\overline{X}$ = 22.14; p < .001). As before, this effect was qualified by a significant three-way interaction of self-esteem, locus of causality, and valence of outcome (p < .001). The results for the positive-outcome items replicated the earlier findings: high self-esteem subjects viewed the internal causes for positive events as more probable ($\overline{X}$ = 32.40) than did moderate ($\overline{X}$ = 31.54) or low ($\overline{X}$ = 29.02) self-esteem subjects. The results for the negative-outcome items, however, were somewhat different from those obtained earlier. In the first study, when asked to *choose the single most probable cause* for the negative outcomes, high self-esteem subjects chose external causes more often than did moderate or low self-esteem subjects. In this study, when asked to *rate the probability of each cause,* high self-esteem subjects rated both internal ($\overline{X}$ = 20.42) and external ($\overline{X}$ = 24.61) causes as less probable than did moderate ($\overline{X}$s = 22.01; 25.66) and low ($\overline{X}$s = 23.96; 26.71) self-esteem subjects. These results suggest that high self-esteem subjects not only tend to externalize the cause of negative outcomes to a greater degree than moderate and low self-esteem subjects; they also tend to see such outcomes (in terms of their possible causes) as generally more improbable.

With regard to sex effects, the results of the second study revealed attributional differences between males and females that were strikingly similar to the differences found between high and low self-esteem subjects. These were evidenced in a three-way interaction of sex, causal locus, and valence of outcome ($p <$ .001). For positive-outcome items, males rated the internal causes as more probable ($\bar{X}$ = 32.02) than did females ($\bar{X}$ = 30.95), while rating the external causes as less probable ($\bar{X}$ = 27.89) than did females ($\bar{X}$ = 29.23). For negative-outcome items, males saw both internal ($\bar{X}$ = 20.53) and external ($\bar{X}$ = 24.93) causes as less probable than did females ($\bar{X}$s = 23.58; 26.35). Thus, for both types of outcomes, the pattern of results for males resembles that of high self-esteem subjects, and the pattern of results for females resembles that of low self-esteem subjects. For positive outcomes, males at each self-esteem level rated the internal causes as more probable than the corresponding females rated them; conversely, for negative outcomes, males at each self-esteem level rated all causes as less probable than the corresponding females rated them. A retrospective look at the data from the first study revealed a similar pattern of sex differences, but these results attained only a conventional (.05) level of significance and not the more stringent (.001) level we had adopted a priori. The sex differences in these studies cannot be regarded as artifacts due to the correlation of sex and self-esteem; the two variables were effectively "blocked" in both designs such that the means and variances of self-esteem at each of its three levels were virtually identical for both sexes.

A final set of results emerged in the second study; these involved the stability factor and essentially replicated the set of findings obtained before. Unstable causes were considered more probable ($\bar{X}$ = 28.62) than stable causes ($\bar{X}$ = 25.32), and this main effect was again qualified by two-way interactions with locus of causality ($p <$ .001) and with valence of outcome ($p <$ .001). For outcomes perceived as internally caused, unstable causes ($\bar{X}$ = 29.33) were considered more probable than stable causes ($\bar{X}$ = 24.26); for outcomes perceived as externally caused, the perceived probability of unstable ($\bar{X}$ = 27.90) and stable ($\bar{X}$ = 26.38) causes differed very little. Similarly, though the causes of negative outcomes were much more likely to be perceived as unstable ($\bar{X}$ = 27.02) than as stable ($\bar{X}$ = 20.81), the causes of positive outcomes were just as likely to be perceived as stable ($\bar{X}$ = 29.83) than as unstable ($\bar{X}$ = 30.22).

Taken collectively, the results of the second study are very similar to those of the first, and both sets of data are consistent with the results of studies we have reviewed earlier. We may summarize the results as follows:

1. With respect to self-esteem differences, the data indicate that high self-esteem subjects are inclined to take credit for positive outcomes by ascribing them to internal causes. In contrast, low self-esteem subjects tend to take less credit for positive outcomes by attributing them relatively more often to external causes. For negative outcomes, high self-esteem subjects either make external attributions or else rate all causal factors (and, by implication, the outcomes

themselves) as improbable, whereas low self-esteem subjects tend to "blame" themselves for such outcomes by making relatively more internal attributions to explain them.

2. With respect to sex differences, the data indicate that males at each level of self-esteem exhibit attributional tendencies resembling those of high self-esteem subjects, whereas the attributional tendencies of females at each level resemble those of low self-esteem subjects.

3. With respect to the stability variable, the data indicate a general tendency for subjects in all groups to choose unstable causes more often than stable causes for the outcomes described, and this tendency is especially pronounced for outcomes that are negative and/or perceived as internally caused.

Interpretations of Studies 1 and 2.   There are essentially two major theoretical interpretations that could be used to account for the three patterns of results listed. First, some or all of these results may be valid representations of the typical kinds of cause—outcome relationships that the subjects have experienced. Second, some or all may reflect relatively invalid, reality-distorting "biases" in the attribution process. These "biases" may be either: (1) learned biases that have been acquired through socialization; or (2) motivated biases that derive either from the need to maintain a stable self-conception (Heider, 1958) or from the need to maintain a feeling of control over one's outcomes (Kelley, 1971).[3] Although we see later that it is nearly impossible to eliminate one or another of these alternative explanations, it is important from a theoretical standpoint that they be viewed as conceptually distinct. We therefore discuss each of the three patterns of results in terms of these alternative interpretations.

Looking first at the self-esteem differences, the difficulty of deciding between the two alternative interpretations becomes immediately apparent. On the one hand, it is possible that the typical cause—outcome relationships experienced by high self-esteem subjects are *genuinely different* from those experienced by low self-esteem subjects. For example, if high self-esteem subjects possess — in addition to their high self-esteem — other socially desirable traits, skills, and abilities that low self-esteem subjects do not possess, their positive outcomes may in fact be more generally attributable to internal factors, whereas the positive outcomes of low self-esteem subjects may not. By the same reasoning, negative outcomes may in fact not be as probable for high as for low self-esteem subjects, especially to the degree that they are internally caused (see Orvis, cited in Weiner, 1974, p. 19). On the other hand, it is possible that the typical cause—outcome relationships experienced by high and low self-esteem subjects are not really different but are *merely perceived differently* because of biases affecting the attributions of

---

[3]Attributions may also be "biased" by restrictions on the available information or by limitations of the information-processing capacities of the individual (see Miller & M. Ross, 1975; L. Ross, 1977; Fischhoff, 1976). As the relevance of these forms of bias to our data would seem to be rather tenuous, we will not take the space to discuss them here.

subjects in one or both groups. Given this type of explanation, however, we still have two viable possibilities confronting us: (1) that these attributional biases have been acquired by subjects — through instruction, imitation, etc. — as a product of the socialization process; or (2) that these biases have not been learned through socialization but instead have emerged psychodynamically as defense mechanisms by means of which subjects attempt to maintain a stable self-conception or a feeling of control over their outcomes.

Of these two forms of attributional bias, the hypothetical processes involved in the first are fairly straightforward: the individual is taught (either explicitly or implicitly) by parents, teachers, peers, and other socializing agents that (s)he is causally responsible for good outcomes but not for bad ones, or vice versa. Presumably, this learning eventually becomes internalized as a general predisposition or attributional style that then influences a wide range of attributions the person makes. The continued use of this attributional style may be the direct cause of corresponding changes in the person's level of self-esteem, or the two variables may be dynamically related such that a change in either variable may effect a parallel change in the other.

The hypothetical processes involved in the second form of attributional bias are a bit more complicated and require some elaboration. Although recent reviews by Miller and M. Ross (1975) and by L. Ross (1977) have provided a number of arguments for caution in assuming that motivational biases in attribution really do exist, Heider (1958) has proposed a fairly convincing psychological basis for them in his assertion that an attribution will not be acceptable to the person unless it "makes sense" by satisfying two major criteria. To make sense, the attribution must not only provide a plausible cause of the outcome to be explained but must also be congruent with the person's view of himself and his "cognitive expectations about connections between motives, attitudes, and behavior, etc., [pp. 172—173]." This reasoning suggests that the entire "naive psychology" of the person may have to be taken into account and that the organizing principles of balance and attribution should not be seen as divorced from each other but should instead be regarded as dynamically related within the person's life space (see also "An Interview with Fritz Heider," pp. 14—17, in *New Directions in Attribution Research*, Vol. 1). A person's attributions must tend, therefore, to be consistent with the self-concept, and vice versa, if a relatively stable view of the self and its relation to the world is to be maintained.

This assumed need for stability in one's view of the self and its relation to the world is most plausibly evidenced in the case of high self-esteem subjects. Their characteristic attributional style — internalizing positive outcomes and externalizing or denying the probability of negative outcomes — could easily be interpreted as serving an ego-defensive function. The second aspect of this attributional style has much in common with the ego-defensive mechanism of projection postulated by Freud (1920), and the combination of both aspects appears to be well-suited to reinforcing the person's high level of self-esteem (by causally linking the self to only positive, successful outcomes) and protecting it from damage (by dissociating the self from negative outcomes and, hence, from "failure" or "blame"). Of course, the extreme and indiscriminate overuse of the two aspects of this style, either singly or in combination, may be neurotic, unrealistic, and probably maladaptive in the long run (the narcissistic braggart or taker-of-undeserved-credit may epitomize the extreme use of the first aspect; our ego-threatened letter writer, with her apparently compulsive need to project failure and blame to external sources, may characterize an extreme use of the second). In the short run, however, or when used in moderation over the long haul, the two aspects of this attributional style may be complementarily employed in a relatively healthy way to maintain and defend a favorable self-conception (see Bradley, 1978; and Chapter 4, by Snyder, Stephan, & Rosenfield, this volume). They may also support "an internal locus of control orientation [p. 23]" that provides a basis for the continuation of control attempts on the part of the individual (Kelley, 1971).

The attributional style most characteristic of low self-esteem subjects — internalizing negative outcomes but not positive ones — could also be interpreted as ego-defensive in the sense that it helps them to stabilize their conception of self with respect to a changing and often capricious environment. It is clearly *not* ego-defensive, however, in the sense that it helps them maintain feelings of self-worth. On the contrary, the extreme use of the two aspects of this attributional style may be profoundly debilitating (as exemplified by the depressed, chronic self-blamer described by Beck). It may also rob the individual of the motivation to attempt to control his outcomes, because all of his rewards are increasingly perceived to be externally controlled, and all of his attempts at internal control are increasingly perceived to result only in "failure." Thus, in exchange for some stability in his self-conception, the individual may have to pay a very heavy price in terms of low self-esteem and a reduced capacity to deal effectively with his environment.

Although this motivational bias interpretation is consistent with Heider's (1958) criteria for "acceptable" attributions, it should be noted that the alternative interpretations — that the obtained differences either reflect a learned bias or correspond to real differences in causal structure — are also compatible with these criteria. We are presently unable to decide which of these explanations best accounts for the attributional differences in self-esteem we have just reported, and this same difficulty must cloud our interpretation of the reported sex differences as well. To be specific, it is possible either that the causal structures of the positive and negative outcomes typically experienced by the females in our studies are in fact: (1) different from those experienced by the males; or (2) are merely perceived differently because of attributional biases. As before, these attributional biases may be derived either from social learning or from the need to maintain a stable self-image and/or sense of control. Although we lean strongly toward the position that the sex differences in attribution reflect biases rather than real differences in causal structure, the present data do not allow us to decide this important issue one way or the other. However, recent data by Dweck and her associates do provide some relevant evidence suggesting that sex differences in attribution may derive in part from the nature and specificity of the evaluative feedback that male and female children receive in the classroom (see Chapter 6, by Dweck and Goetz, this volume, for a fuller discussion of this issue).

Similar ambiguities are encountered when we attempt to interpret the effects involving the stability variable. If these effects reflect veridical differences in causal structure, unstable causal factors may in reality account for more of the subjects' typical outcomes than stable causal factors. The first of the obtained interaction effects suggests the further possibility that unstable internal causes may actually be responsible for more of the subjects' outcomes than stable internal causes. This notion is quite plausible when one considers that, according to Rosenbaum's (1972) causal taxonomy, the class of unstable internal causes includes relatively transient intentions, short-term states of effort and exertion, and variations in mood, emotion, skill, fatigue, physical health, etc. Personal outcomes may often vary in response to short-term "feeling states" such as these. Finally, negative outcomes may typically be due more to unstable than to stable factors not only because they tend to be uncommon and somewhat unpredictable (Kanouse & Hanson, 1971) but also because people typically attempt to eliminate or avoid the more stable, negative-outcome-inducing factors that are located in the environment and to eliminate or suppress analogous factors (undesirable dispositions or traits) that are located in the self.

It is also plausible, however, that these stability effects do not reflect veridical differences in causal structure but are due instead to learned or motivated biases in the attribution process. Such attributional preferences may, in effect, be socialized in the individual, or they may arise psychodynamically out of the individual's need to maintain at least the illusion of control (see Langer, 1975; Langer & Roth, 1976; Wortman, 1976). Of the obtained effects involving the stability variable, the tendency to attribute negative outcomes to unstable rather than to stable factors may be particularly amenable to the second of these interpretations.

In summary, the data from the first two studies yielded a number of replicable effects indicating that sex and self-esteem are related to specific patterns of attributional preference. The data also revealed some interesting preferences with respect to the stability variable posited by Weiner et al. (1971). Unfortunately, for none of these sets of data could we determine whether the obtained effects were due to real differences in the cause—effect structures actually experienced by subjects or to learned or motivated biases in the attribution process. Despite this lack in our understanding of the origin of these various attributional preferences, however, the obtained results clearly seemed to offer some promising leads for future research. In particular, the strong and consistent effects for self-esteem suggested the intriguing possibility that because distinct attributional styles are associated with distinct levels of self-esteem, changes in a person's attributional style might be associated with corresponding changes in self-esteem. It was this possibility that our third study was designed to explore.

Study 3:
Changes in Attributional Style and Self-Esteem

Given the fairly striking differences in attributional style between low and high self-esteem subjects, we began to wonder if it might be possible to effect a positive change in self-esteem by teaching the attributional style most characteristic of high self-esteem subjects to subjects with low self-esteem. In an attempt to test this notion, we designed an experiment for which 60 subjects were selected on the basis of pretest data indicating that they were not only low in self-esteem but also showed the attributional style most characteristic of low self-esteem subjects — internalizing negative outcomes but failing to internalize positive ones.[4] Of these 60 subjects, 45 were randomly assigned to one of three experimental conditions to receive attributional retraining. The remaining 15 subjects were employed as nonretrained controls. The subjects in all four conditions were roughly conterbalanced according to sex.

The 45 subjects in the experimental conditions were contacted by telephone and asked to participate for extra course credit in a project that would extend over a 5-week period. The subjects were not informed of the bases for their selection and were never told during any part of the study that we had any knowledge of their self-esteem scores or that self-esteem was a variable of any interest to us. Instead, as a cover story for the experiment, each subject was told that the project in which (s)he had been enlisted was designed to help refine and improve an "attribution questionnaire" that the authors were developing. The experimenter (MAL) began the session by providing a general description of attributional approaches in terms of their concern with the kinds of causes people ascribe to explain the events they experience. She then explained that an attribution questionnaire had been written and tested by the psychology department and that this questionnaire contained descriptions of a number of hypothetical events for which subjects were asked to assess the probability of certain alternative causes. However, because of some dissatisfaction with the current version of the questionnaire, a 5-week project was being conducted in which a group of students were asked to help generate some new items for a revised version of the questionnaire. These new items would potentially be more credible and meaningful to a college population because they would be based on real situations that students had experienced. [By this point in the cover story, the

---

[4]Because there were too few +E−I subjects available to complete the entire design, each of the four conditions contained equal numbers of +E−I and +=−I subjects matched by sex.

subject had usually recognized the questionnaire the experimenter was discussing and had interrupted her to acknowledge that (s)he had in fact completed it earlier in the semester.]

The experimenter then asked the subject to help to generate possible items for the revised questionnaire by drawing on his or her own experience in "real situations" with "real causes." Using response forms provided by the experimenter, each subject was instructed to record — for each of the 5 weeks — three positive outcome events and three negative outcome events that he or she had actually experienced during that week, along with a list of possible causes for each event. Subjects were instructed to list as many plausible causes as they could think of for the events described, but the *types* of causes they could list were restricted according to their experimental condition. Subjects in Group 1 were asked to list only internal causes for their positive outcome events and only external causes for their negative outcome events. (The meaning of "internal" versus "external" causes was clarified for subjects in the initial session and, if necessary, in subsequent ones.) Subjects in Group 2 were allowed to list any types of causes for their positive outcomes but were required to list only external causes for their negative ones. Subjects in Group 3 were given instructions the reverse of those given to subjects in Group 2; they were asked to provide only internal causes for their positive outcomes but could list any types of causes for their negatives ones. It was explained that the restrictions imposed were necessary because we were interested in only certain types of cause—event combinations. None of the subjects expressed any skepticism about the rationale for or the appropriateness of the task they were requested to do.

It should be clear that what we were really asking the subjects to do was to practice making attributions that were somewhat inconsistent with their assessed attributional style. The attributions that they practiced either partially (Groups 2 and 3) or totally (Group 1) resembled the attributional style most characteristic of high self-esteem subjects. In no way was it implied that the required attributions were somehow better or more appropriate than the ones subjects might have generated spontaneously; nor was it implied that we wanted subjects to consciously adopt or internalize the attributional style on which they were patterned. Rather, it was hoped that by simply attending to and practicing these patterns of attributions, the subjects would tend to perceive such causes as more salient in the future and attend to them more readily. Presumably, they would actually come to see them as more plausible causes for their own life outcomes and change their attributional styles to accord with these new perceptions (for sources relating attention to attribution see Arkin & Duval, 1975; Duval & Hensley, 1976; Duval & Wicklund, 1973; Regan & Totten, 1975; Storms, 1973; Taylor & Fiske, 1975; Wegner & Finstuen, 1977).

Each week the subjects turned in to the experimenter a response form containing written descriptions of the six event-outcomes and their accompanying lists of possible causes. At the end of the 5-week period, the 45 experimental subjects and the 15 control subjects were contacted and asked to come in for a final session (the first session for the controls).[5] In order to obtain in this final session a posttest measure of the subjects' self-esteem in a manner that would not cause them to associate this measure with the one taken earlier or with the current experiment, we arranged for an unfamiliar experimenter to approach each subject in the waiting room a few minutes prior to his or her scheduled session. The second experimenter explained that because his own subject had not shown up, he was willing to offer our subject 50 cents to complete a 10-minute question-naire. The first experimenter always arrived on the scene in time to grant the subject permission to help the second experimenter if (s)he so desired, and all but 1 of the 60 subjects agreed to comply with his request.

Following their completion of the self-esteem scale, each subject proceeded to the scheduled session with the first experimenter. At this session, the experi-mental subjects turned in their final response form, and all subjects were retested with the same attributional style questionnaire they had completed earlier in the semester. A partial debriefing and question-and-answer session followed in which subjects were tested for suspicion and given some limited insight into the true purposes of the study (for ethical reasons, the self-esteem aspect of the research was again omitted from this discussion).

Our hypothesis was that subjects in Group 1 would evidence more "positive" change in attributional style and self-esteem than subjects in Groups 2 and 3 and that subjects in the latter groups would in turn evidence more change than the controls. However, despite our attempt to create impactful, involving manipula-tions, and despite the extreme care with which the experimenter conducted the study, these predictions were not confirmed. There were no significant dif-ferences among the four conditions for the pretest—posttest comparisons of attributional style and self-esteem.

The general inability of our experimental manipulations to effect the predicted changes was somewhat disheartening, but an internal analysis of the data offered at least a ray of hope for our hypothesis. When the subjects, regardless of condi-tion, were dichotomized on the basis of their responses to the second attribution measure, a pretest—posttest analysis indicated that subjects whose attributional style changed in the direction of the typical high self-esteem pattern showed a significantly greater increase in their self-esteem ($\bar{X}$ = 6.1 scale points) than those whose attributional style showed no or reverse change relative to this pattern ($\bar{X}$ = 2.3 scale points; $p <$ .03, one-tailed test). Although this finding does not answer the question of whether a change in attributional style is sufficient to *cause* a change in self-esteem, it at least informs us that changes in one variable are associated with corresponding changes in the other.

When considered in the cool light of retrospective reflection, the failure of our experimental manipulations to influence attributional preferences in this

---

[5]The 15 control subjects were not asked to develop lists of possible causes for personal outcomes during the 5-week period; hence, our control condition was not as comparable to the experimental groups as a "placebo"-treatment control would have been.

study is perhaps not too surprising. Our subjects were asked only to practice an alternative attributional style. They were given no reason to believe that the new style was more appropriate than their old one, nor were they given any motivation to make them want to change. In fact, during the debriefing sessions, the majority of the subjects appeared to be quite resistant to the notion that their attributional style may have been (or should, in principle, be) changed. Many of them saw their own style as appropriate and "modest," in contrast to the +I–E style characteristic of high self-esteem subjects, which these low self-esteem subjects often referred to disparagingly as "immodest" or "egotistical." In short, their own attributional style appeared to be a relatively stable and important aspect of their overall personality structure, just as we have proposed. Lacking the belief that the new attributions were more appropriate than the old ones (see Valle & Frieze, 1976), and having little or no motivation to change, they would not be easily induced to adopt a new style.

The possible importance of making the new attributions appear more credible and desirable than the old ones is suggested by a comparison of our study with a conceptually similar experiment conducted by Dweck (1975). In Dweck's experiment, the effects of successful attributional retraining were demonstrated for a group of grade-school children who were characterized as evidencing the "learned helplessness" syndrome (Crandall, Katkovsky, & Crandall, 1965; Dweck & Reppucci, 1973; Klein, Fencil-Morse, & Seligman, 1976). When compared to a nonretrained group of matched controls, children who had learned to deal with their performance-related failure by substituting an internal unstable attribution ("I didn't try hard enough") for an internal stable attribution ("I can't do it") showed marked improvements in task performance. During the course of the attributional retraining, the experimenter – a highly credible adult authority figure – implicitly discounted internal stable factors as the true causes of the children's failure by explicitly making internal unstable attributions instead. Because effort is not only an unstable factor but a controllable one as well (Barnes, Ickes, & Kidd, 1977; Ickes & Kidd, 1976; Rosenbaum, 1972; Weiner, Russell, & Lerman, this volume), substituting a lack-of-effort attribution for a lack-of-ability attribution presumably facilitated the adoption of a control orientation by the subjects in Dweck's experimental group. Similar results were obtained in a conceptual replication of this study by Chapin and Dyck (1976).

Dweck's (1975) experiment is important in its implication that to successfully intervene to change an individual's attributional preference, any of a number of conditions [discounting the old attribution(s) while actively endorsing the new one(s), establishing a high level of communicator credibility, intervening when the subjects are still young and impressionable, working directly with the subjects in the specific situation in which cognitive motivational deficits are apparent] may eventually prove to be necessary, if not sufficient. Though the establishment of such conditions may be necessary and desirable from a strictly therapeutic standpoint, however, establishing most of them in the context of our research would have been fatal from an experimental standpoint because of the possible "demand characteristic" artifacts they would have introduced into the design. Given a college-age population and our desire for a degree of methodological rigor, we may have already pushed to its reasonable limit the compromise between experimental impact and experimental control.

In addition to providing insight into some of the practical and experimental problems associated with changing attributions, a second reason why Dweck's experiment is important is that it reminds us that the pragmatic aspect of attributional preference is (or should be) of central interest. The fact that self-esteem and attributional style covary may be interesting, but if such measured beliefs predict nothing but other measured beliefs, they are of little consequence to the study of behavior. Our recognition of this concern led us to decide that before we made any further attempts to intervene to change these cognitions, it would be more profitable for us to first attempt to assess the independent effects of self-esteem, sex, and attributional style on behavioral indices of performance. The rationale for examining the behavioral effects of these variables independently was straightforward: Although studies we have reviewed earlier suggested that differences in performance are associated with differences in sex and self-esteem, the data from our first two studies indicated that these individual difference variables were very probably confounded with the subjects' attributional style. It was not clear, therefore, whether either self-esteem or sex would have any effect whatsoever on performance when attributional style was varied independently. It was to provide some clarity (or at least some data) on this point that our fourth study was conducted.

Study 4:
Independent Effects of Self-Esteem, Sex, and
Attributional Style on Performance

In this experiment, the performance of subjects was monitored both prior to and following an experimentally induced failure experience. We chose to study behavioral reactions to failure not only because of our belief that failure has a stronger impact than success on self-esteem but also because our earlier data had indicated that failure outcomes were those for which the attributional responses of high and low self-esteem subjects were most clearly divergent. The task situation, then, was one in which the possibility of obtaining performance effects for self-esteem were presumably very good, assuming that such effects do in fact occur.

Forty subjects, who were preselected for the study on the basis of individual difference data collected earlier, were assigned to eight groups that represented a complete factorial design comprised of three two-level factors: self-esteem (high vs. low, based upon a median split), sex (male vs. female), and attributional style for failure (internal vs. external). Care was taken to block the groups on the three variables so that the means and standard deviations of the blocked variable(s) were comparable for each of the test-variable groups. This assignment procedure allowed us to assess the independent effects of self-esteem, sex, and attributional style by eliminating the possibility that any of these potential sources of variation would be confounded with one or both of the others.

Standardized anagrams were chosen as the task stimuli to be used to test the subjects' performance. When used in an experimental performance setting, anagrams have a number of distinct advantages:

1. They are perceived by subjects as tests of intelligence and therefore elicit a high level of involvement and motivation.
2. They are easily administered on a trial-by-trial basis.
3. Dependent measures of both speed and errors can readily be obtained.
4. Perceived success and failure can be controlled by the presentation of solvable versus insolvable anagrams, and it is not readily apparent to the subject that the experimenter is controlling the outcome.

The procedure we employed can be subdivided into three conceptually distinct stages. In the first stage, each subject was seated in a small work booth facing a panel that contained a switch. The switch activated a timer in an adjacent control room that was used to record the elapsed time for each trial. After giving the subject some general instructions about the anagram task, the experimenter asked the subject to warm up for the test set of anagrams by first completing a short "practice" set. Each of the four anagrams in this practice set was solvable and of moderate difficulty. By means of an intercom, the experimenter signaled the subject from the control room when it was time to begin working each one and signaled again for the subject to stop working as soon as 90 seconds had elapsed. If the subject was able to solve the anagram is less than the time allotted, (s)he signaled the experimenter immediately by flipping the switch on the panel.

Following this practice phase, which was really included in the procedure to provide a pretest or baseline measure of the subject's performance, the experimenter communicated to the subject that the test set of anagrams would now be presented. The presentation of the 13 anagrams in this test series actually constituted the last two stages of the procedure. The middle stage of the procedure was an experimentally induced "failure" that all subjects were forced to experience due to the fact that only two of the first nine anagrams in the test set were really solvable. The last stage of the procedure was based on the presentation of the last four anagrams in the test series. This final set of anagrams actually comprised the posttest measure of the subject's performance. Because these anagrams were all solvable and of moderate difficulty, they were comparable to those used in the practice (pretest) set. In fact, the order of presentation of the two sets was counterbalanced so that the pretest set for one subject was used as the posttest set for another, and vice versa. Performance measures of speed and accuracy were taken for all of the pre- and posttest trials.

A pretest–posttest analysis of the speed (average time to completion) and error (total number incorrect) data revealed no significant differences for self-esteem on either measure ($Fs < 1$). However, as Fig. 5.1 indicates, there were pronounced differences for groups separated on the basis of attributional style. The performance of subjects who were predisposed to internalize their negative or "failure" outcomes became significantly slower ($p < .025$) and less accurate ($p < .005$) following their unsuccessful attempt to solve the insolvable anagrams. No impairment of performance was noted, however, for subjects predisposed to externalize their negative outcomes ($Fs < 1$). This finding is important insofar as it suggests that performance deficits that previously were attributed to self-esteem may really be due to attributional style. Thus, a crucial element underlying these deficits is apparently *not* low self-esteem per se, but the attributional tendency to internalize one's failures. It would probably not be appropriate to conclude from these data that such deficits have a strictly cognitive, as opposed to an affective, basis, but the data are certainly quite consistent with Weiner et al.'s assertion in Chapter 3 of this volume that attributions may provide a necessary precondition for any affective influences that may be operative.

Given this pattern of results, we might expect that attributional style would also account for the variations in performance that have previously been attributed to sex, especially in light of the findings of our first two studies. This, however, was clearly *not* the case. The obtained sex differences were of the same magnitude as those obtained for attributional style and were in the direction predicted by previous research: The performance of the female subjects was definitely impaired by the failure experience, becoming significantly slower ($p < .05$) and less accurate ($p < .005$), whereas the performance of the male subjects was essentially unimpaired by failure ($Fs < 1$).

An examination of Fig. 5.1 reveals that the independent effects of attributional style and sex are additive in nature. In other words, there is a performance deficit associated with internalizing failure and a separate performance deficit associated with being female. Thus, the disruption of performance by a failure experience is least evident for males who externalize failure and most evident for females who internalize it. Because attributional style and sex appear to independently contribute to these differences, how can these two effects be independently accounted for?

Realistically, any of a number of factors may be mediating the attributional style effect, and one can only speculate at this point about their relative importance. One possibility is that subjects who attribute failure internally may experience decreased motivation following a failure experience and simply quit trying. Another possibility is that subjects who internalize failure may make negative self-attributions (of incompetence, stupidity, etc.) that interfere with the attentional aspect of task performance by generating states of anxiety,

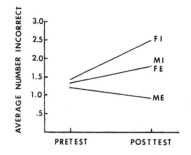 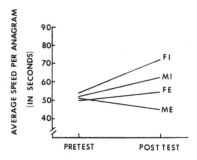

FIG. 5.1.   Measures of performance on a standard anagram task prior to (pretest) and following (posttest) an experimentally induced "failure" experience (Study 4). ME =   males who externalize failure; MI = males who internalize failure; FE = females who externalize failure; FI = females who internalize failure.

evaluation apprehension, and increased self-concern (Michenbaum, 1972; Sarason, 1973; Wine, 1971). In short, a number of attentional, cognitive, and/or emotional factors may be contributing to this effect, either singly or in combination. We give some additional attention to these factors in a moment, when we consider the similarities between the responses of clinically depressed patients and the responses of subjects who internalize negative outcomes as part of their attributional style.

In our attempt to interpret the obtained sex effect, we can benefit from the process of elimination, because our design precluded the possibility that this effect can be attributed to confounding differences in either attributional style or self-esteem. This leaves us in the unhappy position, however, of having to propose that females in general — regardless of their attributional style and self-esteem — may be socialized to react to failure or frustration in an achievement situation by essentially giving up (see Chapter 6, by Dweck & Goetz, this volume). If this explanation is correct, it is hardly surprising that women in this culture are generally less productive than men, because society has burdened them with an internalized motivational handicap that they must somehow learn to overcome if they are to succeed. Quite possibly, the present findings may have some important implications for our view of women in this culture and for our developing awareness of the unrealized potential that they possess.

ATTRIBUTIONAL STYLE, SELF-ESTEEM, SEX, AND DEPRESSION

Self-Esteem and Attributional Style Differences In Depression

Taken collectively, the present findings may have implications not only for our understanding of phenomena related to sex-role identification but for our understanding of clinical depression as well. When the relationships among attributional style, self-esteem, and depression are examined, it becomes evident that major areas of overlap are lowered self-evaluation and self-blame. Moreover, it is the cognitive aspects of these phenomena that have been receiving increasing emphasis in the literature on depression. For example, Lichtenberg (1957) and Megles and Bowlby (1969) stress that depressed patients' feelings of hopelessness typically stem from their belief that they will be incapable of attaining goals because of their incompetence. Their depression appears to deepen when the self-blame engendered by such thoughts becomes generalized from a limited number of specific situations to create a pervasive cognitive style. Similarly, Beck (1967) has hypothesized that the cognitive elements of depression are really more central to the disorder than the emotional factors. He contends that the patients' low self-esteem and thoughts of hopelessness generate their negative moods, in contrast to the generally held notion that the reverse is true. In his study of the verbalizations of depressed patients, Beck found frequently recurring themes of negative self-evaluation, self-blame, expectations of future incompetence, and hopelessness about themselves, their world, and their future.

The covert verbalizations or "self-talk" of depressed patients also suggest that they have developed a general predisposition or bias to interpret situations negatively, even when this response would seem to be totally inappropriate. The depressed patient we referred to at the beginning of this chapter provides a prime example of how a person might construe any situation — even the most neutral and commonplace — as providing evidence of his inadequacy. Beck (1967) gives another example of a young man who placed a similar construction on any situation in which another person seemed indifferent to him: "If a passerby on the street did not smile at him, he was prone to think he was inferior [p. 231]."

In general, depressed patients appear to base their self-appraisals upon a magnification of any failures (or supposed failures) and a minimization of favorable outcomes or attributes. They also make unfavorable social comparisons that contribute heavily to their feelings of inferiority. When comparing themselves to other members of their reference group, depressed patients consistently rate themselves as inferior in intelligence, status, productivity, attractiveness, or financial security. As all of these perceptions are typically associated with the inability or unwillingness to consider alternative interpretations that may be more reality-based, depressed patients exhibit a consistent attributional distortion that Beck (1967) refers to as "a bias against themselves [p. 234]." This attributional bias appears to be associated with their low self-esteem, as evidenced in a study by Laxer (1964). He found that hospitalized patients who exhibited a depressed mood but high other-blame expressed less negative self-evaluation than patients whose depressed mood was combined with a high degree of self-blame.

The self-esteem level and attributional style of clinically depressed patients appear to be essentially similar to those of the normal but low self-esteem subjects who were studied in our research. The data from the clinical literature on depression and from our own studies of attributional style indicate that both groups evidence negative self-evaluation and a general bias toward self-blame. As with our normal, low self-esteem subjects, the use of this characteristic attributional style by depressed patients may be determined in a number of ways. Specifically, this self-blaming tendency may either be a valid representation of the typical cause—outcome relationships experienced by depressed individuals, or it may reflect learned or motivated biases in the attribution process itself.

The foregoing explanations may eventually prove to have some heuristic value in developing the etiology of depression. For example, in some cases, depressive self-blame may be a veridical reflection of causal structure (the person really *does* mess things up habitually; he really *is* incompetent, inferior, etc.). In other cases, depressive self-blame may represent the person's reality-distorting attempt to maintain a state of "balance" between his causal attributions and his low self-esteem (Heider, 1975; Ickes & Harvey, 1977) or between his causal attributions and his perceived locus of control (Kelley, 1971). Finally, a combination of these factors may be operative such that depressive self-blame serves the function of a parsimonious "grand theory" — an integrated and internally consistent self-schema (Markus, 1977) that accounts veridically for at least some of the events in the person's life and can be used with perceptual selection and distortion to account (nonveridically) for others.

But regardless of how the self-blaming tendency becomes incorporated as a pervasive attributional style, this cognitive aspect of depression is important only as part of a more general syndrome that has affective and behavioral aspects as well. Data on the behavioral aspects of the syndrome indicate that depressives tend to have slower reaction times and less consistency in their responses than normals and neurotics (Friedman, 1964; Hall & Stride, 1954; Huston & Senf, 1952; Martin & Rees, 1966). They also tend to underestimate their performance on a number of tasks and/or overestimate their degree of debilitation and may do so even when they are performing adequately (Colbert & Harrow, 1968; Friedman, 1964; Loeb, Beck, & Diggory, 1971). In general, the behavioral aspects of the syndrome are essentially those characteristic of the "learned helplessness" syndrome and include lessened response initiation and slower learning of appropriate instrumental responses after conditions of uncontrollable aversive stimulation (Seligman, Klein, & Miller, 1976; Seligman & Maier, 1967).

Although learned helplessness responses have been demonstrated in humans (Hiroto, 1974; MacDonald, 1946; Racinskas, 1971; Thorton & Jacobs, 1971), it is interesting to note that recent work with human subjects has connected such responses to the kinds of attributions the subjects make to account for some aversive outcome or event. The study by Maracek and Mettee (1972), and the one by Dweck (1975) that we have previously described, are quite relevant in this regard, as are additional studies by Klein, Fencil-Morse, and Seligman (1976), Weiner and Sierad (1975), and Wortman, Panciera, Shusterman, and Hibscher (1976). Taken in sum, the results of these studies strongly indicate that supplying the subject with a new attribution about the cause of his outcome will have a pronounced affect on his subsequent motivation and performance, but only if the new attribution is acceptable to the subject in terms of the two criteria for acceptability specified by Heider (1958, pp. 172–173).

These data suggest that in humans, at least, the learned helplessness syndrome may be precipitated by negative self-attributions that undermine motivation and/or disrupt ongoing performance by increasing anxiety, evaluation apprehension, self-concern, etc. Viewed from this perspective, the learned helplessness syndrome bears a marked resemblance to the "exacerbation" syndrome described by Valins and Nisbett (1971) and Storms and McCaul (1976) and to the "belief creating reality" syndrome described by Snyder and his colleagues (Snyder & Swann, in press; Snyder, Tanke, & Berscheid, 1977). In all of these cases, attributions channel the person's subsequent perceptions and behaviors in a manner that leads to the apparent confirmation of the very attributions that engendered them. Through the spiraling processes of selective perception and behavioral confirmation, and process continues — locking the person into an increasingly vicious cycle in which "belief begets reality begets belief . . ." [Snyder & Swann, in press] .

Sex Differences in Depression

As a final note, we would like to point out that the striking similarity between the individual difference variables of sex and self-esteem in their relationship to attributional preference and performance is again apparent in their relationship to clinical depression. A report of a 1970 survey of the differential utilization of psychiatric facilities by men and women in the United States indicated that approximately twice as many women as men were treated for depression in that year (Cannon & Redick, 1973). Similarly, the data from a more recent and broad-scale review that examined reported rates of depression in men and women in a number of different countries for the last 40 years indicated that the 2:1 ratio is a fairly stable estimate of the degree to which the incidence of depression is greater for women than for men in Western culture (Weissman & Klerman, 1977). The authors of the review concluded that these sex differences "are, in fact, real and not an artifact of reporting or health care behavior [p. 109] ."

These findings are obviously quite consistent with the other data we have reviewed in this chapter. Collectively, all of the relevant evidence appears to implicate a cognitive-motivational syndrome of performance deficit, learned helplessness, and depression — a syndrome to which individuals with low self-esteem, and females in general, are most likely to be susceptible. We believe that enough of the pieces of this particular empirical puzzle are now available that the picture they form is reasonably clear to any who may care to look. To what extent future events may alter this picture remains to be seen.

SUMMARY

The results of the series of studies we have just described appear to fill some major gaps in the puzzle that interrelates the variables of sex, self-esteem, attribution, and performance. The first two studies indicated that attributional style may be an important aspect of the individual's personality — one that is integrally related to such central features of the self-conception as gender and self-esteem. The third study suggested that although attributional style may be relatively stable and not easily modified, changes in this aspect of self are associated with corresponding changes in the level of self-esteem. A final study indicated that debilitated task performance following a "failure" experience is related to differences in attributional style and gender but is not related to differences in self-esteem. Although the data from these studies offered little insight into the means by which various attributional styles develop, the striking similarity of these data to those obtained in the study of depression suggests a possible continuum relating low self-esteem normals to clinically depressed patients. It further suggests that attributional processes may be of central importance in the individual's capacity to establish and maintain a state of mental health.

ACKNOWLEDGMENTS

We would like to acknowledge our thanks to John Harvey, Robert Kidd, and Bernard Weiner for comments on an earlier draft of this chapter. We are also grateful to Herbert M. Lefcourt, who provided us with lists of anagrams scaled according to their level of difficulty; to Ben Harris, who brought the reports of sex differences in depression to our attention; and to Aaron Beck and Abigail Van Buren, who gave us permission to reprint the long quotations used as examples in the chapter's introduction. The preparation of this chapter was supported in part by a grant to the first author from the Graduate School of the University of Wisconsin.

REFERENCES

Arkin, R. M., & Duval, S. Focus of attention and causal attributions of actors and observers. *Journal of Experimental Social Psychology*, 1975, *11*, 427–438.

Barnes, R. D., Ickes, W., & Kidd, R. *Effects of the perceived intentionality and stability of another's dependency on helping behavior: A field experiment.* Manuscript submitted for publication, 1977.

Battle, E. Motivational determinants of academic task persistence. *Journal of Personality and Social Psychology*, 1965, *2*, 209–218.

Beck, A. T. *Depression: Clinical, experimental, and theoretical aspects.* New York: Hoeber, 1967.

Bem, S. L. The measurement of psychological androgyny. *Journal of Consulting and Clinical Psychology*, 1974, *42*, 155–162.

Bradley, G. W. Self-serving biases in the attribution process: A re-examination of the fact or fiction question. *Journal of Personality and Social Psychology*, 1978, *36*, 56–71.

Broverman, I. K., Vogel, S. R., Broverman, D. M., Clarkson, F. E., & Rosenkrantz, P. S. Sex-role stereotypes: A current appraisal. *Journal of Social Issues*, 1972, *28*, 59–78.

Cannon, M. S., & Redick, R. W. *Differential utilization of psychiatric facilities by men and women: United States, 1970.* Statistical note #81, Biometry Branch, Survey and Reports Section, 1973. (Available from National Institutes of Mental Health, 5600 Fishers Lane, Rockville, Md. 20852.)

Chapin, M., & Dyck, D. G. Persistence in children's reading behavior as a function of N length and attribution retraining. *Journal of Abnormal Psychology*, 1976, *85*, 511–515.

Colbert, J., & Harrow, M. Psychomotor retardation in depressive syndromes. *Journal of Nervous and Mental Disease*, 1968, *145*, 405–419.

Coopersmith, S. *The antecedents of self-esteem.* San Francisco: Freeman, 1967.

Crandall, V. Sex differences in expectancy of intellectual and academic reinforcement. In C. P. Smith (Ed.), *Achievement-related motives in children.* New York: Russell Sage, 1969.

Crandall, V., Katkovsky, W., & Crandall, V. Children's beliefs in their own control of reinforcements in intellectual-academic achievement situations. *Child Development*, 1965, *36*, 91–109.

Cruz Perez, R. The effects of experimentally induced failure, self esteem and sex on cognitive differentiation. *Journal of Abnormal Psychology*, 1973, *81*, 74–79.

Deaux, K. Sex: A perspective on the attribution process. In J. Harvey, W. Ickes, & R. Kidd (Eds.), *New directions in attribution research* (Vol. 1). Hillsdale, N.J.: Lawrence Erlbaum Associates, 1976.

Deaux, K., & Emswiller, T. Explanations of successful performance on sex-linked tasks: What is skill for the male is luck for the female. *Journal of Personality and Social Psychology*, 1974, *29*, 80–85.

Deaux, K., & Farris, E. *Attributing causes for one's own performance: The effects of sex, norms and outcomes.* Unpublished manuscript, Purdue University, 1974.

Deaux, K., White, L., & Farris, E. Skill vs. Luck: Field and laboratory studies of male and female preferences. *Journal of Personality and Social Psychology*, 1975, *32*, 629–636.

Duval, S., & Hensley, V. Extensions of objective self-awareness theory: The focus of attention-causal attribution hypothesis. In J. Harvey, W. Ickes, & R. Kidd (Eds.), *New directions in attribution research* (Vol. 1). Hillsdale, N.J.: Lawrence Erlbaum Associates, 1976.

Duval, S., & Wicklund, R. A. Effects of objective self-awareness on attribution of causality. *Journal of Experimental Social Psychology*, 1973, *9*, 17–31.

Dweck, C. S. The role of expectations and attributions in the alleviation of learned helplessness. *Journal of Personality and Social Psychology*, 1975, *31*, 674–685.

Dweck, C. S., & Reppucci, N. D. Learned helplessness and reinforcement responsibility in children. *Journal of Personality and Social Psychology*, 1973, *25*, 109–116.

Feather, N. T. Some personality correlates of external control. *Australian Journal of Psychology*, 1967, *19*, 253–260.

Feather, N. T. Change in confidence following success or failure as a predictor of subsequent performance. *Journal of Personality and Social Psychology*, 1968, *9*, 38–46.

Feather, N. T. Attribution of responsibility and valence of success and failure in relation to initial confidence and task performance. *Journal of Personality and Social Psychology*, 1969, *13*, 129–144.

Feather, N. T., & Simon, J. G. Reactions to male and female success and failure in sex-linked occupations: Impressions of personality, causal attributions and perceived likelihood of different consequences. *Journal of Personality and Social Psychology*, 1975, *31*, 20–31.

Fischhoff, B. Attribution theory and judgment under uncertainty. In J. Harvey, W. Ickes, & R. Kidd (Eds.), *New directions in attribution research* (Vol. 1). Hillsdale, N.J.: Lawrence Erlbaum Associates, 1976.

Fish, B., & Karabenick, S. A. Relationship between self-esteem and locus of control. *Psychological Reports*, 1971, *29*, 181.

Fitch, G. Effects of self-esteem, perceived performance and choice on causal attributions. *Journal of Personality and Social Psychology*, 1970, *16*, 311–315.

Freud, S. *A general introduction to psychoanalysis.* New York: Liveright Publishing Corp., 1920.

Friedman, A. S. Minimal effects of severe depression on cognitive functioning. *Journal of Abnormal and Social Psychology*, 1964, *69*, 237–243.

Gilmore, T., & Minton, H. Internal vs. external attribution of task performance as a function of locus of control, initial confidence and success–failure outcome. *Journal of Personality*, 1974, *42*, 159–174.

Glass, D., & Singer, J. *Urban stress.* New York: Academic Press, 1972.

Hall, K. R. L., & Stride, E. Some factors affecting reaction times to auditory stimuli in mental patients. *Journal of Mental Science*, 1954, *100*, 462–477.

Heider, F. *The psychology of interpersonal relations.* New York: Wiley, 1958.

Heider, F. *On balance and attribution.* Paper presented at the Symposium on Social Networks, Dartmouth University, 1975.

Hiroto, D. S. Locus of control and learned helplessness. *Journal of Experimental Psychology*, 1974, *102*, 187–193.

Huston, P. E., & Senf, R. Psychopathology of schizophrenia and depression. I. Effect of amytal and amphetamine sulfate on level and maintenance of attention. *American Journal of Psychiatry*, 1952, *109*, 131–138.

Ickes, W., & Barnes, R. D. The role of sex and self-monitoring in unstructured dyadic interactions. *Journal of Personality and Social Psychology*, 1977, *35*, 315–330.

Ickes, W., & Barnes, R. D. Boys and girls together – and alienated: On enacting stereotyped sex roles in mixed-sex dyads. *Journal of Personality and Social Psychology*, in press.

Ickes, W., & Harvey, J. *Fritz Heider: A biographical sketch.* Paper presented at CHEIRON, The International Society for the History of The Behavioral and Social Sciences, 9th Annual meeting, University of Colorado at Boulder, June 11, 1977.

Ickes, W., & Kidd, R. An attributional analysis of helping behavior. In J. Harvey, W. Ickes, & R. Kidd (Eds.), *New directions in attribution research* (Vol. 1). Hillsdale, N.J.: Lawrence Erlbaum Associates, 1976.

Ickes, W., & Layden, M. A. Unpublished data, 1975. (Available from Dr. William Ickes, Dept. of Psych., Univ. of Wisconsin, W. J. Brogden Bldg., 1202 W. Johnson St., Madison, Wisc. 53706.)

Kanouse, D. E., & Hanson, L. R. Negativity in evaluations. Morristown, N.J.: General Learning Press, 1971.

Kelley, H. H. Attribution theory in social psychology. In D. Levine (Ed.), *Nebraska Symposium on Motivation.* Lincoln, Neb.: University of Nebraska Press, 1967.

Kelley, H. H. Attribution in social interaction. Morristown, N.J.: General Learning Press, 1971.

Klein, D. C., Fencil-Morse, E., & Seligman, M. E. P. Learned helplessness, depression and the attribution of failure. *Journal of Personality and Social Psychology*, 1976, *33*, 508–516.

Langer, E. J. The illusion of control. *Journal of Personality and Social Psychology*, 1975, *32*, 311–328.

Langer, E. J., & Roth, J. Heads I win, tails it's chance: The illusion of control as a function of the sequence of outcomes in a purely chance task. *Journal of Personality and Social Psychology*, 1976, *32*, 951–955.

Laxer, R. M. Relation of real self-rating to mood and blame and their interaction in depression. *Journal of Consulting Psychology*, 1964, *28*, 214–219.

Lichtenberg, P. A definition and analysis of depression. *Archives of Neurology and Psychiatry*, 1957, *77*, 519–527.

Loeb, A., Beck, A. T., & Diggory, J. Differential effects of success and failure on depressed and nondepressed patients. *Journal of Nervous and Mental Disease*, 1971, *152*, 106–114.

MacDonald, A. Effects of adaptation to the unconditioned stimulus upon the formation of conditioned avoidance responses. *Journal of Experimental Psychology*, 1946, *36*, 1–12.

Maracek, J., & Mettee, D. Avoidance of continued success as a function of self-esteem, level of esteem certainty and responsibility for success. *Journal of Personality and Social Psychology*, 1972, *22*, 98–107.

Markus, H. Self-schemata and processing information about the self. *Journal of Personality and Social Psychology*, 1977, *35*, 63–78.

Martin, I., & Rees, L. Reaction times and somatic reactivity in depressed patients. *Journal of Psychosomatic Research*, 1966, *9*, 375–382.

Megles, F. T., & Bowlby, J. Types of hopelessness in psychopathological process. *Archives of General Psychiatry*, 1969, *20*, 1313–1320.

Michenbaum, D. H. Cognitive modification of test anxious college students. *Journal of Consulting and Clinical Psychology*, 1972, *39*, 370–379.

Miller, D. T., & Ross, M. Self-serving biases in the attribution of causality: Fact or fiction? *Psychological Bulletin*, 1975, *82*, 213–225.

Mischel, W., Zeiss, R., & Zeiss, A. Internal and external control and persistence. *Journal of Personality and Social Psychology*, 1974, *29*, 265–278.

Monson, T. C., & Snyder, M. K. Actors, observers, and the attribution process: Toward a reconceptualization. *Journal of Experimental Social Psychology*, 1977, *13*, 89–111.

Montanelli, D. S., & Hill, K. T. Children's achievement expectations and performance as a function of two consecutive reinforcement experiences, sex of subject, and sex of experimenter. *Journal of Personality and Social Psychology*, 1969, *13*, 115–128.

Morse, S., & Gergen, K. Social comparison, self-consistency and the concept of self. *Journal of Personality and Social Psychology*, 1970, *16*, 148–156.

Phares, E. J. *Locus of control in personality*. Morristown, N.J.: General Learning Press, 1976.

Platt, J., Eisenman, R., & Darbes, A. Self-esteem and internal–external control: A validation study. *Psychological Reports*, 1970, *26*, 162.

Racinskas, J. R. *Maladaptive consequences of loss or lack of control over aversive events*. Unpublished doctoral dissertation, Waterloo University, Ontario, Canada, 1971.

Regan, D. T., & Totten, J. Empathy and attribution: Turning observers into actors. *Journal of Personality and Social Psychology*, 1975, *32*, 850–856.

Rosenbaum, R. M. *A dimensional analysis of the perceived causes of success and failure*. Unpublished doctoral dissertation, University of California, Los Angeles, 1972.

Ross, L. The intuitive psychologist and his shortcomings: Distortions in the attribution process. In L. Berkowitz (Ed.), *Advances in experimental social psychology* (Vol. 10). New York: Academic Press, 1977.

Rotter, J. B. Generalized expectancies for internal vs. external control of reinforcement. *Psychological Monographs*, 1966, *80*, 1–28.

Ryckman, R., & Rodda, W. Confidence maintenance and performance as a function of chronic self-esteem and initial task experience. *Psychological Record*, 1972, *22*, 241–247.

Ryckman, R. M., & Sherman, M. F. Relationship between self-esteem and internal–external control for men and women. *Psychological Reports*, 1973, *32*, 1106.

Sarason, I. G. Test anxiety and cognitive modeling. *Journal of Personality and Social Psychology*, 1973, *28*, 58–61.

Seligman, M. E. P., Klein, D. C., & Miller, W. R. Depression. In H. Leitenberg (Ed.), *Handbook of behavior modification and behavior therapy*. Englewood Cliffs, N.J.: Prentice–Hall, 1976.

Seligman, M. E. P., & Maier, S. F. Failure to escape traumatic shock. *Journal of Experimental Psychology*, 1967, *74*, 1–9.

Shrauger, J., & Rosenberg, S. Self-esteem and the effects of success and failure feedback on performance. *Journal of Personality*, 1970, *38*, 404–417.

Snyder, M. K., & Swann, W. B. Behavioral confirmation in social interaction: From social perception to social reality. *Journal of Experimental Social Psychology*, in press.

Snyder, M. K., Tanke, E. D., & Berscheid, E. Social perception and interpersonal behavior: On the self-fulfilling nature of social stereotypes. *Journal of Personality and Social Psychology*, 1977, *35*, 656–666.

Spence, J. T., Helmreich, R., & Stapp, J. Ratings of self and peers on sex role attributes and their relations to self esteem and conceptions of masculinity and femininity. *Journal of Personality and Social Psychology*, 1975, *32*, 29–39.

Solley, C. M., & Stagner, R. Effects of magnitude of temporal barriers, type of goal and perception of self. *Journal of Experimental Psychology*, 1956, *51*, 62–70.

Storms, M. D. Videotape and the attribution process: Reversing actors' and observers' point of view. *Journal of Personality and Social Psychology*, 1973, *27*, 165–175.

Storms, M. D., & McCaul, K. D. Attribution processes and emotional exacerbation of dysfunctional behavior. In J. Harvey, W. Ickes, & R. Kidd (Eds.), *New directions in attribution research* (Vol. 1). Hillsdale, N.J.: Lawrence Erlbaum Associates, 1976.

Taylor, S. E., & Fiske, S. T. Point of view and perceptions of causality. *Journal of Personality and Social Psychology*, 1975, *32*, 439–445.

Thornton, J. W., & Jacobs, P. D. Learned helplessness in human subjects. *Journal of Experimental Psychology*, 1971, *87*, 369–372.

Valins, S., & Nisbett, R. E. Attribution processes in the development and treatment of emotional disorders. Morristown, N.J.: General Learning Press, 1971.

Valle, V. A., & Frieze, I. H. Stability of causal attributions as a mediator in changing expetations for success. *Journal of Personality and Social Psychology*, 1976, *33*, 579–587.

Wegner, D. M., & Finstuen, K. Observer's focus of attention in the simulation of self-perception. *Journal of Personality and Social Psychology*, 1977, *35*, 56–62.

Weiner, B. *Achievement motivation and attribution theory*. Morristown, N.J.: General Learning Press, 1974.

Weiner, B., Frieze, I., Kukla, A., Reed, L., Rest, S., & Rosenbaum, R. M. *Perceiving the causes of success and failure*. Morristown, N.J.: General Learning Press, 1971.

Weiner, B., & Sierad, J. Misattribution for failure and enhancement of achievement strivings. *Journal of Personality and Social Psychology*, 1975, *31*, 415–421.

Weissman, M. M., & Klerman, G. L. Sex differences and the epidemiology of depression. *Archives of General Psychiatry*, 1977, *34*, 98–111.

Wetter, R. *Levels of self-esteem associated with four sex role categories*. Paper presented at the 83rd annual meeting of the American Psychological Association, Chicago, August 30, 1975.

Wine, J. Test anxiety and direction of attention. *Psychological Bulletin*, 1971, *76*, 92–104.

Wortman, C. B. Causal attributions and perceived control. In J. Harvey, W. Ickes, & R. Kidd (Eds.), *New directions in attribution research* (Vol. 1). Hillsdale, N.J.: Lawrence Erlbaum Associates, 1976.

Wortman, C., Panciera, L., Shusterman, L., & Hibscher, J. Attributions of causality and reactions to uncontrollable outcomes. *Journal of Experimental Social Psychology*, 1976, *12*, 301–316.

Wylie, R. C. *The self concept*. Lincoln, Neb.: University of Nebraska Press, 1961.

II ATTRIBUTION AT THE INTERPERSONAL LEVEL

At the interpersonal level, attributional processes lose none of their importance but gain considerably in their complexity. This added complexity is due largely to the fact that behavior occurring in a social context may be influenced not only by the individual's own attributions but by the real and imagined attributions of other people as well. It is, perhaps, somewhat surprising that attribution theorists and researchers have not given more attention to the interactive nature of inference making in social relationships and to the special problems that arise when one person makes attributions about another's attributions. Indeed, given the social pyschological orientation of most of the people who work in the area of attribution, the relative neglect of these issues may appear to be quite glaring. Any apparent reluctance to tackle these issues may be somewhat more understandable, however, when one considers the obvious complexity of the problems raised at the interpersonal level and the need for an adequate understanding of attributional processes at the personal level as the basis for these more extensive inquiries.

But if attributional problems in a social context are particularly complex, they are also particularly deserving of study. Increasingly, theory and research in the area suggest that attributions, publicly expressed, provide the common currency with which individuals negotiate their conceptions of social or even physical reality. As the various chapters in this section will indicate, attributions play a major role in the interrelated processes of socialization, social perception, and social influence.

Some striking evidence of how self-attributions acquired during the socialization process can affect subsequent behavior is presented by Carol Dweck and Therese Goetz in their chapter, "Attributions and Learned Helplessness." After noting that in earlier research "helpless" children were found to differ from "persistent" children in the kinds of attributions they made to account for failure, the authors describe how Dweck and her colleagues explored some developmental antecedents of these differences in a series of interlocking field—observational and laboratory—experimental studies. The data they present bear on the interesting paradox that girls appear to develop a helpless orientation more readily than boys during the grade-school years, even through evidence suggests that girls consistently receive higher grades, more praise, and less criticism from their teachers. By observing and coding the actual evaluative feedback given to boys and girls in fourth- and fifth-grade classrooms, Dweck and her colleagues determined that for girls negative feedback was almost exclusively directed at the intellectual aspects of their work; whereas, for boys, it was most often directed at the nonintellectual aspects of their work or at their conduct. This and related investigations strongly indicate that the source and specificity of negative feedback — rather than its absolute or proportional amount — may be the critical factors leading to the development of a learned helplessness orientation in children.

Not only do attributions figure prominently in the development of an orientation to achievement settings, they are also quite important in shaping the development of conceptions of morality. In his chapter, "Attribution, Socialization, and Moral Decision Making," Richard Dienstbier discusses two major issues relating to the role of attribution in moral decision making and moral behavior. The first issue concerns the question of how a person defines a situation as one calling for a moral decision. Data from a series of studies are presented that indicate that fairly subtle aspects of a situation can influence the perception of its moral relevance, even when the act in question (for example, cheating) is ostensibly immoral across all situations in which it occurs. The second issue concerns the question of how the level and attributed cause(s) of a person's arousal in a temptation situation affect the likelihood that a behavioral transgression will take place. The results of a number of studies by Dienstbier and his colleagues are reported relevant to this question; in general, they suggest that arousal attributed to internal injunctions against the prescribed behavior will be sufficient to inhibit it, whereas arousal attributed to a morally irrelevant source will not. In a final section, Dienstbier explores the implications of arousal-distribution processes for moral induction, the use of punishment during socialization, and the etiology and treatment of psychopathy.

Dienstbier's chapter reintroduces a theme that has been evident throughout many ofthe previous chapters of this volume — that attributions and affective responses are complexly interrelated. This theme is extended and systematically developed by Dennis Regan in "Attributional Aspects of Interpersonal Attraction." His review of the relevant research in the area of attraction suggests that "attribution and attraction seem mutually interrelated in complex ways [p. 229]," although empirical studies often obscure this interdependence by imposing the forced separation of these factors into independent and dependent variable categories. Indications of their interdependence emerge, however, when Regan attempts to conceptually integrate findings in the literature that reflect three distinct kinds of effects: (1) effects of causal attributions about another's behavior on liking for the other; (2) effects of self-attributions on liking for the other; and (3) effects of the degree of attraction to the other on attributions about the other's behavior. Regan argues that although some of this empirical evidence is methodologically weak or of dubious external validity, the three lines of research generally converge to demonstrate the centrality of attributional processes in interpersonal attraction. His suggestions regarding the type of work that is most needed in the area of attraction should prove to be of value in guiding the course of future research.

In some situations, attribution may be enmeshed in the complex psychological dynamics of an individual's sometimes desperate attempts to understand and thereby more adequately cope with trying personal problems. Perhaps no situation represents this class of situations better than that involving conflict and separation in close relationships. In their chapter, "Attribution in the Context of Conflict and Separation in Close Relationships," John Harvey, Gary Wells, and Marlene Alvarez report data deriving from research on people's use of attributions in nonmarital close relationships involving conflict and in marital relationships involving recent separation. Their data suggest that sex differences in attributional perspective, divergence in attributions, and — importantly — the lack of understanding of divergence may be related to conflict in the relationships of young couples. Divergent perspectives also appear to be involved in the attributions of recently separated individuals; however, Harvey, Wells, and Alvarez's data suggest that many other attributional tendencies are operative in separation as well. Individuals in this situation report increased and intensive cognitive activity — often of a justificatory and defensive nature — in attempting to understand what went wrong in their relationships.

The final chapter of this section is an attempt to extend Kelley's (1967) notions regarding the transmission of attributional information as a primary means of social influence. In "Attributional Strategies of Social Influence," Avi Gottlieb and William Ickes propose that influence is typically mediated by a change in the attributor's perception of a given event or entity. They suggest that though such a change is sometimes effected by means of an emotional appeal or by virtue of the personality and characteristics of the communicator, it is most commonly effected by the provision of attributional information of the types specified in Kelley's (1967) model — consensus, consistency, and distinctiveness. Following a review of Kelley's model and an examination of the implicit assumptions he makes about the role of attributions in social influence, the authors propose a more differentiated model of the processes by which the various types of attributional information are related to attitude and behavior change. Their discussion of the major aspects and implications of the model addresses such issues as the overlap in the three types of attributional information, their relative effectiveness in altering perceptions of socially versus physically defined events or entities, and the strategic use of such influence "tactics" as interpolation, extrapolation, differential salience, and informational contrast.

6 Attributions and Learned Helplessness

Carol S. Dweck
Therese E. Goetz
University of Illinois

Learned helplessness in achievement situations exists when an individual perceives the termination of failure to be independent of his responses. This perception of failure as insurmountable is associated with attributions of failure to invariant factors, such as a lack of ability, and is accompanied by seriously impaired performance. In contrast, mastery-oriented behavior — increased persistence or improved performance in the face of failure — tends to be associated with attributions of failure to variable factors, particularly to a lack of effort. One would think that persistence following failure would be related to one's level of ability or to one's history of success in that area. Yet our research with children has shown that, compared to achievement cognitions, these variables are relatively poor predictors of response to failure.

In this chapter we examine the role of attributions in determining the response to failure of learned helpless and mastery-oriented children. First we review research that establishes the link between attributions and reactions to failure and that documents the nature of the performance change occasioned by failure. Next we explore the generality of individual differences in helplessness, specifically sex differences, and present findings that indicate how these individual differences develop. In addition, we show how attributions can mediate the generalization of failure effects to novel achievement and academic situations and demonstrate how this phenomenon can account for individual differences in particular academic areas such as sex differences in verbal and mathematical achievement. We then describe current research on the applicability of this learned helplessness analysis, developed in intellectual-achievement failure situations, to children's responses to failure in social situations. Finally, we present evidence that helpless and mastery-oriented children differ not only in the attributions they report when asked — in less structured situations — in the timing of their achievement-related cognitions and, indeed, in the role played by causal attributions.

ATTRIBUTIONS, HELPLESSNESS, AND RESPONSE TO FAILURE

Learned helplessness was first investigated systematically in animals by Seligman and Maier (1967), who found that subjects who were pretreated with unavoidable, inescapable shock in one situation subsequently failed to avoid or escape from shock in a different situation in which control animals readily learned the avoidance contingency. In contrast to the normal animals, whose behavior following shock was characterized by intense activity, these animals tolerated extreme amounts of shock passsively. Few attempts were made to prevent its recurrence. Even with forced exposure to the contingency between respond- and shock termination, many trials were required before all of the animals began responding reliably on their own (Seligman, Maier, & Geer, 1968). The authors proposed that the animals exposed to prior, inescapable shock learned that the probability of shock termination given a response was equal to the probability of shock termination given no response — or, that shock termination and responding were independent. In other words, in the same way that organisms can learn about contingencies, they can learn about the absence of contingencies. Moreover, this learning can generalize to similar situations, seriously decreasing the probability of attempting instrumental responses and therefore of recognizing the presence of a contingency when one in fact exists.

Analogous divergent behavioral patterns are apparent when children are confronted with failure in intellectual problem-solving situations. Some children tend actively to pursue alternative solutions when they encounter failure, often to a greater extent than prior to failure. The performance of others, however, undergoes marked deterioration, with some children becoming literally incapable of solving the identical problems they solved with relative ease only shortly before. Our first question, then, was whether this behavioral parallel between the experimental animals in the Seligman and Maier studies and the children in problem-solving situations is accompanied by parallel cognitions. Do children who tend to give up in the face of failure tend to see the remedy as beyond their control, to see the probability of success following failure as negligible whether they respond or not? To answer this question, a study was conducted (Dweck & Reppucci, 1973) in which a series of problems was administered to children by two experimenters, one of whom gave soluble problems, the other insoluble ones.[1] Problems from the success and the failure experimenters were randomly interspersed. After a number of trials, however, the failure experimenter began to administer soluble problems, ones that were virtually identical to some that had been administered by the success experimenter earlier. A surprisingly large number of children failed to solve these problems, despite the fact that they were motivated to solve as many problems as they possibly could (they earned tokens, redeemable for highly attractive prizes, for correct solutions), despite the fact that they had solved similar problems from the success experimenter, and despite the fact that they continued to show a rather large practice effect on problems administered by the success experimenter.

On the basis of their performance on the soluble problems from the failure experimenter compared to the analogus problems from the success experimenter, the children were split at the median into two groups: those who failed to solve the problems or who showed the greatest increases in solution times versus those who tended to maintain or improve their performance. It should be emphasized that these two groups had not differed on initial performance. If anything, among the females, those who subsequently showed the most deterioration had shown superior performance under success. What did distinguish the two groups were their attributional patterns, that is, their characteristic ways of explaining their intellectual academic successes and failures (see Weiner, 1972, 1974).

Children's attributions were assessed by means of the Intellectual Achievement Responsibility Scale (Crandall, Katkovsky, & Crandall, 1965), a forced-choice attribution questionnaire in which achievement situations with positive and negative outcomes are depicted and the child selects the one of two alternatives that best describes how he would explain that outcome. One of the alternatives always presents an external factor as the cause of the success or failure, whereas the other alternative presents an internal factor, either one's ability or one's effort, as the cause.

We found that children who persisted in the face of failure placed significantly more emphasis on motivational factors as determinants of outcomes. Attributions of failure to lack of motivation imply that failure is surmountable through effort, a factor that is generally perceived to be under the control of the individual. Children whose performance deteriorated tended more than

---

[1]In this and in subsequent studies the participants were late grade-school-age children (grades four to six). Throughout the research great care was taken to ensure that every child left the experimental situation feeling that his performance had been commendable. For example, in a typical study following the failure trials, children were given mastery experiences, were assured that they had conquered a difficult task more quickly than most, and were told that they had done so well there was no need to complete the remaining problems. Thus the procedure incorporated what is essentially persistence training.

persistent children to place the blame for their failures on largely uncontrollable external factors rather than effort. When they did take responsibility for failures, they were relatively more likely than the persistent children to blame their lack of ability. Both attributions to external factors and attributions to lack of ability imply that failure is difficult to overcome, particularly within a given situation where transformations in, say, the teacher's attitude or in one's ap-attributions — minimizing the role of effort — was more characteristic of girls than boys.

Given that helpless children emphasize the unchangeable nature of failure and de-emphasize the role of effort in overcoming failure, it is not surprising that their performance suffers. But terms like "decreased persistence," "reduced effort," or "impaired performance" can encompass a multitude of behaviors. What then is the precise nature of the performance decrement that takes place when failure occurs? We know that helpless children do not merely slow down and take greater care, since their error rate, as well as their response latency, increases with failure. Do they simply withdraw when failure occurs and begin to respond randomly? Do they at first attempt sophisticated alternative solutions but abandon them more quickly than the more persistent children? Do they, under the pressure, slip a notch or two to less mature problem-solving strategies that are easier to execute but less efficient and less likely to yield correct answers? Or, does the sophistication of their strategies undergo a gradual erosion as they experience successive failures until they become incapable of problem solving? By the same token, one might ask what accounts for the often improved performance of mastery-oriented children following failure. Again, citing an undifferentiated construct like "effort" is somewhat unsatisfying as an explanation. After all, one would hardly expect a knotty intellectual problem to yield to the same kind of exertion as a demanding physical task. Clearly, "heightened effort" must be broadened to include alterations in task strategy in addition to increases in speed, concentration, force, and the like.

In order to examine the precise nature of the performance change exhibited by helpless and mastery-oriented children under failure, children were given a task that allowed us to monitor moment—to—moment changes in their problem-solving strategies (Diener & Dweck, 1978, Study I). Children, categorized as helpless or mastery-oriented on the basis of their tendency to neglect or emhasize effort as a determinant of failure, were first trained on eight soluble discrimination-learning problems. For each problem they were shown pictures on cards, two at a time, that differed in three respects: the shape of the form depicted (e.g., circle or triangle), the color of the form (e.g., red or green), and the symbol that appeared in the cente of the form (e.g., star or dot). One of the six stimulus values (e.g., red) was correct for the whole problem, and the child's task was to discover which one by utilizing the feedback ("correct" or "wrong") that was provided. On the first training problem feedback was given after every choice, but by the seventh problem children received feedback only after every fourth response. From the child's choices in the block of four trials before feedback was given, we could infer his hypothesis. From the sequence of hypotheses over blocks of trials, we could infer his problem-solving strategy. Strategies could be ordered in terms of maturity, the most sophisticated being the one that on the average yields the solution most quickly and the least sophisticated being one that can never lead to problem solution and that is exhibited most often by the youngest children.

All children in the study were able to employ problem-solving strategies that were successful in reaching the correct solutions on the training problems. On all measures of performance on the training problems — sophistication of strategy, trials to criterion on each problem, number of hints required to reach criterion on the training problems, efficiency of feedback utilization — helpless and mastery-oriented children were virtually identical.

When children reached criterion on the eighth training problem, four insoluble test problems were administered. In essence, every fourth trial the child was told that his choice was incorrect. (The number of trials administered on the test problems was limited, so that the feedback could conceivably be veridical.) Changes in problem-solving strategy over the four failure problems were monitored. The mastery-oriented children not only were able to maintain the sophistication of the problem-solving strategy they had displayed earlier, but, to our surprise, an appreciable proportion (26.3%) of them actually began using even more mature strategies — those typical of older children. It would appear that in response to the limited number of trials they were now allowed, they attempted and perfected a more advanced form of problem solving than they appeared to be capable of before failure. Helpless children, in contrast, showed a steady regression in strategy across the failure problems. By the second trial, 37.9% had already abandoned strategies that could lead to solution, and by the fourth trial, 68.9% were failing to show any sign of a useful strategy. Of these, two-thirds showed the repeated choice of a single stimulus value regardless of feedback and one-third showed alternating choice of the left and right stimulus, regardless of its nature. Although some girls displayed a sudden dramatic decline in strategy use near the beginning, most of the children displayed a more gradual regression over the course of the failure trials. None showed any evidence of attempting more sophisticated strategies.

After the test problems, children were asked to generate attributions for their failures: "Why do you think you had trouble with these problems?" Over half of the helpless children cited a lack of ability (e.g., "I'm not smart enough") as the cause. None of the mastery-oriented children offered this explanation. Instead they tended to cite effort and other potentially surmountable factors. Again, despite the equivalent initial proficiency and successes of the helpless and mastery-oriented children, the cognitions they entertained about their failures differed and their performance over failure trials became progressively divergent. In fact, by the time helpless children made their postfailure attributions to lack of ability, they could perhaps find evidence for this in their performance on the later trials.

Thus helplessness, the perception of failure as uncontrollable, does indeed predict responses to failure. But does it cause them? If so, we reasoned, then it should be possible to alter children's responses to failure by altering their attributions for failure (Dweck, 1975). Specifically, helpless children taught to attribute their failures to a lack of effort — as mastery-oriented children do — should become more able to cope with failure effectively. To test this possibility, a number of extremely helpless children were identified and were assigned to one of two relatively long-term treatment procedures. All of the children showed the attributional pattern indicative of helplessness on the Intellectual Achievement Responsibility Scale, and, on a questionnaire that directly pitted attributions of failure to lack of effort against attributions of failure to lack of ability, all showed a strong tendency to endorse the lack of ability alternative. Moreover, every one evidenced severely impaired performance following the occurrence of failure, with some being unable to recover baseline performance for several days after a relatively mild failure experience.

In order to assess precisely the effects of failure on the problem-solving performance of the helpless children before and after treatment, stable baselines of speed and accuracy were established on math problems. When the baseline performance had stabilized, failure trials were interpolated between the sets of problems the children had been solving daily. The decrease in the speed and accuracy on the problems that followed the failure problems (compared to the identical problems on the previous day) was used as the index of the disruptive effects of failure.

Following this assessment of their reactions to failure, the children were given one of two treatment procedures in a different situation. Half of the children received only success experiences in the treatment situation, a procedure recommended by advocates of what might be called the "deprivation theory" of maladaptive responses to failure. This position holds that poor reactions to failure stem from a lack of confidence in one's abilities, which in turn stems from a scarcity of success experiences in a given area. If such children, then, could be supplied with the missing success, their confidence would rise and would bolster them against the negative effects of failure. Indeed, there is evidence to suggest that reported expectations of success are correlated with persistence in the face of failure (e.g., Battle, 1965; Feather, 1966). Moreover, the treatment was one that highlighted the contingency between the child's efforts and his successes.

The other half of the children received attribution retraining. Although here, too, success predominated, several failure trails were programmed each day. On the occasions when failure occurred the child's actual performance was compared to criterion performance and the failure was explicitly attributed by the experimenter to a lack of effort. Thus the children in this group received direct instruction in how to interpret the causes of their failures. Both treatments were carried out for 25 daily sessions. At the middle and the end of training, children were returned to the original situation, and the effects of failure on their performance were again assessed.

By the middle of training, all of the children in the attribution retraining condition showed improvement in their response to failure, although all still showed some impairment when their postfailure performance was compared to that of the previous, prefailure day. However, by the end of training, none of the children showed any appreciable impairment, and unexpectedly most of them showed improvement in performance as a result of failure. According to the investigator (Dweck), who tended to eavesdrop during the testing, several of the children, upon encountering failure, were heard to mutter such things as "I missed that one. That means I have to try harder." In addition, when the effort versus ability attribution measure was readministered by an individual unconnected with the study, children in the attribution retraining treatment showed significant increases in their tendency to emphasize effort over ability as a determinant of failure.

The children in the success-only treatment, as well as the proponents of the "deprivation model," fared far more poorly. This group showed no improvement on the midtraining and postraining failure tests and no change on the attribution measures. Some of the children even showed a tendency to react somewhat more adversely to failure than they had before the start of this treatment. (Of course, children in this group were subsequently given attribution retraining.) Thus even though the performance of these children had been showing steady improvement during training and during the nonfailure days of testing, failure remained a cue for continued failure, and they remained incapable of dealing with it competently. Just as level of proficiency did not predict mastery-oriented responses to failure, a history of success did not do so either. In contrast, intervention at the level of failure attributions can essentially eliminate the deleterious effects of failure.

Why does prior mastery fail to predict future mastery attempts? Why is success not enough? To begin to answer this question, we turn to the issue of how helplessness develops.

THE DEVELOPMENT OF HELPLESSNESS

We have studied the development of helplessness· in the context of sex differences in children's attributions for and responses to failure — a context in which an intriguing paradox exists. It is oftened assumed that girls learn to blame their abilities for failure because teachers and other adults view them as less competent than boys and somehow convey this to them. This seems to be far from the case. If anything, in grade school it is the girls who are receiving this information that they are the ones who possess the ability; hence, the paradox.[2]

Girls, on the average, are far more successful than boys in the academic arena during the elementary school years.They receive consistently higher grades (e.g., McCandless, Roberts, & Starnes, 1972) and regularly outscore boys on tests of reading achievement (see Asher & Markel, 1974). In addition, they receive less criticism from teachers (see Brophy & Good, 1974) and are, in fact, more highly regarded by teachers on almost every conceivable dimension: skills, motivation, personal characteristics, conduct, and more (Coopersmith, 1967; Digman, 1963; Stevenson, Hale, Klein, & Miller, 1968). One would hardly call this discrimination against females. What is more, girls themselves think that teachers believe them to be smarter, that teachers believe they work harder, and that teachers like girls better (Dweck, Goetz, & Strauss, 1977).

Yet despite this record of success and this largely benign environment, girls show far greater evidence of helplessness than boys when they receive failure feedback from adult evaluators. Girls place less emphasis than boys on motivational factors as determinants of failure and are more likely than boys to attribute failure feedback to a lack of ability (Dweck & Reppucci, 1973; Nicholls, 1975). In line with this, they are also more likely than boys to show decreased persistence or impaired performance when failure occurs, when the threat of failure is present, or when the evaluative pressure on a difficult task is increased (Butterfield, 1965; Crandall & Rabson, 1960; Dweck & Gilliard, 1975; Nicholls, 1975; Veroff, 1969). This occurs even on tasks at which girls have demonstrated their ability or have even outperformed boys.

Boys, in spite of their poorer grades and the greater criticism they receive, and despite the lower esteem in which they are held by teachers, respond quite differently to failure feedback from adults. They tend to attribute it to controllable or variable factors. In line with this, they tend to confront failure with improved performance or increased persistence and to seek out tasks that present a challenge. Boys have also been found to credit success to their abilities more readily than girls (Nicholls, 1975).

Some have argued that this differential response to failure stems from boys' and girls' discrepant socialization histories (e.g., Barry, Bacon, & Child, 1957; Crandall, 1963; Veroff, 1969). Boys, it is said, have been trained to be independent and to formulate their own standards of excellence against which to judge the adequacy of their performance. Thus, this line of reasoning continues, when a boy receives negative feedback, he can accept it or reject it depending on how it matches his own assessment. Girls, however, are believed not to develop independent standards and therefore to remain more dependent upon external

---

[2]This is not to deny that later in their academic careers girls may indeed encounter these attitudes.

evaluation. They consequently look to the feedback of others to assess their performance and evaluate their abilities.

If this position were correct, one would expect the typical sex differences in responses to failure to have wide generality. For example, although the research showing the sex differences has always been conducted with adult evaluators, one would expect the difference to remain relatively constant regardless of who delivered the feedback. Boys' internal standards, after all, should withstand variations of this sort. Yet we know that in the late grade-school years, peers become increasingly important to boys as sources of evaluative feedback (Bronfenbrenner, 1967; Hollander & Marcia, 1970). Perhaps, then, boys do not view adult criticism as indicative of their abilities, but would view such feedback from peers as reflecting their level of competence.

In research designed to test this hypothesis (Dweck & Bush, 1976), children attempted several trials of a task and received failure feedback from either a male or female, adult or peer evaluator. As in past research, when feedback was provided by an adult evaluator, particularly a female adult, girls showed greater helplessness than boys; they were more likely to attribute their failures to a lack of ability and to show impaired performance in the face of failure. Interestingly, there was a tendency for boys to blame the female adult evaluator more than girls did and for those who did so to show impaired performance. However, most boys attributed their failures to a lack of effort and persisted under failure. With peer evaluators (and particularly the male peer), the pattern was essentially reversed. Boys were the ones who saw the failure feedback as indicative of their abilities and the ones whose performance suffered. In fact, boys receiving negative feedback from male peers showed the most impairment of any group. Girls in this condition tended to ascribe the feedback to a lack of effort and to show significant improvement in their performance under failure.

These findings suggest that it is not boys' and girls' general socialization histories that determine response to failure, but rather their specific histories with particular agents. This, in turn, implies that in order to learn how feedback acquires different meanings for the two sexes, one should analyze the pattern of evaluative feedback they experience from the various evaluators. Because adults are the major evaluators in all academic environments, this analysis was undertaken in grade-school classrooms and the findings were then corroborated in a laboratory experiment (Dweck, Davidson, Nelson, & Enna, in press).

Of particular interest were the ways in which negative feedback was used — how much it was used compared to positive feedback, what it was typically used for, the specificity with which it was used — and the attributions teachers made for children's intellectual failures. It would be expected, for example, that negative feedback from an evaluator who is typically more negative than positive, whose feedback typically refers to nonintellectual aspects of behavior, whose criticism is used diffusely for a wide variety of referents, and who attributes failure to lack of motivation, would *not* be interpreted as indicating a lack of ability. It would be expected that negative feedback from an agent who is generally positive, who uses feedback quite specifically to refer to intellectually inadequate aspects of performance, and who does not attribute failure to a lack of effort, would more readily be attributed to a lack of ability. These, basically, were the patterns hypothesized to occur for boys and girls in the classroom.

Trained observers coded every instance of evaluative feedback given by fourth- and fifth-grade teachers to their students during academic subjects. The observers noted whether the feedback was positive or negative and recorded the class of behavior for which the feedback was given — either conduct, nonintellectual aspects of academic work (e.g., neatness), or intellectual aspects of academic work (e.g., correctness of answer). Observers also noted when teachers made explicit attributions for children's successes or failures.

The results have helped to resolve the paradox of how the more favorable treatment of girls can lead to their denigration of their competence, to helplessness, and to lessened ability to cope with failure. They also show how boys can learn to discount failures or see them as unrelated to their abilities. But first a brief indication of ways in which girls and boys did not differ. They did not differ in the absolute amount of feedback for *intellectual aspects of work* (i.e., average per boy and average per girl within classroom) or in the portion of this feedback that was positive, negative, or absent. Therefore, sex differences do not appear to be related to the amount or proportion of success and failure feedback for intellectual performance.

However, when one looks at the feedback in general and at the feedback for intellectual quality of work within the context of all feedback, striking sex differences become apparent. First, negative feedback to boys was overall, far more frequent. Since negative outcomes that are in accord with environmental forces can plausibly be attributed to them (see Enzle, Hansen, & Lowe, 1975; Kelley, 1971), boys may attribute failure feedback to the teacher's attitude. Second, negative feedback was used in a more diffuse and a more ambiguous fashion vis-á-vis the intellectual quality of boys' work. Past research has clearly shown that feedback used in a nonspecific manner to refer to a wide variety of nonintellectual behaviors comes to lose its meaning as an assessment of the intellectual quality of the child's work (Cairns, 1970; Eisenberger, Kaplan, & Singer, 1974; Warren & Cairns, 1972). In fact, negative feedback for boys was used *more often* for conduct and nonintellectual aspects of work than it was for the intellectual quality of their academic performance (67.5% and 32.5%, respectively).

One might argue that although conduct feedback may convey to the child something about the teacher's values or attitudes toward himr or her, such feedback can easily be discriminated from feedback for work-related matters and would not seriously affect the information value of feedback addressed to the child's work. In contrast, intellectual and nonintellectual aspects of work occur simultaneously and if feedback is typically used for both, then the basis for the feedback or its referent on any given occasion is more likely to be ambiguous. However, even if conduct is excluded from the analysis and we look only at work-related feedback, still 45.6% of the feedback for boys' work referred to intellectually irrelevant aspects of their performance. That means that almost half of the criticism that boys got for their work had nothing to do with its intellectual adequacy. Instead this feedback referred to such things as neatness, instruction-following, and style of response delivery — "form" rather than "content." Finally, looking at the explicit attributions teachers made for children's intellectual failures, we find that teachers attributed boys' failures to lack of motivation eight times more often than they did girls'. In short, when boys are given failure feedback by adults they can easily view it as reflecting something about the evaluators' attitude toward them or as being based on an assessment of some nonintellectual aspect of their work. When they do see it as referring to the intellectual quality of their work, they can attribute the failure to a lack of motivation.

In striking contrast, girls received relatively little negative feedback for conduct, and the vast majority (88.2%) of the negative feedback they received for their work referred specifically to its intellectual aspects. Thus, since the teacher is generally positive toward girls (they also got more praise than boys), they are less apt to see criticism as reflecting a negative attitude toward them. Second, since feedback is used for them in a very specific fashion for intellectual aspects of work, girls are not as likely to see the assessment as being based on an evaluation of nonintellectual qualities. Finally, since teachers view girls as highly motivated and girls themselves concur in this assessment, they cannot attribute their failures to a lack of motivation. They may have little choice but to view the negative feedback as an objective evaluation of their work and to attribute their intellectual failures to a lack of ability.

Thus the two sexes differ widely in the degree to which negative feedback serves as a valid indicant of the intellectual ability displayed in their academic performance. The results for positive feedback, although not as striking, were essentially the opposite. For work-related praise, 93.8% was contingent upon the intellectual quality of work for boys, but only 80.9% for girls, suggesting that positive evaluation for boys may be more indicative of competence than it is for girls.

These patterns of negative and positive feedback to the two sexes were both consistent across classrooms and rather general across children within classrooms. Are teachers simply reacting to the different behavior of the two sexes, or are they instead reacting differentially to similar behavior from the two sexes? The answer is probably a bit of both. Although it is clear that boys are often more disruptive, less neat in their work, and less motivated to perform well in the elementary school years, there is also some evidence that they tend to be scolded more often and more severely than girls for similar transgressions (Etaugh & Harlow, 1973).

In terms of trying to understand how the use of negative feedback determines its meaning for boys and girls, however, this question may not be critical. We have shown (Dweck et al., in press) that any child exposed to the contingencies that boys and girls are exposed to in the classroom will interpret the feedback accordingly. These contingencies can serve as direct and powerful causes of children's attributions. We have taken the "teacher-girl" and the "teacher-boy" contingencies of negative feedback that we observed in the classroom and have programmed them in an experimental situation.

Specifically, on an initial anagram task with mixed success and failure trials, children received negative feedback that either: (1) referred exclusively to the correctness of their answers (like girls in the classroom), or (2) referred sometimes to correctness and sometimes to neatness (like boys in the classroom). There were two teacher-girl groups — one matched to the teacher-boy group on number of intellectual (correctness) criticisms and the other matched on total number of criticisms. All children next performed a second task (a digit—symbol substitution task) at which they failed on the initial trials and received standardized task) at which they failed on the initial trials and received standardized feedback from the same experimenter. They were then given a written question that asked them to attribute the failure feedback on this second task to one of the three factors described: ability, effort, or the experimenter.

As predicted, most of the children in the teacher—boy condition (75%) did *not* view the failure feedback on the second task as reflecting a lack of ability. Instead, insufficient effort was the alternative that was most frequently endorsed. Although an attempt was made to assure the children of the anonymity of their choices, only two of the 60 children in the study cited the evaluator as the cause of their failure. However, both of these children were in the teacher—boy condition.

In sharp contrast, children in both the teacher—girl conditions overwhelmingly interpreted the failure feedback they received as indicating a lack of ability. Only 25% of these children ascribed their failures to a lack of effort. There were no differences between male and female subjects in their attribution choices in any of the conditions.

These findings clearly indicate that regardless of sex, children who receive failure feedback that is solution-specific (and for which no alternative explanation is provided) are far more likely to regard subsequent failure feedback from that agent as indicative of their ability than are children who receive failure feedback that is often solution-irrelevant. It appears then that the pattern of feedback observed in the classroom for teacher—boy versus teacher—girl interactions can have direct effects on children's interpretations of their failures.

THE GENERALIZATION OF HELPLESSNESS

What are the implications of these individual differences in attributions for the generalization of failure experiences to new situations? To the extent that one's perceived cause of failure remains in effect in a new situation, then one will view past outcomes as predictive of future ones (cf. Brickman, Linsenmeier, & McCareins, 1976). In this way, an individual's causal attributions might serve as mediators of failure effects from one situation to others involving new tasks or new evaluators.

For girls, attributions of failure to a lack of ability on a task or in an academic area imply that when presented with a similar task in the future, the past outcome is relevant to subsequent ones and, in this case, would presage a poor future. To the extent that one encounters similar academic subjects throughout school, girls' earlier condemnations of their ability will continue to be applicable. Thus girls' attributions of failure to ability may discourage continued "testing" of the environment in future grades both because similar tasks may mediate generalization of the effects of past failures and because it is unpleasant to conclude that one lacks ability despite renewed effort.

For boys, however, although blaming the teachers' attitudes or biases may impair motivation and performance in the immediate situation, by blaming the evaluator they can maintain their belief in their ability to succeed. Therefore, when the agent changes, as when they are promoted to the next grade or attend a new school, they can discount their past failures and can approach the situation with renewed effort. Thus boys' attributions of past failures to the agent may encourage testing of the environment when the agent changes. Moreover, attributions of past failures to a lack of effort imply that when in the future one cares to succeed one can at that time begin to apply oneself.

When one considers success feedback, the picture becomes even clearer. To the extent that girls' successes are not viewed by them as indicative of their ability but are attributed to the beneficence of the agent or to intellectually irrelevant aspects of their work, then past successes should not be seen by them as predictive of future success with a new agent. For boys, however, past successes have been achieved despite an "inhibitory" environmental force (Kelley, 1971) the teacher — and are therefore more apt to be chalked up to their abilities. Thus in new school situations with a new teacher, boys will see their successes more than their failures as indicative of ability and as predictive of future performance.

This analysis — postulating attributions as mediators of the effects of past outcomes — provides a mechanism with which to explain the commonly found sex differences in expectancy of success. Crandall (1969) presents a good deal of evidence that girls underestimate their chances for success relative to what their past performance in similar situations would warrant. boys, on the other hand, are more likely to overestimate their chances of future success relative to their past accomplishments. For example, girls have been found to predict lower grades for themselves than boys do for themselves, even when they have received equal or higher grades than boys in the past. This effect does not appear to be due to sex typing or social desirability of responses. It is plausible to assume that in formulating an expectancy, one will focus on those past outcomes that are most indicative of what is likely to occur in the situation at hand. If boys focus on their successes and girls on their failures, this would yield overestimation and underestimation, respectively. In fact, Crandall presents data to suggest that when feedback is mixed or inconsistent, girls tend to weight the negative aspects and boys the positive.

This analysis further predicts that the sex discrepancy in achievement expectancies would be maximal at the beginning of a school year, when the subject matter (ability areas) remain roughly constant but the teacher (evaluator) has changed. Under these circumstances one would expect boys' expectancies to rise dramatically, but girls' perhaps to decline. However, as the year progresses and both boys and girls learn that this year's teacher is similar to last year's — for example, in attitudes, criteria, grading practices, and feedback patterns — the gap in expectancy should diminish.

Two studies were designed to investigate the hypothesis that sex differences in attributions mediate the generalization of prior failure experiences to new situations. Essentially, one would predict that no change in the situation or changes in factors perceived to be irrelevant to past failures should lead to persistence of failure effects, whereas changes in factors that are viewed as causes of failure should encourage recovery from failure.

In the first, laboratory study, children worked on a task (on which both neatness and accuracy were said to be important) and received failure feedback from an adult female experimenter after each of the first four trials. Expectancy of success was monitored prior to each trial. For the fifth trial, children experienced one of four conditions: a new task, a new evaluator, both a new task and a new evaluator, or no change from the previous four trials.[3] The recovery of the child's expectancy on the fifth trial served as the index of the perceived relevance of the change to his future outcomes. This procedure is analogous to that employed in studies of habituation, in which a stimulus is repeatedly presented until the response to it has diminished. The original stimulus and the novel stimulus are then administered during testing. The magnitude of the recovery of the response reflects the perceived novelty of the testing stimulus.

---

[3]Pilot work had ensured that both tasks employed elicited identical mean initial expectancies, that boys and girls did not differ in their initial expectancies, and that ability, effort, and evaluator attributions for failure were all perceived as plausible candidates.

In the present study, it was predicted that changes in only those factors to which boys and girls attribute their failures would promote a recovery of expectancy by creating the perception of the situation as a new one. Specifically, since boys tend to attribute failure to teacher-like agents more than girls do, it was expected that boys' expectancies would recover significantly more than girls' when the evaluator was changed. This prediction was confirmed. Not only did girls show no recovery, but their expectancies declined from Trial 4 to Trial 5 much as they did in the condition in which nothing was altered.

Since girls tend to attribute their failures to a lack of ability more than boys do, it was predicted that their expectancies would recover more than boys' when a new task was introduced. The perception of oneself as lacking ability should no longer be valid if the task on which one failed has been eliminated — provided that the ability one blamed was task-specific. Boys' recovery should be limited by the continued presence of the same evaluator. The results indicated that although both sexes recovered significantly in this conditions, girls by no means regained their initial level of confidence and in fact did not recover any more than boys did. More striking, however, was our finding that even when both the evaluator and the task were varied — making it a largely new situation — girls' expectancies did not show complete recovery. Boys', naturally, did.

Of course it is possible that girls were simply more reluctant to state high expectancies again given the fate of their original predictions. However, these findings suggest that perhaps failure feedback leads girls to consider themselves lacking in an intellectual ability that goes beyond the particular task at hand. In this way, they may be transferring a failure experience from one domain to another by applying a general label to their perceived deficit. It is interesting to note that we started with a situation in which boys and girls confronted a new task with equivalent expectancies and ended with a situation in which boys and girls confronted a new task with the typical sex difference in expectancy.

The second study tested the hypothesis that sex differences in academic achievement expectancies would be maximal at the beginning of a school year at which time the ability areas remain similar but the evaluator is different. Prior to their first and second report cards of the year, over 300 fourth-, fifth-, and sixth-grade children were given questionnaires that assessed how well they expected to do on their upcoming report cards. Although girls had received significantly higher grades than boys on their final report cards the previous year, girls predicted significantly poorer performance for themselves than boys predicted for themselves on their first report cards. Needless to say, girls then recieved higher grades on their first (and second) report cards. By the second report card the gap had closed, but girls still did not give higher predictions than boys, which is what accurate estimates from both would yield.

The results from these studies provide strong support for the view that individual differences in attributions result in differences in the generalization of failure effects. The results also suggest that failure effects may have more of a long-term and cumulative effect for girls than for boys and provide a way of understanding why, despite their early advantage, girls begin to lag behind boys in many achievement areas later on. Even more interesting is the fact that this analysis can account for the development of differential performance by the two sexes in different subject areas.

It has long been a course of curiosity that although girls fairly typically have outperformed boys on tests of verbal achievement, starting at around the junior high or early high school years boys begin to outscore girls on tests of mathematical achievement (Maccoby & Jacklin, 1974). If one analyzes the characteristics of math versus verbal skills as they are acquired over the school years and considers this in conjunction with responses to and generalization of failure experiences, an explanation for the emerging achievement differences becomes clear (although, of course, it does not rule out other potential contributing factors). Once the basic verbal skills — reading, spelling, vocabulary — are acquired, increments in difficulty are gradual. Never again, or rarely, is the child confronted in school with a new unit that puts him at a complete loss or for which a totally new set of concepts and skills must be mastered. Learning to read a new word or to spell it or define it involves fundamentally the same process one has gone through before with old words. New learning is, in a sense, assimilated into a larger, pre-existing body of knowledge.

With math, however, a new unit may involve totally new concepts, the relevance to which of past learning may not be immediately or ever, clear. New units in math often involve quantum leaps as from arithmetic to algebra to geometry to calculus and so on. For a young child even the links between addition and multiplication or between multiplication and division may not be readily apparent. This characteristic of the acquisition of mathematical skills provides numerous opportunities for initial failures and, if one is so inclined, for concluding that one lacks ability. To the extent that one blames a lack of general mathematical ability, the effects of the failures may generalize to all future tasks subsumed under the label "math," resulting in: (1) lowered persistence in the face of difficulties, which are inevitably encountered even by those accomplished in mathematics; (2) avoidance of math courses when that option becomes possible; and (3) perhaps interference with new learning, which in math often requires sustained attention and the maintenance of systematic problem-solving strategies, both of which are hampered by helplessness. Thus, it may be the case that two children who start out with equivalent skills and equivalent "aptitude" in elementary school end up with divergent skills later on. This possibility — that differential achievement can be accounted for by how well the nature of skill acquisition coincides with attributional tendencies — is currently under investigation (by Carol Dweck and Barbara Licht).

Although this research has focused on sex differences in helplessness, the variable "sex," like many demographic variables, simply serves as a convenient way of summarizing a particular learning history. In this case the histories of the two sexes favor different attributional tendencies. However, there are important within-sex differences in these tendencies and one would expect the relationships elaborated here to apply to these individual differences as well.

In the same vein, most of the research to date relating perceptions of causality to responses to outcomes has focused on intellectual achievement situations; yet similar relationships should hold for situations involving social interaction.

HELPLESSNESS AND SOCIAL INTERACTIONS

In the same way that intellectual failures can be met with a variety of responses, so, too, can social rejection. One may attempt a variety of strategies designed to reverse the rejection, or one may respond with behavior that represents a marked deterioration from the previous interaction. For instance, a normally socially facile person may withdraw or resort to hostile retaliatory measures. Do attributions guide selection of coping patterns in social situations in the same way that they appear to in academic–intellectual ones?

The importance of this question is highlighted when it is viewed in the context of past work on children's interpersonal coping and peer relationships. Most previous investigations have virtually ignored the role of perceptions of control and have concentrated instead on cognitive and social skills, with the assumption that problems in coping most be primarily a result of deficits in these skills (Allen, Hart, Buell, Harris, & Wolf, 1964; Baer & Wolf, 1970; Keller & Carlson, 1974; O'Connor, 1969, 1972; Spivack & Shure, 1974). Yet those programs that have focused on skill or overt behavior alone have not reliably promoted or maintained effective peer relationships (see Gottman, 1977). The most effective programs have been the ones that appear to be teaching the contingency between the child's actions and the social outcomes he experiences (see Gottman, Gonso & Schuler, 1976; Oden & Asher, 1977).

Focusing on perception of control over aversive outcomes brings to the fore a number of important possibilities not addressed by earlier approaches. Past researchers have considered only rejected and isolated children, implicitly assuming the more popular ones to be free of potentially serious problems in interpersonal coping. However, just as learned helplessness in achievement situations appear to be unrelated to competence, responses to negative social outcomes may be relatively unrelated to social skills; popular as well as unpopular individuals may have coping problems. For instance, some popular children may interpret their few experiences with social rebuff as indications of permanent rejection not open to change by their actions, parelleling the instance of the "A" student who attributes a low grade to a lack of ability despite all previous evidence to the contrary. Moreover, it may be that some individuals are isolated *not* because they lack social skills or the knowledge of appropriate behavior, but because they fear or have experienced social rejection and view it as insurmountable.

We are currently conducting an investigation to establish the relationship between perceptions of control and causal attributions for rejection.[4] In order to tap causal attributions for social rejection, we have developed a questionnaire depicting a series of hypothetical social situations in which children are either rejected or accepted by same-sex peers. Responses on this measure are being related to responses in a situation in which each child must cope with potential social rejection by a peer of the same sex. In this way, we can determine whether causal attributions that imply difficulty in surmounting rejection are in fact associated with deterioration of social behavior in the face of rejection.

In the questionnaire, children are presented with an instance of rejection and are asked to evaluate a list of reasons the rejection may have occurred. Both the situations and their causes were selected as the most representative of those generated by children in the course of extensive interviews. For example: "Suppose you move into a new neighborhood. A girl you meet does not like you very much. Why would this happen to you?" The reasons include such factors as personal ineptitude, a characteristic of the rejector, chance, misunderstandings, and a mutual mismatch of temperaments or preferences.

Just as failure attributions to some factors, like effort, imply surmountability, so too do rejection attributions to misunderstanding. Similarly, attributing rejection to one's lack of ability (e.g., "It happened because it's hard for me to make friends") implies a relatively enduring outcome, as would an ability attribution for academic failure. Blaming the rejector for interpersonal rejection has the same implications as blaming the evaluator for academic failure. Given the parallel implications of interpersonal and intellectual attributions, one would also expect parallel reactions and generalization effects.

The data from over 100 children tested thus far indicate that individuals differ in consistent ways from one another in the causes to which they ascribe social rejection. In order to test the hypothesized relationships between attributions and responses to rejection, a method of sampling each child's interpersonal strategies both before and after social rejection has been devised. Children try out for a pen pal club by communicating a sample getting-to-know-you letter (conveying the "kind of person" they are) to a peer evaluator. The experimenter then relays to the child the evaluator's decision — in this case, not to admit the child into the club. To assess the effect the evaluation has on subsequent responses, the child is told he has a chance to try again, to send another message to the same evaluator.[5] After a short wait, enthusiastic acceptance is relayed to the

---

[4]This research is the second author's dissertation.

[5]A test for generalization to a new evaluator is also included.

child, and his home address is recorded to place on file for the club (which has actually been set up for this project).

These communications are currently being rated by trained, independent judges along a number of dimensions designed to reflect the child's strategy for attaining acceptance and to reveal alterations in strategy across situations. As with the attribution measure, the data thus far indicate that there are striking differences in strategies for coping with interpersonal rejection. Responses range from mastery-oriented patterns, with more and different strategies used after rejection, to complete withdrawal — about 10% of the children could not come up with a second message to the rejecting evaluator, even after gentle prodding and prompting. Only after the experimenter explicitly attributed the rejection to an idiosyncrasy of the committee member would these children produce a message for a different evaluator, which enabled us to provide acceptance feedback.

The initial data indicate that those children who either gave up or were extremely reluctant to send a message to the rejecting committee member also favored attributions that emphasized the insurmountability of rejection. Those children who were notable for their self-confident responses to the rejection and thoughtful approaches to their second communication emphasized the role of surmountable factors like misunderstandings. Thus, preliminary analysis of the attribution and coping measures suggests that individual differences in attribution are indeed systematically related to responses following rejection in the predicted ways.

HELPLESSNESS AND THE OCCURRENCE OF ATTRIBUTIONS

Researchers in the area of attributions assume that following some discrete event, such as the delivery of evaluative feedback, an attribution is always made (see Dweck & Gilliard, 1975). Individual differences are assumed to occur only in the nature of the attribution that is made at that point. However, we have recently completed research that shows there are also clear individual differences in the timing or occurrence of causal attributions when the situation is less structured (Diener & Dweck, 1978).

When attributions were measured at a prespecified time — either on a preexperimental attribution questionnaire or on a task-specific, postfailure attribution probe — we obtained the typical helpless versus mastery-oriented differences in attributions and task persistence. However, when children's spontaneous, ongoing reports of achievement-cognitions were monitored, we found that attributions were made primarily by helpless children and not by mastery-oriented ones. In this research, described in part earlier, children were trained to criterion on a discrimination-learning task. Following the training problems, failure feedback ("wrong") was begun, and chances in problem-solving strategies were tracked. As noted above, helpless children showed a steady decline in the sophistication of the strategies they employed, whereas nearly all of the mastery-oriented children maintained their strategies; and a number even began using more mature strategies than they had before. These results were replicated in a second study, which was identical in all respects but one: prior to the seventh training trial, children were requested to verbalize what they were thinking, if anything, while performing the task. The instructions gave them license to report anything − from justification for their stimulus selection to plans for their lunch. This procedure allowed us to monitor differences in not only the nature of particular achievement-related cognitions but also in the presence, the timing, and the relative frequency of various cognitions.

All children verbalized freely; and, prior to failure, the verbalizations of the helpless and mastery-oriented children were virtually identical. Almost all of the statements pertained to task strategy and almost none reflected achievement related cognitions. Following the onset of failure, however, a dramatic shift took place. Both groups of children began to report many more achievement cognitions, but what they emphasized differed markedly. The helpless children rather quickly began to make attributions for their failures, attributing them to a lack or loss of ability, and to express negative affect about the task. The mastery-oriented children, in contrast, did not make attributions for failure. Instead they engaged in self-instruction and self-monitoring designed to bring about success. They continued to express a positive prognosis for future outcomes (e.g., "One more guess and I'll have it") and to express positive affect toward the task (e.g., "I love a challenge"). It would appear that despite the feedback of the experimenter, the mastery-oriented children did not consider themselves to have failed. They were making mistakes, to be sure, but they seemed certain that with the proper concentration and strategy they could get back on the track. Thus they dwelled on prescription rather than diagnosis, remedy rather than cause.

One might argue that although mastery-oriented children did not verbalize effort attributions, they were implied by their self-instructions. However, the few attributions that were explicitly verbalized did not seem to fall into any one category. Moreover, when one considers the nature of the task, it is clear that identifying the cause of failure is in this case irrelevant to achieving success. Whether the cause is thought to be insufficient effort, bad luck, increased task difficulty, or lower ability than previously believed, the remedy still would be sustained concentration and the use of sophisticated strategies.

This observed difference in the tendency of helpless and mastery-oriented children to attend to the cause of failure raises a number of interesting questions. Would the mastery-oriented children have attended to the causes of failure sooner if a diagnosis were necessary for the prescription of a remedy? Or, is there among them a subset of children who are too "action oriented" to analyze causal factors systematically? Although helpless children readily concede that they have failed soon after negative feedback begins, at what point do mastery-oriented children define themselves as having failed, when do they attribute, and at what point do they consider terminating their efforts? Are there those among the mastery-oriented who suffer from what may be termed the "Nixon syndrome" — unusually prolonged persistence designed to forestall the admission of failure (cf. Bulman & Brickman, 1976)? For such children, as for helpless children, failure may have highly negative connotations for their competence; yet rather than surrender to it prematurely, they persist past the point of diminishing returns in the belief that, as expressed by Richard Nixon, "You're never a failure until you give up."

In short, in past research experimenters have typically at a predetermined point defined failure for the subject and asked him to make a causal attribution. The current results suggest that there are also important differences in the timing or even the occurrence of attributions. The implications of these differences have yet to be explored.

ACKNOWLEDGMENTS

The research reported here was supported in part by Grant NE G-00-3-0088 from the National Institute of Education to the first author and by Grant ND 00244 from the U.S. Public Health Service.

REFERENCES

Allen, K. E., Hart, B., Buell, J. S., Harris, F. R., & Wolf, M. M. Effects of social reinforcement of isolate behavior of a nursery school child. *Child Development,* 1964, *35,* 511–518.

Asher, S. R., & Markel, R. A. Sex differences in comprehension of high- and low-interest material. *Journal of Educational Psychology,* 1974, *66,* 680–687.

Baer, D. M., & Wolf, M. M. Recent examples of behavior modification in pre-school settings. In C. Neuringer & J. L. Michael (Eds.), *Behavior modification in clinical psychology.* New York: Appleton–Century–Crofts, 1970.

Barry, H., Bacon, M. C., & Child, I. L. A cross-cultural survey of some sex differences in socialization. *Journal of Abnormal and Social Psychology,* 1957, *55,* 327–332.

Battle, E. Motivational determinants of academic task persistence. *Journal of Personality and Social Psychology,* 1965, *2,* 209–218.

Brickman, P., Linsenmeier, J. A. W., & McCareins, A. Performance enhancement by relevant success and irrelevant failure. *Journal of Personality and Social Psychology,* 1976, *33,* 149–160.

Brofenbrenner, U. Response to pressure from peers versus adults among Soviet and American school children. *International Journal of Psychology,* 1967, *2,* 199–207.

Brophy, J. E., & Good, T. L. *Teacher–student relationships.* New York: Holt, 1974.

Bulman, R. J. & Brickman, P. *When not all problems are soluble, does it still help to expect success?* Unpublished manuscript, 1976.

Butterfield, E. C. The role of competence motivation in interrupted task recall and repetition choice. *Journal of Experimental Child Psychology*, 1965, *2*, 354–370.

Cairns, R. B. Meaning and attention as determinants of social reinforcer effectiveness. *Child Development*, 1970, *41*, 1067–1032.

Coopersmith, S. *The antecedents of self-esteem*. San Francisco: Freeman, 1967.

Crandall, V. C. Sex differences in expectancy of intellectual and academic reinforcement. In C. P. Smith (Ed.), *Achievement-related motives in children*. New York: Russell Sage Foundation, 1969.

Crandall, V. C., Katkovsky, W., & Crandall, V. J. Children's beliefs in their own control of reinforcements in intellectual–academic situations. *Child Development*, 1965, *36*, 91–109.

Crandall, V. J. Achievement. In H. W. Stevenson (Ed.), *Child psychology*. The sixty-second yearbook of the National Society for the Study of Education. Chicago: NSSE, 1963.

Crandall, V. J., & Rabson, A. Children's repetition choices in an intellectual achievement situation following success and failure. *Journal of Genetic Psychology*, 1960, *97*, 161–168.

Diener, C. I., & Dweck, C. S. An analysis of learned helplessness: Continuous changes in performance, strategy, and achievement cognitions following failure. *Journal of Personality and Social Psychology*, 1978, in press.

Digman, J. M. Principal dimensions of child personality as inferred from teachers' judgments. *Child Development*, 1963, *34*, 43–60.

Dweck, C. S. The role of expectations and attributions in the alleviation of learned helplessness. *Journal of Personality and Social Psychology*, 1975, *31*, 674–685.

Dweck, C. S., & Bush, E. S. Sex differences in learned helplessness: (I) Differential debilitation with peer and adult evaluators. *Developmental Psychology*, 1976, *12*, 147–156.

Dweck, C. S., Davidson, W., Nelson, S., & Enna, B. *Sex differences in learned helplessness: (II) The contingencies of evaluative feedback in the classroom* and *(III) An experimental analysis*. Developmental Psychology, in press.

Dweck, C. S., & Gilliard, D. Expectancy statements as determinants of reactions to failure: Sex differences in persistence and expectancy change. *Journal of Personality and Social Psychology*, 1975, *32*, 1077–1084.

Dweck, C. S., Goetz, T. E., & Strauss, N. *Sex differences in learned helplessness: (IV) An experimental and naturalistic study of failure generalization and its mediators.* Unpublished manuscript, 1977.

Dweck, C. S., & Reppucci, N. D. Learned helplessness and reinforcement responsibility in children. *Journal of Personality and Social Psychology*, 1973, *25*, 109–116.

Eisenberger, R., Kaplan, R. M., & Singer, R. D. Decremental and nondecremental effects of noncontingent social approval. *Journal of Personality and Social Psychology*, 1974, *30*, 716–722.

Enzle, M. E., Hansen, R. D., & Lowe, C. A. Causal attributions in the mixed-motive game: Effects of facilitory and inhibitory environmental forces. *Journal of Personality and Social Psychology*, 1975, *31*, 50–54.

Etaugh, C., & Harlow, H. *School attitudes and performance of elementary school children as related to teacher's sex and behavior.* Paper presented at the meeting of the Society for Research in Child Development, Philadelphia, March 1973.

Feather, N. T. Effects of prior success and failure on expectations of success and subsequent performance. *Journal of Personality and Social Psychology*, 1966, *3*, 287–298.

Gottman, J. M. The effects of a modeling film on social isolation in preschool children: A methodological investigation. *Journal of abnormal Child Psychology*, 1977, in press.

Gottman, J., Gonso, J., & Schuler, P. Teaching social skills to isolated children. *Journal of Abnormal Child Psychology*, 1976, *4*(2) 179–197.

Hollander, E. P., & Marcia, J. E. Parental determinants of peer-orientation and self-orientation among preadolescents. *Developmental Psychology,* 1970, *2,* 292–302.

Keller, M. F., & Carlson, P. M. The use of symbols modeling to promote social skills in preschool children with low levels of social responsiveness. *Child Development,* 1974, *45,* 912–919.

Kelley, H. H. *Attribution in social interaction.* Morristown, N.J.: General Learning Press, 1971.

Maccoby, E. E., & Jacklin, C. N. *The Psychology of sex differences.* Stanford, Calif.: Stanford University Press, 1974.

McCandless, B., Roberts, A., & Starnes, T. Teachers' marks, achievement test scores, and aptitude relations with respect to social class, race, and sex. *Journal of Educational Psychology,* 1972, *63,* 153–159.

Nicholls, J. G. Causal attributions and other achievement-related cognitions: Effects of task outcomes, attainment value, and sex. *Journal of Personality and Social Psychology,* 1975, *31,* 379–389.

O'Connor, R. D. Modification of social withdrawal through symbolic modeling. *Journal of Applied Behavior Analysis,* 1969, *2,* 15–22.

O'Connor, R. D. Relative efficacy of modeling, shaping, and the combined procedures for modification of social withdrawal. *Journal of Abnormal Psychology,* 1972, *79,* 327–334.

Oden, S. L., & Asher, S. R. Coaching children in social skills for friendship-making. *Child Development,* 1977, *48,* 495–506.

Seligman, M. E. P., & Maier, S. F. Failure to escape traumatic shock. *Journal of Experimental Psychology,* 1967, *74,* 1–9.

Seligman, M. E. P., Maier, S. F., & Geer, J. The alleviation of learned helplessness in the dog. *Journal of Abnormal and Social Psychology,* 1968, *73,* 256–262.

Spivak, G., & Shure, M. B. *Social adjustment of young children.* San Francisco.: Jossey–Bass, 1974.

Stevenson, H. W., Hale, G. A., Klein, R. E., & Miller, L. K. Interrelations and correlates in children's learning and problem solving. *Monographs of the Society for Research in Child Development,* 1968, *33*(7, Serial No. 123).

Veroff, J. Social comparison and the development of achievement motivation. In C. P. Smith (Ed.), *Achievement-related motives in children.* New York: Russell Sage, 1969.

Warren, V. L., & Cairns, R. B. Social reinforcement satiation: An outcome of frequency or ambiguity? *Journal of Experimental Child Psychology,* 1972, *13,* 429–260.

Weiner, B. *Theories of motivation.* Chicago: Markham, 1972.

Weiner, B. *Achievement motivation and attribution theory.* Morristown, N.J.: General Learning Press, 1974.

7

Attribution, Socialization, and Moral Decision Making

Richard A. Dienstbier
University of Nebraska, Lincoln

The position developed in this chapter resembles that taken by Aristotle more than that taken by most modern psychology theorists, for our discipline has tended to approach moral behavior and thought as if two quite different topics were at issue. Modern psychological approaches to moral behavior have tended to focus almost entirely on the emotional mediation of avoidance and upon avoidance behavior, whereas moral judgment theorists have tended to isolate themselves from both emotion and behavior, concentrating instead on verbal expressions of moral judgments. Aristotle suggested a more integrated view — that effective moral training should first involve the young child's emotional dispositions, so that the child learned to love and hate correctly; later when maturation allowed reasoning to emerge, there would be a "symphony between habituated preferences and what reasoning shows to be good" (Fortenbaugh, 1975, p. 49). In an attempt to account for that "symphony" between habituated preferences and reasoning in the realm of morality, this chapter's theoretical development deals with the interplay between emotions and cognitions about emotions, using emotion attribution theory concepts.

The initial issue of this chapter centers upon what constitutes a moral decision, and whether moral decisions and situations should be defined on the basis of specific significant behavior by the actor (e.g., cheating, stealing, killing) or upon a more complex basis. Research is presented that supports the more complex approach by demonstrating that when moral values are made salient, cheating (usually thought to be clearly morally relevant) either increases or decreases, depending on certain specific attributes perceived in the temptation context. It is concluded that a complex attributional analysis of situations from the individual's viewpoint is very important in establishing the relevance of moral values to decision making in various contexts.

The largest portion of the chapter deals with research indicating that causal attributions about the sources and meaning of emotional arousal determine the impact of that emotional response in moral decisions and behavior for adults and children. Finally, the effectiveness of various socialization techniques for both normal and psychopathic individuals is analyzed, using emotion attribution concepts.

DEFINING MORAL SITUATIONS

What is morality? What constitutes a moral decision? What unique attributes must a situation appear to have before moral values and constructs are elicited? The issue is sufficiently complex that it is not solved by simply defining (as has Turiel, 1977) certain classes of behavior, such as helping others in distress, as arbitrarily relevant to morality, whereas other classes, such as not handling a forbidden toy, are not morally relevant.

For the purpose of this paper, a moral situation is defined as one that evokes significant guilt or shame when transgression occurs; normally, this will mean that a significant transgression of one's values has occurred or may occur. Moral self-control, on the other hand, will be defined as behavior where one resists temptation in a moral situation in spite of the perception that detection of transgression is *not* possible. One test of whether any situation is a moral one would be to determine whether the rate or frequency of moral behavior changed (appropriately) as moral values were made increasingly salient; the first section of this chapter addresses that issue. A second test of whether morality is perceived as relevant to a situation would be achieved by manipulating emotional arousal or attributions about arousal to ascertain whether moral behavior increased or decreased; the second and larger half of this chapter presents research using such techniques.

Whether an individual responds to a situation as a moral dilemma depends primarily upon the cultural, social, and individual context, for categories of be-havior defined as morally relevant by one culture, society, or by one individual may not be similarly defined by others. Within the culture of "middle America," for example, situations involving theft, or others being hurt, may be perceived by most people as clearly calling for moral decisions, whereas less consensus as to moral relevance would exist for some other issues (e.g., sexuality or "hard work"). But even for a single individual it is likely that the temptation to perform a specific behavior in one situation might be considered a moral dilemma, where-as in another situation that same behavior would not be regarded as morally relevant. Consider the following example: A college student with the need and opportunity to cheat attributes the quality of unfairness to the test and the testing context. In that example, attributing the quality of "unfairness" might reduce the perception of moral relevance for the act of cheating, resulting in increased cheating. It should be apparent that a great variety of situational attributes could potentially affect the degree to which moral values and constructs are perceived to be relevant to the choices in a given situation. Such attributions about the meaning of each situation will undoubtedly affect the attributions about the nature and meaning of any emotional arousal experienced; attributions of both types will then influence the individual's behavior.

A series of studies are presented to illustrate the point that specific attributions made about temptation situations (rather than a specific behavior) determine whether an individual will respond to a temptation situation as a moral dilemma. Specifically, the research demonstrates the importance of attributes about the victim of cheating in making moral values relevant or irrelevant to cheating. The logic of the research is simple: If making moral values or norms salient for research subjects increases their tendency to cheat on an exam, then the act of cheating is not being perceived as morally wrong; on the other hand, if cheating decreases when morality is made salient — compared to a control group for whom morality is not made salient — the dilemma of whether or not to cheat is indeed a moral one. Besides being based on common sense, this approach is based on the research of Schwartz and his colleagues (Schwartz, 1976), who demonstrated increased helping behavior after altruistic values were made salient. Schwartz used the term *norm activation* to suggest that the norms governing altruistic behavior were already well internalized but were potentiated or activated by a technique that reminded subjects of those norms. In this paper, I shall continue to use the term *value* to mean norms that have been "internalized."

In the first study of the five-study series (Dienstbier, Kahle, Willis, & Tunnell, 1977) done with the able assistance of Gil Tunnell, we initially attempted to explore aspects of emotion attribution theory that will be discussed in a later section of this chapter. However, the study led us down a theoretical path not initially intended. Freshmen male college students were recruited for a study on verbal skills and were subjected to a series of tests beginning with a vocabulary test; that test was followed by a reading comprehension test (the independent manipulation) and then with the opportunity to cheat (the dependent measure) on the vocabulary test. Three different versions of the reading comprehension test concerning the Minoan culture were constructed: The first versions described their high moral principles based upon personal values and guilt (the internal manipulation). The second version presented similarly high moral principles based upon social norms and shame (the external manipulation). The third version dealt with the high level of achievement of the Minoans (the control manipulation). As confirmed by an elaborate postexperimental questionnaire (such as used in all the reported studies), the later cheating situation was perceived by the subjects as free from possible detection. It was hypothesized that the moral values that were made salient would be effective in reducing cheating on the vocabulary test for the two groups of subjects who read about moral principles, in contrast to the (higher) cheating frequency of control subjects. [The reading comprehension passages were based upon an article by Davies (1969) on the achievements of the Minoans.]

In order to induce motivation to cheat on the vocabulary portion of the test, the subjects were told that typical college freshmen scored at least 18 right on the 30-item, multiple-choice test and that their performance on the test would be compared with their future college performance, since the "Board of Psychologists" developing the test thought that vocabularly scores would correspond with grades in certain courses. Subjects were seated in a row of booths so that the 6 to 10 subjects present for any one session could not see each other; cheating occurred as answer changing during a "delay" period after the test was over when the answers were available. Since subjects were told that the machine-gradable answer paper for the vocabularly items must be marked darkly and be free of sloppy erasures, they had an excuse to use pencils during the "delay" period. However, they were warned (as in all the cheating research) *not* to change any answers. [The procedure for detecting cheating using pressure-sensitive paper is described in detail elsewhere (Dienstbier & Munter, 1971).] The results were clearly contrary to the hypothesis; the control-condition subjects cheated less (21% of the 48 subjects cheated) than subjects in the other two conditions (33% of the 43 internal-condition subjects cheated, and 43% of the 44 external subjects cheated). In other words, the activation of moral values by the internal and external reading comprehension passages increased rather than prevented cheating – a totally unexpected finding. An informal statistical analysis (Fisher Exact Test, two-tailed) was undertaken, indicating a significance level of $p = .06$ between the control condition cheating rate and the combined rate for the internal and external conditions. It was not until after that effect was replicated with diverse procedures and materials that the finding was regarded as credible.

Since the results in that initial research were contrary to expectations, a second study was undertaken with different materials, again in an effort to study differences between cheating rates of the internal vs. external subjects, and with the control condition included merely for additional information. That study, done with the aid of Keith Willis, was carried out with freshmen female subjects and vastly revised independent manipulation ("reading comprehension") materials, though the vocabularly testing part and the cheating opportunity remained essentially the same as for Study 1. Subjects were told that since they all had similar psychology backgrounds (being all in their first basic course), the reading comprehension test would deal with psychology topics (giving no one an unfair advantage). After reading and being tested on a passage on the sensation of touch, the independent manipulation passage was read. Internal- and external-condition subjects read about the development of conscience. The external passage indicated that the final developmental step was completed when the individual experienced shame when detection for transgression was thought to be possible. The internal passage concluded that conscience was developed when guilt was experienced even when detection was impossible. Control subjects read about the development of immediate memory. A replication of the results of Study 1 was achieved, with almost identical cheating rates in the internal and external conditions (32% of the 56 subjects in both conditions cheating) and with that rate considerably higher than that for the control subjects (14% of the 29 subjects cheating, $p = .05$, Fisher's exact test, one-tailed).

With the unexpected results of both Studies 1 and 2, where increased cheating followed the activation of moral values, a more thoughtful analysis was undertaken of the situation as it was perceived by the subjects. In Studies 1 and 2, following procedures developed for other research (Dienstbier & Munter, 1971; Dienstbier, 1972), subjects had been told that the vocabularly test was an experimental test being developed by a board of psychologists, and they were asked to put their social security number of their answer sheet, "just in case the board developing this test decides to compare your vocabularly test performance with your later grades in certain courses – so far they have found that most successful college students score at least 18, as freshmen, on this 30-item test; a chance score is 10."

I began to suspect that the subjects perceived the research as basically unfair and that this unfairness could be offset by their cheating on the vocabulary test to achieve a score somewhat higher (and closer to 18) than they had achieved fairly. Furthermore, cheating against an anonymous "board of psychologists" might have seemed too impersonal a jesture to be morally relevant. Certainly similar logic is often used by individuals dealing with large institutions such as their government at tax time, insurance companies, etc. More cheating after moral value activation could be due to the subjects being able to make a clearer definition of morality following the reading of the conscience-oriented passages, so that their present situation could be more clearly defined as nonmoral. To test this notion, a third study was undertaken with materials and procedures *identical* to those used in Study 2, except that Keith Willis, who had acted as the experimenter in Study 2, remade our tape-recorded instructions in his voice and claimed that this research was for his Ph.D. dissertation. References to the "board of psychologists" were changed to "I" in the instructions pertaining to the vocabulary test, and subjects were told: "Don't change any answers. I really need accurate data for my thesis, or I'll have to spend another semester trying." It was believed that those changes would cause subjects to redefine the situation as a moral one, with cheating an immoral behavior, so that moral value activation via the initial reading comprehension passages would result in reduced cheating. That prediction was realized with 12% of the 69 subjects in the two experimental groups cheating, compared with 29% of the 34 control-condition subjects ($p = .03$, Fisher's exact test, two-tailed). The difference in cheating patterns between Studies 2 and 3, run in successive semesters but with the same experimenter and format except as described above, was statistically significant ($p < .005$ using an interaction test for arc-sine transformed percentages, two-tailed, developed by

Langer and Abelson, 1972). The results of Study 3 were essentially replicated in a study to be reported below.

To buttress our confidence that the complete reversal in cheating pattern between Studies 2 and 3 was really due to the perception that cheating in Study 3 was perceived to be morally wrong, especially after reading about morality, I conducted a fourth study with Lynn Kahle. In Study 4, subjects participated in either the Study 2 or the Study 3 version of the procedure, except that when the opportunity to cheat occurred, we asked subjects to record their level of resentment toward either the "board of psychologists who developed the vocabulary test" in the Study 2 version or toward "Lynn Kahle who developed the vocabulary test" in the Study 3 version. Subjects were given the opportunity to respond on a 5-point scale with possible answers from "none" (scored 1) to "extreme" (scored 5). Those subjects in Study 2 (with the board of psychologists) were significantly more resentful than subjects in the Study 3 version (means of 1.98 vs. 1.43, $F = 7.08, p < .01$). Furthermore, as predicted, whereas subjects in the internal and external conditions of Study 2 were considerably more resentful than subjects in the comparable conditions of Study 3 (means of 2.13 vs. 1.47, $t = 2.77, p < .01$, two-tailed), resentfulness of control-condition subjects did not differ significantly between the studies (mean of 1.67 vs. 1.35, $t = .89$, NS). The pattern of data therefore supports the contention that although subjects resented the "board of psychologists" considerably more than the graduate-student researcher, their resentment was accentuated by the reading of the morally relevant reading comprehension materials.

The research presented above (along with a similar study presented below to explore other emotion-attribution issues) demonstrates that specific attributions that people make about potential moral dilemmas determine whether moral values (and by inference, avoidance-mediating emotions such as guilt and shame) are relevant to one's behavioral choices. Making morality salient in the research above either increased or decreased the number of subjects who saw the situation as morally relevant and who resisted the temptation to cheat.

Unfortunately, this research does not provide an easy key to the analysis of what situations will be perceived as moral ones for people in the real world or for research subjects in other settings. Instead, I am able to suggest only a research technique. When the researcher assumes that individuals will bring to bear certain values or constructs in response to the research situation, one method for checking that assumption is to preactivate those values or constructs to determine whether appropriate behavior is stimulated.

In the following section, in contrast, the manipulations were not relevant to the activation of moral values, so that we may assume that the subjects in the various research conditions were equally likely to perceive the moral implications of their behavior (often cheating) and were equally motivated to cheat. The issue of the following section and of the research described in it is how the quantity and quality of emotional arousal and experience affected moral self-control.

EMOTIONAL AROUSAL AND MORAL BEHAVIOR

The larger half of this chapter will deal with those emotion attribution processes that occur when an individual faces temptation to transgress against moral values. The first issue, addressed through the research reviewed immediately below, is whether the avoidance of behavior that has potentially painful (physically or psychologically) consequences is facilitated by heightened emotional arousal.

Studies by Lykken (1957) and by Schachter and Latané (1964) demonstrated that psychopathic criminals, who are often described as without fear or guilt, failed to avoid errors punished with shock in a lever–maze learning situation. (At each "choice point" in the "maze" there was a correct response, an incorrect and shocked error, and several possible wrong options not resulting in shock.) The demonstration by Schachter and Latané that psychopaths would learn to avoid shocked errors when pre-aroused with an adrenalin injection (contrasted with a placebo injection) provided significant evidence for the conclusion that nonaroused psychopaths disregarded the shock because of a lack of normal emotional response to that punishment. Other work by Schachter and Ono (Schachter & Latané, 1964) demonstrated that college students would cheat more under tranquilizer (chlorpromazine) conditions than under placebo conditions. Those findings gave considerable support to the traditional view that avoidance behavior (including the avoidance of behavior seen as immoral) is mediated, or at least facilitated, by emotional arousal. It is only a short conceptual step to the normally accepted idea that moral transgressions are resisted in part through the aid of conditioned fear, anxiety, shame, and guilt, an analysis akin to the two-factor avoidance theory advanced by Mowrer (1950). The two-factor avoidance theory states that avoidance is facilitated by conditioned emotional responses that increase as the individual approaches the forbidden behavior and decrease (providing reinforcement) as the individual avoids or retreats from that behavior. That theory provides our starting point.

EMOTION ATTRIBUTION AND MORAL BEHAVIOR

Schachter and Singer (1962) demonstrated that emotional arousal artificially induced by adrenalin injection could be experienced as different emotional states (anger or euphoria), depending on the different attributions made in different experimental contexts. The application of that idea to our avoidance formulation, as developed above, would lead to a theoretical elaboration. Identical levels of emotional arousal, resulting from a conditioned emotional response to (planned) transgression behavior, might lead to very different reactions depending upon the attributions made about the causes or meaning of that emotional response. Actually, this statement represents little more than the imposition of attribution-theory language on an idea occasionally articulated by socialization theorists. In the words of punishment researchers Walters and Parke (1967), it may be that "social conformity is not facilitated by arousal per se but only by emotional responses that are associated with, or are contingent on, certain kinds of cognitive structuring [p. 220]." The first study in the next series of research was an attempt to ascertain whether the attributions an individual made about emotional arousal naturally induced in a temptation situation could be manipulated to alter the frequency of tempted behavior substantially. Since other research had clearly demonstrated that the *quantity* of arousal was important in avoiding punished errors and temptation, this initial research posed the question of whether *qualitative* differences in emotional states under constant arousal conditions would yield similar results. This question, if answered affirmatively, would prove far from academic, for it could provide one of the keys to understanding how cognition plays a role in modulating the impact of emotional inhibition without immediately changing the quantity of emotional arousal. In nonmoral contexts, we are used to viewing such emotional–cognitive interactions. Even if we are slightly acrophobic, for example, we know that if we can retain our belief that a roller coaster ride is safe, and if the emotional arousal is not overwhelming, we may experience the arousal state as positive excitement rather than negative fear. In our attribution language system, we would say that the attribution of the cause of the emotional experience to the speed and fast stimulation changes rather than to danger allows the emotional arousal to be experienced as noninhibitory; we continue to seek out rather than to avoid the roller coaster.

In the first study of this series, done with the aid of Pamela Osborne Munter (Diestbier & Munter, 1971), I attempted to study the impact of different emotion attributions on cheating. Recruited for a study on the impact of a vitamin supplement on vision, subjects ingested a placebo pill and participated in a vocabulary testing exercise like that described previously, while waiting for the pill (supposedly containing the vitamin supplement) to take effect. The vocabulary test was introduced as a delay task, but subjects were told that the board of psychologists would probably get in contact with them if they scored less than 20. Subjects were then exposed to the autokinetic illusion to verify that the pill really did have psychoactive effects and were subsequently given an opportunity to cheat on their vocabulary test, although they were warned (as in all cheating research), "Do *not* change any answers." At the beginning of the study and immediately prior to the opportunity to cheat, half of the subjects were told to anticipate possible side effects of autonomic arousal (e.g., heart acceleration, face flushing, etc.), and half were told to expect benign side effects irrelevant to autonomic arousal.

In that study 49% of the 47 subjects who had anticipated arousal side effects cheated, compared to 27% of the 48 control (nonarousal) subjects ($p < .05$, by Fisher's exact text, one-tailed). But the entire effect was due to the male subjects. Follow-up research (Dienstbier, 1972) indicated that the effect held for women, too, provided that the threat for failure was made less intense; it was concluded that either the women experienced the threat of Study 1 as more attention-preoccupying than did the male subjects, so that they did not attend to the more subtle aspects of the study pertaining to the pill side effects, or that they were unable to attribute their higher emotional arousal to the pill. The hypothesis underlying those studies on cheating was confirmed, with those subjects who could attribute their emotional arousal to the placebo pill apparently experiencing their arousal differently, so that it did not play an inhibitory role as they considered cheating.

Of course there are alternative explanations of this type finding — explanations that do not assume that the vital process is the attribution of arousal to the pill. For example, there is a possibility that merely inducing the subject to attend to the arousal side effects might somehow cause the associated emotion to become reduced or dissociated from the temptation dilemma. In a second study discussed in the 1972 article, that question was experimentally addressed. In a four-condition study, half the subjects received no placebo but were told that they were in a "control" condition. It was explained to those "control" subjects that after the "real" subjects received the vitamin-related drug, we assess drug effects through a side effects checklist that those "real" subjects fill out. Therefore, it was explained, the possibility exists that suggestions induced by the questions on the checklist might result in the side effects being reported by the "real" subjects when they really are not experienced. To ascertain whether this checklist suggestion effect was real, "control" subjects were told that they would be run through exactly the same study as the "real" subjects, except that they would receive no pill. This is exactly what was done. Half of the "control" subjects were given drug information and a checklist concerning the arousal side effects, while the second half received the benign side effects description and checklist. "Control" subjects were asked to "look into themselves conscientiously, just as 'real' subjects did," to assess whether they experienced the symptoms listed on their side effects lists. The other two conditions of the study received the "arousal" and "benign" placebo pills in conditions that essentially replicated the Dienstbier and Munter (1971) study. In the "control" conditions, merely attending to arousal side effects did not facilitate cheating, but the facilitory effect of the arousal placebo in the replication conditions was observed. The statistically significant interaction helped to establish that the hypothesized emotion—attribution process may be the best logical explanation for the increased cheating with the arousal placebo.

The success of the studies described above led to interest in working with the attributional processes that would be employed by children as they faced temptation. I speculated that those attribution processes would be shaped by the socialization practices of the parents, since different socialization techniques would provide different emotion-attribution information. For example, if physical punishment followed stealing, would the punishment itself become so salient and obvious a source of the child's emotional discomfort that the child would attribute emotional discomfort to the punishment rather than to the stealing?

If so, in future temptation (to steal) situations, would the conditioned emotional response be attributed to the punishment rather than to the act of stealing? If, in the former case, the emotion was experienced as fear of punishment, would the emotion be as effective in preventing stealing than if equally intense emotions in anticipation of stealing were attributed to the act of theft itself? It was hypothesized that the externally attributed emotional discomfort (to the punishment rather than to the act) would be less effective in inhibiting transgression when the situation was perceived as detection-free. That formulation is untestable for ethical reasons; but the hypothesized ineffectiveness of external emotion attributions is testable if external attributions are induced through verbal rather than physical techniques.

Studies of Self-Control with Children

The research that was subsequently undertaken to explore those questions was therefore restricted by practical and ethical considerations to testing whether children would react differently in a detection-free situation after they had been given the attributions that previous emotional discomfort (from a failure in a similar situation) was due to being "found out for doing the wrong thing" (external) as compared to attributing their emotional arousal simply to "having done the wrong thing" (internal). If merely giving different verbal attribution information in an experimental context substantially altered subsequent behavior in a "detection-free" situation, evidence would be provided to support the conclusion that like the college freshmen in the cheating studies, the emotional attributions children make are relevant to moral behavior and extremely susceptible to influence; hence such processes could be an important key to understanding socialization differences.

In the placebo research with adults, emotion attributions were manipualted in a manner quite dissimilar from real-life experiences. This use of techniques not encountered outside of the laboratory was undertaken in order to establish that attribution manipulations could induce differential cheating. The shift to research with children, however, represented an examination of emotion-attribution at quite a different level from that in the placebo—cheating research with adults, for the emotion attributions that were induced with children are closer to potential real-life manipulations.

In the first two studies (Dienstbier, Hillman, Lehnhoff, Hillman, & Valkenaar, 1975) of the series with second-grade children, a three-part procedure was developed in which different children were first asked to watch a slot car that supposedly had a problem. After their failure at that watching task was dicovered by the experimenter, their emotional responses were attributed to being found out (external) or to having done the wrong thing (internal). The children were subsequently left alone, watching the car for a period of 12 minutes with their attention to the task timed after the situation was changed so that they believed that they could not again be detected for not watching the slot car. Since this procedure is presented in detail elsewhere (Dienstbier et al., 1975), it will be presented here but briefly and only the results from the second of the first two studies will be considered.

In Study 2 of the research with children, second-grade identical and same-sex fraternal twins were used as subjects, with one of each twin pair in each of the two experimental conditions; Steve Slane and Gail Zimmerman worked diligently with me in Study 2. The subject population was chosen since environmental and genetic differences (and hence error variance) are nearly equated between conditions, permitting a sensitive matched-pairs analysis of the results; differences between the means of the two experimental conditions would not be altered by this choice of subjects, however.

In the first phase of Study 2, the randomly chosen participating twin was told that the slot car he was to watch was very old and that it might break if it fell off the track. It was explained to each twin that recent problems with the slot cars included the cars occasionally going too fast so that they jumped the track but that when that was about to happen a "trouble light" on the side of the "control" (observation) booth would go on. The child was then to push in the level indicated on the side of the booth to prevent the "accident." It was explained that the man who fixed the toys would soon be called by the experimenter but that he could not fix the slot car track unless it was well "warmed up." The child was then asked to help by watching the slot car run on the track during the experimenter's absence to "call the toy man"; the experimenter then left the room. Immediately following the child's transgression to the set criterion (not watching for 6 continuous seconds), the slot car was made to jump the track and the "trouble light" was turned on; the slot cars were controlled by an observer hidden in a one-way observation booth in the toy room. The experimenter, having been signaled by the observer, returned immediately after the slot car accident and accompanied the child to another room (allowing the observer to remain blind). The experimenter then became aware of the first twin's condition and delivered the independent manipulation. Except for the attribution message itself, the language and emphasis of the two manipulation scripts was identical. The excerpt below is a brief portion of the external manipulation script with the italicized parts indicating the phrases added to modify the internal script to external:

"I bet you feel a little bad now that *I found out* the car fell off. I've seen other kids feel bad when *someone found out* they weren't able to do exactly what they were supposed to do. When the other kids who tried to watch the car couldn't, they felt bad *when I found out too,*" etc.

At the end of the independent manipulation, delivered in a gentle and warm manner, the child and experimenter returned to the toy room. The child was told that since he/she knew how to start and control the new "unbreakable" slot car, that during the ensuing period with the experimenter locked out (and with no key), "even if it (the slot car) did fly off the track, I'd never find out, since you could put it back yourself and I can't come right in with the door locked." After establishing a "secret knock" so that the child would know when the experimenter had returned, the child was left "alone" in the locked (from the outside) room for 12 minutes of slot-car watching. At the end of the 12-minute period, all children were praised lavishly for their apparently excellent performance and exchanged with their opposite-condition co-twin.

The average time of transgression (not watching the slot car) was 322 seconds for the external-condition twins, compared to 177 seconds for the internal condition children ($t = 8.0$, $p < .001$). Aside from the mean superiority of the watching behavior of the internal twins, a remarkable data pattern emerged; for each of the six twin-pairs run, the external twin transgressed approximately twice as much as the internal twin, a finding not dependent upon which condition was run first.

An analysis of the procedure from the child's standpoint may be useful in putting the powerful data of Study 2 in perspective. First, when the first slot car jumped the track, followed almost immediately by the entrance of the experimenter, the suddenness of all that happened seemed to leave most children quite surprised; most subjects displayed emotional arousal as evidenced by agitation and attempts to remedy the situation and/or to provide excuses for the accident. Second, although the experimenter acted warmly and kindly toward the child, he was a male (hence less often encountered in this type of situation by second-grade children), and he was a relative stranger (hence somewhat unpredictable and arousal provoking). In response to some combination of those factors and the experimenter's warm delivery of the independent manipulation, most children responded quite positively by apparently trying to pay attention. Finally, the task was a new one, with which the children had not had extensive previous experience, so that they were (presumably) somewhat more suggestible than if a more mundane task had been used. Since most factors, as described above, differ from typical socialization situations in the home or school, with the differences favoring more potential impact for our manipulation, it should not be expected that a similar level of single-trial success would be achieved with similar attribution manipulations in natural settings.

Although most early moral training is undertaken by socialization agents in circumstances of the child's heightened arousal following transgression (as in Study 2), I was curious if similar results could be achieved in an "innoculation" paradigm — a situation of reduced emotional arousal with emotion-attribution information given prior to any transgression or error. As the idea developed of exploring emotion-attribution procedures under conditions of reduced emotional arousal, the question of whether the arousal concept was necessary at all was confronted. An analogy to perceptual-motor functioning seems appropriate. In perceptual-motor tasks, when a skill such as driving a car is well learned, we take account of and respond to extremely small variations in directions, guiding the vehicle accordingly but usually without an awareness that is consciously articulated. It may be that we similarly learn to take account of even slight variations in autonomic arousal levels, responding and "correcting course" with (often) minimal or no articulated awareness. This belief in the influence of even slight arousal was clearly anticipated by William James (1890): "The changes are so indefinitely numerous and subtle that the entire organism may be called a sounding-board which every change of consciousness, however slight, may make reverberate. . . . Every one of the bodily changes, whatsoever it be, is felt acutely or obscurely, the moment it occurs [p. 450]." Although these observations do not conclusively establish the necessity of the arousal concept, they do encourage the continued use of the concept, especially in light of the earlier cited research. That is, it is apparent that most research subjects did not undergo strong arousal in the studies of Schachter and Ono (in Schachter & Latané, 1964) and those previously cited cheating studies with adults by my associates and me; yet the subjects responded to a tranquilizer in Schachter's research and emotion-attribution information about the (placebo) pill in our research as if their emotional arousal were quite significant.

The third study in this series with children, undertaken with the aid of Gil Tunnell and Cynthia Gallentine, was conducted to ascertain whether the emotion-attribution procedure would yield similar results under conditions of reduced emotional arousal and when given prior to any temptation or failure.

Data were gathered from 10 pairs of same-sex fraternal and identical twins, 2 pairs of which were in first grade, with the remaining 8 in second grade. As in Study 2, the randomly chosen participating twin was instructed in the necessity of running the slot car to warm it up and on the procedures for preventing a slot-car accident.

But we eliminated any personal failure experience and attempted to minimize the emotional arousal of our subjects by telling them about a child who helped the toy room experimenter "yesterday" who did not do a good job watching; the slot car therefore jumped the track. Together the experimenter and child discovered the slot car to be broken — a result that supposedly could only happen to such a "very *old*" slot car.

The experimenter and the child then left the toy room (to aid in keeping the observer blind) to find a "*new*" slot car that the child would watch. During that absence the experimenter became aware of the (first) twin's condition and administered the independent manipulation, which was similar to that used in Study 2. The external passage below, representing the first two sentences of the independent manipulation, has the phrases italicized that represent those added to change the internal script to the external version:

"When the car fell off before, yesterday, when that other kid was supposed to be watching it, he felt very bad *when I found out.* I know he did, because he told me how bad he felt *that I found out* the car fell off."

Upon returning to the toy room, the child was told that although this *new* slot car could not break if it flew off the track, it would be bad for it to jump the track many times; constant attention to the car and trouble light would therefore be required to prevent accidents. The locked toy room door and secret knock portions were similar to Study 2. To make sure that subjects understood that their conduct was free from detection, they were told: "Remember, this is a new car and so it won't break even if it does fly off, and you know how to put it on and how to start it up. And the door will be locked so I can't come right in. So even if it did fly off the track, I'd never know." The experimenter then left the room, remaining out for 12 minutes, during which the child's attention to the constantly running slot car was monitored by the hidden observer.

Mean seconds transgressing (not watching) for the internal-condition twins were 188, compared with 346 seconds for the external-condition co-twins (t = 2.54, sig $<$ .02, one-tailed).

Although Study 3 demonstrated that an inoculation procedure under apparently lower arousal levels could be similar in power to the Study 2, procedure utilizing prior failure, an even more convincing demonstration would be effected if children could be induced to make internal (vs. external) attributions through the use of materials not directly relevant to the dependent measure task. That is, one problem for the unambiguous interpretation of these studies with children is that it is possible that the internal manipulations drew more attention to the dependent measure task, or somehow made that task seem more important, etc.

Subsequent research with children was designed to answer the question of whether stories with internal or external emotion-attribution themes would differentially effect behavior in a difficult ("detection-free") watching task that would be presented to the children as unrelated other research. Although the initial research, conducted with the aid of Arleen Lewis, Charles Kaplan, Max Lewis, and Keith Willis, was successful in demonstrating superior watching performance from second-grade children after listening to internally oriented stories (heroes feeling good or bad due to their own performance, contrasted to external stories with feelings dependent upon others knowing), subsequent studies in our lab failed to replicate that effect; and I feel no confidence in our initial positive findings.

The positive outcome of such studies would have relevance beyond that of allowing greater certainty in the interpretation of the research with children. If it could be demonstrated that general internal emotion-attribution information given in a nonarousing context can be applied by individuals in specific moral situations, then the potential for developing moral training curriculae based upon this potentially powerful technique would be established.

Fortunately, research with college-age adult subjects has been successful in demonstrating this effect. The first series of research reported in this chapter (Dienstbier et al., 1977) utilized "reading comprehension test" materials to induce subjects to accept either internally or externally oriented emotion-attribution beliefs about the functioning of conscience. But differences between those two manipulations were not found in the three studies reported; the only differences that consistently emerged were between the passages relating to morality (both internal and external) and the control condition material (dealing, in Studies 2 and 3, with memory). Fortunately for understanding the problems of Studies 2 and 3, a manipulation check question was included in the test questions; the question asked experimental condition subjects to indicate what the "final step" was in the development of morality. One of the two possible answers was externally oriented toward parental punishment, whereas the other was internally oriented toward one's own moral principles. The question was answered correctly by only 62% of the subjects in Study 3. Since chance alone would yield a base rate of 50% correct, it was apparent that differences between the internal and external conditions were only weakly established, with approximately 24% of the subjects *knowing* the right answer (and half of the remaining 76% of the subjects guessing the correct option, yielding the 62% correct figure).

To strengthen the internal vs. external manipualtion in Study 5, those passages were completely rewritten. In all other critical ways, Study 5 was identical to Study 3, being supposedly the thesis research of the graduate student assistant; this study was conducted with the able assistance of Lynn Kahle. It was predicted that with the distinction between the internal and external emotion-attribution passages much better defined, the external condition and control condition cheating rates would be closer, with the internal condition cheating rate being significantly less than the other two.

Freshmen women subjects were recruited exactly as in Studies 2 and 3 to participate in a study "involving verbal testing." Those subjects who read the one-page, internally oriented passage on "The Development of Moral Self-Control" read that

> Even if the child has never been scolded or punished by parents, the child may begin to experience emotional tension when considering the violation of moral rules about things such as lying, cheating, or stealing . . . the individual will resist temptation to avoid the emotional tension even though no one else may ever know of the transgression . . . as we mature, the pleasure which we anticipate from knowing that we have acted morally correct remains a strong motivating force in helping us to be strong in the face of temptation. . . . Research has demonstrated that often very strong feelings of emotional tension result from individuals violating their own moral values, even though other people important to them do not know of those violations.

Comparable passages from the externally oriented passage were as follows:

> After being scolded or punished a number of times by parents or others, the child begins to experience emotional tension when considering the possibility of being found violating the moral rules about things such as lying, cheating, or stealing . . . the individual will resist temptation to avoid the emotional tension which is tied to the risk of being found out . . . as we mature, the pleasure which we anticipate from others knowing that we

have acted morally correct remains a strong motivating force in helping us to be strong in the face of temptation . . . research has demonstrated that often very strong feelings of emotional tension result when other people who are important to us discover and confront us over violations of moral values.

In addition to the differences in the internally and externally oriented passages as sampled above, at the end of each of the passages a "study question" was added to further involve the subject in thinking about morality in the manner described in each passage. Subjects were told that if they thought carefully about the passage and the question it would help them in the later reading-comprehension questions. Internal condition subjects were asked, "Can you think of any time recently when you were confronted with a moral choice but resisted transgression and remained strong due to feelings of emotional tension associated with your knowledge that the transgression violated your own moral values?" External subjects read a similar question ending with ". . . due to feelings of emotional tension associated with you knowing that other people might find out that you had transgressed?"

As in the previous studies of this type, after answering the first set of reading-comprehension questions, the subjects were given the right answers to the vocabulary test during the "delay" period and an excuse to use their pencils. As in Study 3, they were told that in order to avoid having to work an extra semester on his doctoral dissertation research, the researcher needed accurate data, etc.

The manipulation check indicated that a much better distinction between the independent manipulation passages concerning morality was achieved, with 83% of the subjects choosing correctly the internally or externally oriented option to the manipulation check question following the delay period, indicating that 66% *knew* the correct answer. As always, suspicious subjects were eliminated via a detailed, funnel-type (Page, 1971), postexperimental questionnaire.

As predicted, whereas only 15% of the 66 internal-condition subjects cheated, 30% of the 66 external subjects and 31% of the 61 control-condition subjects cheated. By chi-square tests, the frequency of cheating in the internal condition is significantly different from that of the external condition (chi-square = 4.31, $p < .05$) and significantly different from the control condition (chi-square = 4.60, $p < .05$).[1]

---

[1]It is of some interest to note, because social psychology is currently involved in discussions of the impact of social psychological knowledge on social change, that the independent manipulation in this study consisted of passages discussing two psychological theories, each having some claim to validity. That is, the external passage was essentially a presentation of the two-factor theory of avoidance (avoiding temptation to avoid previously conditioned emotion responses), whereas the internal manipulation was essentially the theory of self-control presented in this paper and previously by Dienstbier et al. (1975). Differential cheating rates were achieved, then, by reading two different current psychological theories.

RESEARCH SUMMARY AND GENERAL THEORETICAL ISSUES

It is useful now to consider the implications of all the emotion-attribution research reviewed so far for the purposes of summary and in order to explore that research for general conclusions not obvious when viewed piecemeal. The initial research with college students was presented to establish the importance of emotion-attributions in making moral decisions, using a placebo pill manipulation that had very little apparent mundane or real-world realism. On the other hand, the research with conceptually similar phenomena with children presented manipulations that had far more mundane realism, such that it would not be difficult to speculate productivity about direct applications of that research to real-life socialization problems. But in changing significantly both the age of subjects and the form of the manipulation between the initial adult and the subsequent child research, we had only the theoretical bridge of emotion-attribution theory to join those two quite different research series. The final study presented above, using the reading comprehension approach to independent variable manipulation, substantially reinforces that conceptual bridge, for the manipulation content (e.g., "the individual will resist temptation to avoid the emotional tension even though no one else may ever know of the transgression") was purposefully designed to resemble that of the research with children (e.g., "He told me he would have felt bad even if I didn't find out"). The major value of that study, however, is the finding that adults were able to apply that general emotion-attribution information about morality to a specific cheating situation. Although that finding would be more useful if it could have been demonstrated with young children as well, it is likely that our failure to demonstrate such a general-to-specific transition with children was due to working with children who were too young (second grade).

Another theme emerges from the research program as a whole other than the apparent convergence of support for the emotion-attribution analysis. It does appear that children and young adults alike are remarkably susceptible to external influences in the attributions they make about the causes and meanings of their emotional arousal.

How Attributional Processes Affect Behavior

There are apparently three ways in which the attribution of emotion to an irrelevant source may affect the impact of that emotional arousal on subsequent behavior. Any or all of these effects may influence the role of emotional arousal in resistance to temptation.

The major emphasis derived largely from the research presented above has been that emotional arousal attributed to an irrelevant source simply becomes less effective in influencing behavior than if the source of the emotional arousal is experienced as relevant. For example, one may make attributions, which are essentially cognitive, that an emotional response is due to acrophobia on a high bridge or due to fear of detection in obviously safe situations. Although the quality of the emotional experience may change little or not at all under such circumstances, if the emotional arousal is of low or moderate intensity, it may be relatively easy and logically consistent to override the avoidance tendency motivated by the emotional state. This response somewhat resembles the classical psychoanalytic defense of "denial," but it should be increasingly more difficult to accomplish with increasing levels of negative emotion. The ineffectiveness of the arousal (placebo) pill manipulation with women in the initial emotion-attribution study discussed previously was probably due in part to this phenomenon of increased difficulty in acting against the emotion when the emotion is intense. Similarly, Nisbett and Schachter (1966) found that under high-fear conditions, subjects would not attribute shock-induced arousal to a placebo pill with "arousal" properties, though such an effective attribution to the placebo (allowing more shock tolerance) under low-fear conditions did occur.

The second way in which the attribution of emotion to "irrelevant" sources may attenuate the impact of that emotion is through the quality of the emotion changing as a result of that reattribution. This result is probably more likely when a mixture of situational cues is present, some of which make positive and some negative emotions relevant. As suggested in the example used early in this chapter, the roller coaster may prove very exciting, even to the mildly acrophobic rider, if positive cues (other people laughing, the fast motion, etc.) are present *and* if the negative emotional state is perceived to be due to irrelevant sources (e.g., the rider perceives the danger as nonexistent, so the fear is due to personal and irrelevant deficiencies). Which of the cues are responded to will depend in part upon the chronic dispositions and needs of the individual. For example, the intense need for excitement believed to be common to the psychopath (Hare, 1970) may dispose such an individual to regard cues associated with negative emotional states (objective danger, etc.) as irrelevant (Yochelson & Samenow, 1976). If negative cues are suddenly perceived as relevant, however, the positive emotional state may quickly revert to a negative experience. This re-experiencing of a different emotional state due to focusing upon selective cues has been termed "drive displacement" by Epstein (1967), who noted that novice skydivers often experience outbursts of anger prior to jumping, while steadfastly denying the experience of fear. Perhaps Frued observed these processes in developing the "reaction-formation" concept.

Finally, there is evidence that if the source of emotional arousal is seen as irrelevant, the emotional response will extinguish more quickly than if the emotion is seen as situationally appropriate. Loftis and Ross (1974) demonstrated that an emotional response would extinguish more quickly with continued presentation of the conditioned stimulus if subjects believed that the response was not due to the presented stimulus. This phenomenon is also quite likely to be observed mainly at low emotional arousal levels, for case histories are numberous of phobic individuals failing to experience extinction of their emotional reactions to stimuli that they knew (cognitively) to be "irrelevant."

SOCIALIZATION

The various attributional messages that are potentially imparted by different socialization techniques can lead to different behavioral results following temptation through any or all of the three processes described above. Additionally, other variables are relevant in socialization, complicating prediction even further. Thus while each socialization technique may lead to different attributions about emotional arousal, each may be applied at varying levels of intensity, creating differing arousal levels. Additionally, each child has a different ability to generate and sustain emotional arousal, and children will certainly differ in their readiness and ability to make attributional inferences from the same situation. The characteristics of the socialization agent, the background variables of immediate environment (physical and social), and the relationships that exist between the socialization agents and child will also influence both the level of emotional arousal generated by any socialization interaction and the child's attributions about the cause and meaning of that emotional response. All these variables could interact in determining the effectiveness of any approach to the socialization of moral self-control for each individual. Such complexity must lead to extreme caution in assuming the validity of statements about the *overall* superiority of specific socialization techniques; it is not surprising that research on the success of most techniques has been quite inconsistent (Hoffman, 1970). Although the potential interactions of relationship, individual differences, and socialization techniques are far too numerous and complex for a comprehensive analysis in this chapter, I will discuss some of the more significant factors and relationships. Many of these analyses are admittedly highly speculative, and it will be obvious that much research remains to be done before even moderate certainty is likely.

The ability to generate and sustain emotional arousal may be one of the most important individual difference variables relating to socialization success. Such differences, often labeled "temperment," have been identified by many researchers (e.g., Cattell, 1965; Buss, 1975; Martin, 1971) as due more to heredity than to environmental variations (within American society). Most of the literature on socialization relates to that majority of children who come from the broad middle of the continuum of ability to generate and sustain emotional arousal and to those children who come from the high end, appearing anxious and overstimulated. This is simply because for such children, normal socialization works; whereas with psychopathic children — who evidence a lack of inhibiting emotional reactions — little socialization or therapeutic success is generally reported. Even within this normal group, the ability to generate and sustain arousal related directly to the level of socialization, as indicated by the recent work of Waid (1976). That work demonstrated that for a subject population recruited through the Rutgers newspaper, greater skin conductance in response to a loud tone corresponded to higher levels of socialization as indicated on the CPI Socialization Scale. This phenomenon has been previously noted between normal and psychopathic populations, with less skin conductance by the psychopathic group in response to noxious stimulation (Hare, 1970). There are two reasons for this more effective socialization of emotionally reactive children; obviously, such children would experience more inhibiting arousal when facing temptation, but it is also likely that less severe parental responses were needed (and therefore utilized) in socialization, with the result that the child would be more likely to attribute emotional arousal to the act of transgression, rather than to the parental confrontation.

Induction.   As defined by Hoffman (1970), induction involves the explanation to the child of why certain behaviors are right or wrong; it is, according to Hoffman's review of socialization, a most effective technique for inducing moral self-control. An emotion-attribution analysis would support and explain that conclusion as follows: Following transgression and confrontation, even when interaction between parent and child will proceed on a verbal level, some arousal is quite likely, especially for the child who is higher in emotional reactivity. This arousal will be due to the child's recognition that the parent (or other socialization agent) is unhappy with the child, that the child has transgressed against a standard that may already be partially socialized (perhaps merely at the cognitive level), and that some inconvenience or pain to the child may be forthcoming. But by asking the child to focus upon the harm done by the act of transgression, the child is induced to attribute that emotional discomfort — whatever the real cause — to the transgression. When confronting later temptation, the emotional discomfort resulting from the contemplation of transgression should be experienced as relevant and should play a role in aiding the resistance to that temptation. Hoffman has convincingly presented the case that induction is the most effective technique for the development of self-control; I would suggest that this would be true for children in the middle to upper range of emotional responsiveness; the likely ineffectiveness of induction for emotionally unresponsive children will be elaborated below.

Punishment.   The literature on morality and socialization indicates that when moral self-control is defined similarly to the manner used previously in this paper (resistance to temptation under circumstances of perceiving that detection for transgression is impossible), that punishment is ineffective — appearing to be negatively related to morality (Maurer, 1974). On the other hand, some research has suggested that love withdrawal or "psychological" punishment is positively related to self-control (e.g., Sears, Maccoby, & Levin, 1957).

Punishment is obviously quite effective as a behavior control technique, either when punishing agents will normally be present during temptation periods (e.g., controlling social problems such as hitting adults) or if the organism is not sufficiently developed cognitively to be able to know that there are detection-free contexts or to recognize them with confidence when they occur. However, its major limitation in inducing moral self-control, as defined above, is that more than any other technique the act of punishment draws to itself the attribution of negative emotional arousal. The child with the stinging hand may find it almost impossible to attribute the resultant emotional state to the act of transgression rather than to the punishment. Yet, it is possible that judicious use of punishment could be useful in eliciting emotional arousal that might (in later similar transgression circumstances) be attributed to the transgression itself rather than to the early (in socialization) punishment, especially in less emotionally reactive children. However, remembering the relationship of misattribution and emotional arousal level, positive results with punishment should hold most for mild punishment of young children when it is delivered in conjunction with internal emotion attributions (e.g., "you feel bad because you transgressed," etc.). With older children who are normal in emotional reactivity, a paradoxical effect is predicted, with the more severely punished child manifesting less resistance to temptation, since such a child could make accurate emotion attributions about the external cause of the emotional arousal. There is some evidence of exactly that effect. In their chapter on the influence of punishment, Walters and Parke (1967) reference their own research in which the impact of punishment intensity and timing interacted for 6- to 8-year-old boys. High-intensity punishment was more effective in eliminating or delaying reaching for a "forbidden toy" in a 15-minute "alone" condition if administered early in the reaching sequence; however, it was the low-intensity punishment that was most effective if punishment was delivered late in the sequence (where it occurs in most natural socialization). They suggested that the late, mild punishment might have served as *a more effective cue*, allowing the association of the disapproved *activity* with disapproval more than would be the case for the high-intensity punishment. My interpretation would suggest that when elicited early in a behavioral sequence, the emotional arousal resulting from punishment may be mistakenly attributed to the act of transgression, since this attribution is facilitated by the temporal contiguity (but even this effect should hold only for relatively young children). On the other hand, once a transgression has been completed, if negative emotional arousal is then induced by a very salient (severe) external cause, the child should be able to clearly differentiate the punishment-induced emotional state from the previous state associated with the transgression itself, facilitating the attribution of the negative emotional state to the punishment. Thus a late, severe punishment should be particularly ineffective for normal children. Emotion-attribution theory can therefore provide an articulate framework for explaining why punishment is negatively associated with the development of self-control when used in the home and classroom.

When punishment is inconsistent, it should be even less effective, since the child will be more able to correctly perceive the external cause of the negative emotional state if it does not always follow identical transgressions. Similarly, if an individual is punished very frequently, perhaps due to low emotional reactivity (hence slow avoidance learning), strict parents, or due to living in a particularly tempting environment, that individual too should learn more accurately the real causes of emotional arousal and hence should be less likely to attribute negative emotional arousal to the act of transgression. One can foresee the possibility of a vicious circle, with the lower emotional reactivity leading to less emotional conditioning, resulting in more punishment, which would in turn lead to more accurate attributions about the source of emotional arousal following transgression; the ultimate result would be that moral development would be substantially inhibited.

When punishment is "psychological" or love-oriented, on the other hand, some authors have suggested that it is effective in the hands of warm parents (i.e., Sears et al., 1957). If there is no specific salient and identifiable parental act of punishment in such cases, it may be easier for the child to misattribute emotional arousal to the act of transgression rather than to the punishment itself. In short, if both physical and "psychological" punishment induce similar arousal levels, the more tangible physical punishment should prove less effective for the development of self-control. When the child perceives the parents to be warm, it would be more difficult to attribute to them and to their act of punishment the negative emotional response from the punishment, making attribution to the transgression itself easier. We might also surmise that warm parents probably tend to use milder punishments and more explanations, facilitating internal attributions.

Psychopathy and Emotional Arousal

In his review of the psychological literature relating to psychopathy, Hare (1970) indicated the Group for the Advancement of Psychiatry has proposed that the term "tension-discharge disorder, impulse-ridden personality" be used in place of the term "psychopath" when applied to children. Hare (1970) described such an individual as one with "little anxiety, internalized conflict, or guilt ... with failure to develop the capacity for tension storage and for the postponement of gratifications [p. 5]." (For this chapter I continue to use the term *psychopath* for young and old.) Not only are psychopaths typically regarded as being "guiltless" and low in capacity for fear and/or anxiety, but as high in their need for stimulation and thrill seeking. We focus first on the implications of the psychopath being unable to develop or sustain emotional arousal from those socialization experiences that would cause negative emotional states in more normal children.

It should be expected, if a lack of powerful and sustained emotional arousal is the root problem of psychopathy and some habitual criminality, that the primary focus of socialization and correctional techniques for such individuals might be through procedures that reliably elicit high levels of emotional arousal (in normal children), provided that those techniques are consistently applied and contingent upon negative behavior. Hare (1970) has suggested that psychopathy "may very well represent the outcome of interactions between the characteristics (possibly congenital) of the child and the socialization techniques employed [p. 101]." He noted that cold, unaffectionate fathers who tend to be strict disciplinarians have fewer psychopathic children and that strict discipline tends also to decrease the incidence of adult psychopathy. But Hare's analysis should hold only if the strict discipline is consistently applied, for the reasons detailed above. Inconsistent, harsh discipline occasionally (noncontingently) applied should lead to opposite results, since accurate (external) attributions become quite likely under such a regime. We would therefore expect that even with strict and consistent discipline that real self-control, as defined herein, would be achieved only slightly by children who are chronically unresponsive emotionally, but that in detection-possible situations such procedures would result in socially acceptable behaviors.

The picture that emerges is of an emotionally unresponsive individual who is likely to become the target of increasingly harsh socialization as parents and others become desperate to achieve positive change. But while that punishment should lead to some conditioned emotional arousal when the psychopath faces temptation, it should be quickly discounted in detection-free situations by the psychopath as irrelevant, since correctly attributed to punishment. This would be particularly true if the harsh discipline administered were inconsistent. There is some evidence that a noncontingent, harsh environment is related to the development of psychopathic behavior (e.g., Hare, 1970; Mandler, 1964; Kohlberg, 1958) and that even though psychopaths do evidence normal levels of autonomic arousal under threatening conditions, they do not experience arousal as inhibiting emotion (Schachter & Latané, 1964).

Let us return to the issue of the understimulated and stimulation-seeking side of psychopathy. Hare has reported that even in controlled laboratory situations, the psychopathic delinquent indicates a greater preference for levels of complexity and novelty (associated with seeking higher arousal levels) than normal controls. In their study of *The Criminal Personality*, Yochelson and Samenow (1976), though reluctant to use the term *psychopath*, described the habitual criminal as almost constantly seeking emotional stimulation, partly through turning to criminal activity; thus crime is not just the logical result of the lack of inhibitory control. Criminal activity may be experienced by the psychopath as a positive emotional stimulant.

Indeed the psychopath's apparent need for criminal activity for its own sake (rather than to remedy conditions of material deprivation, if we accept the thesis of Youchelson and Samenow) provides some support for the idea that the criminal experiences the emotional arousal associated with the risks of crime as a positive emotional experience. That enjoyment of risk is apparently attained through the elimination of the belief that capture is possible, rendering punishment-induced emotional arousal irrelevant. Yochelson and Samenow discuss a process they label "cutoff" in which the chronic criminal convinces himself that all forms of danger are irrelevant to the ensuing crime.

It was suggested above that the misattribution of the cause of (and hence the nature of) certain emotions is maximized under conditions of low or moderate emotional arousal; the process of re-experiencing the emotional state associated with danger should therefore be relatively easy for the psychopath.

This reinterpretation of emotional arousal originally stimulated by punishment and later elicited by the temptation situation does not mean that all emotional states might easily be derived from that initial fear-stimulated arousal. Discrete emotion theorists such as Izard (1971, 1972) and Plutchik (1962) have suggested that certain emotions are somewhat alike in underlying arousal. It should therefore be easier to reattribute or re-experience certain emotional states as certain specific others, since some arousal components may be common to a limited few similar states. (Certain arousal components may be common to all emotional states that have sympathetic nervous system involvement, as emphasized by Schachter and Singer [1962].) Specifically, for the present issue, Hebb (1946) has noted that the arousal common to both fear and excitement allows easy transition between those two emotional experiences.

CONCLUSIONS

Often the imposition of a new theoretical system to a substantive area of venerable interest results in new insights into those traditional issues and problems. Frequently, however, as new systems are embraced by theorists, momentum carries the theory into areas well beyond its range of convenience, as evidenced by claims of insight qualitatively superior to that provided by other theoretical systems. In this chapter I have attempted to avoid both the image and the fact of exaggerated claims; in fact, the major insight developed in this chapter could be that the emotion-attribution approach leads to recognition that the issues of moral socialization are far more complex than traditional moral judgment or avoidance theories would suggest. In place of relatively simple universal insight into moral development and socialization, I offer a theoretical pool muddied by complexity. However, it is possible with continued research and development that emotion-attribution theory will provide a means for establishing some order, to the end that the most important elements of individual differences and of socialization practices may be identified. A more articulate understanding of the interactions of those elements may follow.

With several colleagues (Dienstbier et al., 1975) I have previously attempted to use emotion-attribution concepts to understand the interrelationship of moral judgment, as understood by Kohlberg (1958), with moral behavior approaches as derived from learning theory. Without resurrecting that discussion, I shall elaborate briefly on the process of moral decision making, using emotion-attribution concepts. It was argued, in the paper cited above, that in our society the number of behaviors that elicit moral dilemmas are few, involving (usually) pain inflicted upon others, honesty, and theft or destruction of property. As we mature, it is likely that different attributions about the specific temptation situations will emerge; that is, with social and cognitive development, different elements of an honesty situation may be important in defining the moral relevance of potential action. (The initial reading-comprehension research with college students reflects and illustrates this process.) But emotional responses may arise even in situations where initial attributional processes were not conducive to such response, for it is likely that the conditioned emotional response to such situations will be less differentiated than those attribution processes. (The research by Schachter and his colleagues with adrenalin and with the tranqualizer chlorpromazine reflects the importance of the level of emotional arousal.) It is suggested that one of the ways in which the individual decides which of several behavioral choices are appropriate is through an assessment of the emotional arousal that follows thinking about each behavioral option. As the individual contemplates the behavioral options available, attempting simultaneously to make judgments about how the relevant value should be hierarchically ordered, emotion-attribution processes become important. Although our experience with our subjects suggests that individuals are not aware of making such attributions at a verbally articulate level, the research with adult subjects using placebo pills with "arousal" side effects, the final reading comprehension study with adults, and the research with children all reflect this process.

At each of the levels of decision making described previously, the degree to which emotion is actually elicited, the attributions made, and the control those attributions have over the meaning of the emotional response will depend on the indivudal's capacity to develop and sustain arousal and the past socialization history of the individual as that history influences both the quantity and quality of emotional response developed in specific temptation situations.

REFERENCES

Buss, A. H. *A temperament theory of personality development.* New York: Wiley, 1975.

Cattell, R. B. *The scientific analysis of personality.* Baltimore: Penguin, 1965.

Davies, E. This is the way Crete went. *Psychology Today*, 1969, *3*, 42–47.

Dienstbier, R. A. The role of anxiety and arousal attribution in cheating. *Journal of Experimental Social Psychology*, 1972, *8*, 168–179.

Dienstbier, R. A., Hillman, D., Lehnhoff, J., Hillman, J., & Valkenaar, M. C. An emotion–attribution approach to moral behavior: Interfacing cognitive and avoidance theories of moral development. *Psychological Review*, 1975, *82*, 299–315.

Dienstbier, R. A., Kahle, L. R., Willis, K. A., & Tunnell, G. *Cheating as a function of emotion–attribution and moral value activation: Reactance vs. conformance.* Unpublished paper, 1977.

Dienstbier, R. A., & Munter, P. O. Cheating as a function of the labeling of natural arousal. *Journal of Personality and Social Psychology*, 1971, *17*, 208–213.

Epstein, S. M. Toward a unified theory of anxiety. In B. A. Maher (Ed.), *Progress in experimental personality research* (Vol. 4). New York: Academic Press, 1967.

Fortenbaugh, W. W. *Aristotle on emotion.* New York: Barnes & Noble, 1975.

Hare, R. D. *Psychopathy: Theory and research.* New York: Wiley, 1970.

Hebb, D. O. On the nature of fear. *Psychological Review,* 1946, *53,* 259–276.

Hoffman, M. L. Moral development. In P. H. Mussen (Ed.), *Carmichael's manual of child development* (Vol. 2, 3rd ed.). New York: Wiley, 1970.

Izard, C. E. *The face of emotion.* New York: Appleton–Century–Crofts, 1971.

Izard, C. E. *Patterns of emotions.* New York: Academic Press, 1972.

James, W. *The principles of psychology* (Vol II). New York: Henry Holt & Company, 1890.

Kohlberg, L. *The development of modes of moral thinking and choice in the years ten to sixteen.* Unpublished doctoral dissertation, University of Chicago, 1958.

Langer, E. J., & Abelson, R. P. The semantics of asking a favor: How to succeed in getting help without really dying. *Journal of Personality and Social Psychology,* 1972, *24,* 26–32.

Loftis, J., & Ross, L. Retrospective misattribution of a conditioned emotional response. *Journal of Personality and Social Psychology,* 1974, *30,* 683–687.

Lykken, D. T. A study of anxiety in the sociopathic personality. *Journal of Abnormal and Social Psychology,* 1957, *55,* 6–10.

Mandler, G. Comments on Dr. Schachter's and Dr. Latané's paper. In D. Levine (Ed.), *Nebraska Symposium on Motivation* (Vol. 12), Lincoln, Neb.: University of Nebraska Press, 1964.

Martin, B. *Anxiety and neurotic disorders.* New York: Wiley, 1971.

Maurer, A. Corporal punishment. *American Psychologist,* 1974, *29,* 614–626.

Mowrer, O. H. *Learning theory and personality dynamics.* New York: Ronald Press, 1950.

Nisbett, R. E., & Schachter, S. Cognitive manipulation of pain. *Journal of Experimental Social Psychology,* 1966, *2,* 227–236.

Page, M. M. Postexperimental assessment of awareness in attitude conditioning. *Educational and Psychological Measurement,* 1971, *31,* 891–906.

Plutchik, R. *The emotions: Facts, theories, and a new model.* New York: Random House, 1962.

Schachter, S., & Latané, B. Crime, cognition, and the autonomic nervous system. In D. Levine (Ed.), *Nebraska Symposium on Motivation.* Lincoln, Neb.: University of Nebraska Press, 1964, *12,* 221–273.

Schachter, S., & Singer, J. E. Cognitive, social, and physiological determinants of emotional state. *Psychological Review,* 1962, *69,* 379–399.

Schwartz, S. Normative influences on altruism. In Leonard Berkowitz (Ed.), *Advances in experimental social psychology* (Vol. 10). New York: Academic Press, 1976.

Sears, R. R., Maccoby, E. E., & Levin, H. *Patterns of child rearing.* Evanston, Ill.: Row, Peterson, 1957.

Turiel, E. Social convention and morality: Two distinct conceptual systems. In Herbert E. Howe, Jr. & C. Blake Keasey (Eds.), *Nebraska Symposium on Motivation* (Vol. 25). Lincoln, Neb.: Unibersity of Nebraska Press, 1977, in press.

Waid, W. M. Skin conductance response to both signaled and unsignaled noxious stimulation predicts level of socialization. *Journal of Personality and Social Psychology,* 1976, *34,* 923–929.

Walters, R. H., & Parke, R. D. The influence of punishment and related disciplinary techniques on the social behavior of children: Theory and empirical findings. In Brendan A. Maher (Ed.), *Progress in experimental personality research* (Vol. 4). New York: Academic Press, 1967.

Yochelson, S., & Samenow, S. E. *The criminal personality, Vol. I: A profile for change.* New York: Jason Aronson, 1976.

8 Attributional Aspects of Interpersonal Attraction

Dennis T. Regan
Cornell University

> *I do not love thee, Doctor Fell,*
> *The reason why I cannot tell;*
> *But this alone I know full well,*
> *I do not love thee, Doctor Fell.*
> —Thomas Browne

The entire body of literature on interpersonal attraction, or liking, conducted over the last few decades by social psychologists can be viewed as a rejection of the implications of this children's poem. Although laymen may sometimes feel that their attraction or repulsion toward another person is inexplicable, investigators have searched diligently for reliable antecedents of attraction and have generally attempted to place their findings in an integrative theoretical context. When the research is viewed as a whole, we can discern two broad theoretical perspectives that have dominated it.

One of these perspectives sees man as determined to maximize the rewards and minimize the costs incurred in interaction with others. The social exchange theories of Homans (1961) and Thibaut and Kelley (1959) are well-articulated exemplars of this tradition. The economic model of interaction advanced by these theorists views man as bringing to any interpersonal encounter certain "investments" of time, energy, prestige, ability, and the like and extracting from the encounter certain rewards obtained at varying cost. Those encounters that are most rewarding — where the ratio of rewards to costs incurred is both favorable and more favorable than that obtainable elsewhere — will be preferred, and the individual should most like others who provide him with such outcomes. Lott and Lott (1974) have argued that the determinants of liking, viewed as an anticipatory attitudinal response, can be placed in a stimulus—response framework derived from Hullian learning theory. In their formulation, "liking for a person will result under those conditions in which an individual experiences *reward in the presence of that person,* regardless of the relationship between the other person and the rewarding event or state of affairs [p. 172]."

The other dominant perspective admits that people who reward one are liked but argues that *what is rewarding* for a person varies dramatically as the context of the interaction is altered. For these more cognitive theorists, the outcomes accruing to an individual in interaction have different meanings depending on contextual factors. This fact, it is argued, reduces the usefulness of a theoretical position in terms of rewardingness alone, because the power of even obvious interpersonal rewards (such as favorable evaluations from another) varies from situation to situation (e.g., Mettee & Aronson, 1974). It should be noted that this is often admitted by reinforcement theorists. Lott and Lott (1974) state explicitly that "it is obviously difficult to specify beforehand what will constitute a reward for human beings for whom the same event or objective stimulus can vary in meaning as a result of differences in past experiences, in present motives, and in the context of the situation [p. 173]." The difference seems to be one of emphasis.

It is, of course, the cognitive tradition in which attributional concerns have been most systematically explored. Many of these concerns can be traced to Heider's (1958) original theorizing about the role of causal attributions in affecting interpersonal reactions — and in being affected by them. It is not necessary here to present Heider's ideas in detail; they have been summarized many times, both in these volumes and elsewhere. But basic to his approach is the contention that the perceiver takes an active role in inferring the meaning of an action — the underlying intention or disposition of the actor that it reflects — and that it is this ascribed meaning, arrived at from an inference about the cause of the behavior, that determines reactions to the actor. It is a truism, for example, that we generally react more favorably to people who benefit us than to those who harm us. But in a strong statement of his position, Heider argued that "if *P* is benefitted or harmed, his response will depend *mainly* on the way he interprets these events [p. 276, emphasis added]." In Chapter 10 of *The Psychology of Interpersonal Relations,* Heider offers rich and extensive speculation about the ways in which motives or intentions are likely to be assigned for benefits or harms received and the importance of such attribution of underlying disposition for affecting sentiment toward the other. He also stresses the way in which assigned intention or meaning of the action is itself likely to be affected by prior sentiment toward the other, primarily in ways consistent with his well-known balance principle: given some slippage in the meaning that could be assigned to an action, it will be assigned in a manner consistent with affect toward the actor.

In the essay that follows, we will examine in detail the relationship between causal attributions for another's (and our own) behavior and interpersonal attraction, or liking for that other. But for attributional processes to play any role in predicting attraction to others, they must occur. We might begin by asking when, or under what conditions, people are likely actively to concern themselves with providing a causal explanation for the behavior of another person. Several writers have suggested that the attribution process is typically directed toward certain goals, and in particular toward the goals of understanding, predicting, and exercising control over the social environment (e.g., Kelley, 1972). If it is simply unimportant for a person to understand the intention behind or future implications of another's behavior, he is presumably relatively unlikely (unless asked) to engage in an attributional analysis in the first place.

There has been little systematic exploration of the conditions under which people are likely to become actively concerned with making attributions about another's behavior. But in a recent article, Berscheid and her colleagues (Berscheid, Graziano, Monson, & Dermer, 1976) have made a beginning, explicitly deriving their suggestions from the assumption that attributions are made in order to increase control over significant aspects of the environment. These authors gathered evidence supporting the hypothesis that people are most likely to be concerned with making systematic causal attributions when they are dependent on another person for future social outcomes, when the relationship is likely to be relatively enduring rather than short-lived, and when relatively little is already known about the other. These hypothesized conditions do indeed seem to be those where concerns about future prediction and control of the other's behavior are likely to be strong. And it is worth noting that, in the typical laboratory study of attribution, only the last factor (little is already known about the actor) is present. Subjects do not expect this other (with whom they often do not even interact) to control their future outcomes for any significant length of time. This point is important, because Berscheid et al. suggest that the *outcome* of the attribution process itself may be significantly affected by the importance of the perceiver's motivation to predict and control.

Given that the perceiver is motivated "to find sufficient reason why the person acted and why the act took on a particular form" (Jones & Davis, 1965, p. 220), how does he go about this? The details differ slightly from theorist to theorist, but there is in the literature general consensus on the broad outlines of the attribution process. Making attributions is seen as a largely rational activity in which the perceiver integrates two kinds of information in arriving at an estimate of the locus of causality for an action: the particular behavior of the actor, and the surrounding circumstances. Using this information, he arrives at an estimate of the degree to which the behavior was caused by factors internal to the actor — such as his particular intentions, dispositions, traits, moods, or other individuating characteristics, versus factors external to the actor, particularly situational forces and demands operating on him.

In assigning causality to these two broad classes of factors, the perceiver is thought to integrate the available evidence much like a scientist using John Stuart Mill's method of difference, or the analysis of variance (Kelley, 1967). That is, the perceiver assigns causality to that factor or set of factors that tends to be present when the effect or behavior is present and absent when the effect is absent. In the simplest case, a behavior that always occurred in the presence of a particular environmental factor and never in its absence would be seen as "caused" by that factor. Similarly, an actor's own characteristics would be seen as the cause of a behavior if that actor always (or at least typically) engaged in that behavior, whatever the situation, whereas other actors did not. For even these crude analyses, evidence of covariation of possible cause and effect over time would be extremely helpful in making a rational causal decision. But frequently in real life, and typically in research settings, the perceiver does not have access to information over time. He thus applies a somewhat more simple rule: the more obvious or salient the environmental factors that might have produced the effect (or, the stronger the apparent environmental press), the less likely the action is to be attributed to characteristics of the actor, and — by implication — the less informative it is seen about such characteristics. Kelley (1972) has termed this the "discounting principle: the role of a given cause in producing a given effect is discounted if other plausible causes are also present [p. 8]." The discounting principle further implies that if there exists a strong environmental inhibitory cause — something in the situation that could have been expected to inhibit the observed behavior or cause its opposite — the inference to an internal cause will be even stronger.

From this rather sketchy analysis of attributional processes,[1] one aspect of their relevance for the development and maintenance of interpersonal attraction can be clearly seen. Attributional processes may yield inferences, held with varying degrees of strength and confidence, about the intentions, motives, and dispositional qualities of others. If we value or like others in part because of the desirable qualities we see in them, and if attributional processes determine the nature of our perception of these qualities, then causal attributions about others'

---

[1]It should be explicitly noted here that attribution theorists have long been intrigued by the possibility of self-serving, defensive, and other biasing tendencies in making attributions. In his original paper, Kelley (1967) explicitly discussed several attributional biases; the self-serving nature of the resolution of attributional conflicts in couples has been suggested in the previous volume in this series (Orvis, Kelley, & Butler, 1976); and Jones, in his discussion with Kelley included in this volume, predicts an increase in attention to such "exceptions" to a disinterested, rational processing of information in arriving at attributions. Nevertheless, in considering the relationships between attribution and attraction, the "rational" model has dominated both the conceptual and empirical work.

behavior are likely to be important in determining our attraction to others. In the section that follows, we examine both assumptions: that attributional processes systematically affect our perceptions of the internal and dispositional qualities of others, and that our perception of those qualities affects our attraction to them.

But the relationship between attribution and attraction is certainly more complex than implied in the preceding analysis. Thus far we have focused only on the *effects* of causal attributions for *another's* behavior on attraction. But others not only behave in ways that may affect us and give rise to inferences about their intentions, motives, and dispositions; we also behave toward them, frequently in ways that may bring benefit or harm to them. There is a burgeoning literature on the determinants of attributions about our own behavior — the inferences we make about what caused us to behave in a particular way. These self-attributions (or self-perceptions) may also influence attraction toward another. If, for example, we harm another person under conditions leading us to infer a negative attitude or disposition toward that person, liking for him will be diminished. We shall discuss in detail the effects of self-attributions on liking.

And finally, there are reasons to expect that liking for another will itself influence attributions about that person's behavior. Behavior, as Heider indicated, does not come clearly labeled as to its cause. In forming attributions about another's behavior, we use whatever information seems relevant in arriving at a causal analysis. Where we know nothing about the actor, and observe only one behavioral sequence, we have nothing but the behavior itself (its typicality, extremity, and the like) and the perceived environmental circumstances on which to base an attribution. But when we already know something about the actor's dispositions, and have developed liking or disliking for him, we may arrive at an attribution for his present behavior that is relatively consistent with this prior knowledge. Good actors are expected to do good things, and bad actors are expected to do bad things. Where the nature of the action "fits" what we know about the actor, we should be relatively likely to attribute the behavior internally — to his intentions and dispositions. But where the behavior is at variance with our knowledge of (and liking for) the actor, we should be relatively more likely to seek an explanation in factors external to the actor.

Thus, we shall examine evidence, in sequence, on: (1) the effects of causal attributions for another's behavior on liking; (2) the effects of self-attributions on liking for another; and (3) the effects of attraction on attributions for another's behavior. Throughout, we treat attribution and attraction as if they were separable independent and dependent variables. This provides a convenient framework for summarizing the relevant research, but it does violence to our assumption that in natural settings, attributions and attraction interact in complex and mutually interdependent fashion.

EFFECTS OF ATTRIBUTIONS
ABOUT ANOTHER'S BEHAVIOR
ON ATTRACTION

It was suggested earlier that stronger and more confident attributions about the personal characteristics of another are made when the apparent environmental press toward the behavior is weak and particularly when situational factors could have been expected to inhibit the behavior. Perhaps the classic demonstration of this phenomenon is the "submariner — astronaut" experiment of Jones, Davis, and Gergen (1961). In that study, subjects heard an interview of a student who had been told to play the role of an applicant for a job as either a member of a submarine crew or an astronaut. The characteristics of a person likely to be selected for and successful at each job were also described, such that a submariner was portrayed as a relatively other-directed individual who enjoys interacting with others, whereas an astronaut was expected to be relatively inner-directed, requiring little interpersonal interaction and content to spend long periods of time alone (the study was done before the era of multiple-manned space craft). In the interview, the candidate portrayed himself as either relatively inner-directed or other-directed. Subjects were then asked to estimate his *true* characteristics and indicate the confidence with which they held their estimates.

As predicted, subjects made confident and rather extreme estimates when the candidate's behavior was out-of-role (the submarine candidate describing himself as inner-directed, and the other-directed applicant for a position as an astronaut). We should expect this, because there was no plausible environmental or situational cause for the actor's behavior in these cases. Indeed, there was a plausible environmental inhibitory cause for this behavior, since the candidate in these conditions portrayed himself as *opposite* to what he had been told were the desirable characteristics for the job. But when his behavior was in-role (submariner candidate indicating other-directedness, or inner-directed astronaut candidate), subjects found it thoroughly uninformative about his true characteristics. Subjects in these two conditions did not differ in their estimates of his true nature, and their estimates were given with very little confidence.

Many other studies exist that support the same basic point: When there exists a strong, plausible environmental cause for behavior, that behavior will be found relatively uninformative about the person. A line of investigations initiated by Jones and his colleagues has supported the hypothesis that another's attitudes are inferred from his behavior with less confidence and extremity when he had little or no choice about engaging in the observed behavior. But it should be noted that in many of these studies (e.g., Jones & Harris, 1967; Miller, 1976; Snyder & Jones, 1974), the *direction* of the other's behavior itself affects attitudinal attribution. Although the difference in attributed attitude is typically less extreme than that found if the actors presumably had choice about their behavior, subjects are still likely to infer a difference of attitude between two people who advocate opposite positions on orders from the experimenter. This tendency for the direction of behavior to "swamp the field" sometimes occurs in the literature on attraction as well, as we see later.

There does, then, exist fairly persuasive evidence that, at least for the kinds of dispositional attributes investigated in the research cited, individuals' attributions are consistently affected by the kinds of situational variables highlighted by attribution theory and in particular by the apparent presence or absence of freedom of choice on the part of the actor. We now turn to the question of whether attraction toward others is mediated by, or consistently related to, these attributions.

Perhaps the most direct and intuitively appealing hypothesis is this: if the information available about another's behavior and the surrounding circumstances is such as to foster an attribution to a valued, desired, or "good" personal characteristic, that individual will be liked. The classic research on this hypothesized sequence is the well-known work by Thibaut and Riecken (1955). Because this article is so frequently cited and is so informative about both the plausibility and the difficulties in providing an attributional analysis of liking, it bears close scrutiny.

Thibaut and Riecken conducted two experiments. In each, a naive subject attempted to convince two confederates of the experimenter who differed in status to comply with a request (either to donate blood, or to share a dictionary). Both confederates eventually and simultaneously complied. The authors hypothesized both: (1) that the compliance of the higher-status target would be attributed internally, to his personal characteristics, whereas the compliance of the low-status target would be more likely to be attributed externally, to the pressure placed on him by the subject's request; and (2) that as a consequence of attributing the higher status target's compliance to presumably attractive or favorable "internal" characteristics, liking for him would increase more than liking for the low-status target.

The results are generally cited as supporting the hypothesis, and in a rough way they did. On attributions, in one study (blood donation) subjects were forced to make a choice, assigning internal locus of causality to one confederate and external causality to the other. Overwhelmingly, on this forced-choice item subjects assigned an internal locus of causality to the higher-status confederate. In the same study, change from the beginning to the end of the experiment in liking for the confederates was also assessed, and relative to the lower-status confederate, the high-status confederate (whose compliance was attributed internally) increased in attractiveness. But the liking result was not very strong ($p \leqslant .04$), and the forced-choice item was insufficiently sensitive to provide evidence that liking was mediated by the attributions.

In their other experiment (dictionary sharing), Thibaut and Riecken did not force subjects to discriminate between the confederates in assigning locus of causality, asking them instead to indicate for each the degree to which the compliance was caused by internal and external factors. Five of 21 subjects in fact did not differentially locate cause for the compliance, and an additional four assigned it the "wrong" way (the lower-status confederate's compliance was seen as more internally caused). Thibaut and Riecken did find (after eliminating those subjects who failed to attribute causality differentially to the two confederates) a relative increase in attraction to the confederate whose compliance was seen as more internal. But we cannot tell whether the attributions contributed to the liking, as hypothesized, or whether subjects assigned internal causality to the confederate whom they had come to prefer.

To clarify and extend these findings, Pilkonis (1976) conducted a conceptual replication of the "blood donation" experiment of Thibaut and Riecken. As in their experiment, his confederates differed in apparent status (via an omnibus manipulation of age, education, social class and career aspirations). After hearing the subject's appeal to give blood, both confederates (independently) complied or refused. Measures of attribution, including both a forced-choice item and individual ratings of degree of internal and external causality for the decision, were taken; and subjects' liking for each of the confederates was assessed. The findings on attribution closely matched those of Thibaut and Riecken; by both measures of locus of causality, the compliance of the higher-status confederate was attributed relatively internally. Furthermore, as predicted, the effect of status on attributions differed, depending on whether the confederates had complied or refused. If they had refused, the lower-status confederate was assigned a more internal locus of causality; for him the subject's message should have been a stronger environmental force, and he must then have had stronger internal reasons for refusing to comply. But the liking data failed thoroughly to replicate the pattern found by Thibaut and Riecken. There was no tendency for liking to be consistently associated with attribution. Although the higher-status confederate's compliance was assigned a relatively internal locus of causality, he was not better liked after compliance. However, the direction of the confederate's behavior was associated with liking; compliers were liked better than refusers. Pilkonis concluded that causal attributions, at least in this setting, are a relatively weak determinant of attraction.

There do exist other studies in which conditions that have been shown to affect the attribution process are manipulated and their effects on attraction to the actor assessed. For example, both Nemeth (1970) and Steiner and Field (1960) present evidence that perceivers' attraction toward an experimental accomplice is more powerfully affected if the accomplice appears to have acted freely than if his behavior was constrained by explicit instructions from the experimenter. In both experiments an attributional link to attraction is plausible, although in neither case is it demonstrated conclusively. These studies do increase our confidence somewhat that, at least when experimental manipulations give clear indication about the actor's choice or constraint, interpersonal attraction is sensitive to the kinds of contextual factors that influence causal attributions.

Nevertheless, with the possible exception of the Steiner and Field (1960) study, in none of the experiments cited was there much reason for the subject to be significantly affected by the actor's behavior (his outcomes in no way hinged on the actor's compliance or "helping"). Nor was there much reason for the subject to think that the actor's behavior was uniquely directed at him as an individual. Jones and Davis (1965) have suggested that both these factors — the "hedonic relevance" of another's actions for the perceiver and the "personalism" of the behavior — significantly affect both the attribution process and interpersonal attraction. Other things being equal, they argue that it is particularly those actions that promote or thwart the goals or values of the perceiver (hedonically relevant actions) and those actions that seem uniquely or particularly directed at the perceiver as an individual (personalistic actions) that are likely to give rise to strong, confident attributions about the intentions and dispositions of another. And we have suggested that it is such attributions that are particularly likely to be important determinants of attraction.

Jones and Davis summarized the evidence available in 1965 on the roles of hedonic relevance and personalism in affecting both attributions and evaluations of the other. It was slender then and remains so, but it is intriguing and suggestive. We shall examine a few issues regarding the effects of those variables on reactions to one type of behavior — evaluations received from others. On the assumption that we generally seek to build and maintain a positive conception of ourselves, and that evaluations from others are a significant factor in influencing the self-concept, such evaluations should be particularly hedonically relevant for the recipient. We may assume that in general positive evaluations are preferred to negative ones. But given an evaluation, it is important for the recipient to fathom the intentions of the evaluator, in order to assess the evaluation's veridicality and significance for him. Indeed, there does exist a fair body of evidence suggesting that reactions to another upon receiving such evaluations are influenced by contextual factors that should affect the attribution process (see Mettee & Aronson, 1974, for an extensive review).

In his work on ingratiation, for example, Jones (1964) has indicated the perceiver's need to determine whether a flattering, positive set of evaluations from another represents that person's true opinion or was engaged in for ulterior or manipulative purposes. If such purposes are detected, presumably the target will discount the evaluation and will not be much attracted to the evaluator. An experiment by Dickoff (1961) provides supporting evidence. In her study, subjects receiving a uniformly extreme set of positive evaluations from another indicated more liking for the evaluator if she presumably had an "accuracy set" in giving the evaluations (she was being trained as a clinician and was attempting to give objective judgments) than if she was giving the evaluations because she hoped later to persuade the subject to take part in her own research project (ulterior motive set).

Jones (1964) has also argued that an observer to a flattering interchange is more likely, for a variety of reasons, to perceive "ulterior motives" and discount the flattery than is the actual target of the evaluations. For an observer, of course, the evaluations lack both hedonic relevance and personalism. An interesting experiment by Lowe and Goldstein (1970) suggests an important difference between targets and observers of evaluations in their sensitivity to the "context" of such evaluations. Their experiment was in part a conceptual replication of Dickoff (1961), with two additional factors. Subjects were exposed to an evaluator who gave either highly positive or negative evaluations, and they subsequently learned that the evaluator was a candidate for a research position where her likelihood of being hired depended either on how accurate her evaluations of the target had been (accuracy set) or on how much she had been able to win the target's approval. Furthermore, subjects were either themselves the target of the evaluations or were asked to role-play and pretend that they had received such evaluations.

Lowe and Goldstein overall found most liking for the evaluator in the accuracy-positive condition and least liking in the accuracy-negative condition. But significant interactions between the role-play variation and both the set of the evaluator and the direction of the evaluation indicated that the relevance of the evaluation to the subject importantly affected sensitivity to the independent variables. Involved subjects were more sensitive than role players to the direction of the evaluation; their liking for the evaluator was much more affected ($p <$ .01) by whether the evaluation was positive or negative. But involved subjects were significantly *less* sensitive to the context of the evaluations – the accuracy vs. approval-set manipulation. Consistent with the tendency for the attraction of the involved subjects to display differential sensitivity to the direction vs. the context of the received evaluations, their attributions of the evaluator's social perceptiveness (ability to "figure people out") reflected the direction of the evaluation but not the set of the evaluator. Social perceptiveness of the evaluator was rated high by targets receiving a positive evaluation and low by those receiving a negative evaluation. Observers were much less likely to "assimilate" this ability attribution to the overall hedonic tone of the evaluations.

These results suggest that both the attributional process and attraction to another may obey somewhat different rules depending on the hedonic relevance to (or "involvement" of) the perceiver. The real-world situations to which we wish to generalize the results of our investigations are typically those in which the person is involved with another whose actions are directly relevant to his outcomes (see Berscheid et al., 1976). It is thus clearly important in the future to investigate both attributions and attraction in situations having more direct relevance for the perceiver than has typically been the case.

Finally, we consider the role of personalism. We referred to personalism above as indicating the degree to which another's actions seem uniquely directed at us as individuals. The suggestion by Jones and Davis (1965) is that "action which is both relevant and personal has a direct and dramatic effect on evaluative conclusions about the actor [p. 247]." But, as with hedonic relevance, there is little published work that systematically investigates the effects of personalism on both attribution and evaluation. An elegant study by Potter (1973) was conducted with these aims in mind, and the results are instructive.

Naive subjects arrived in pairs for a study of "impression formation." During the experimental session, they took two "objective" tests of social insight, participated in an interview and a brief "get acquainted" conversation together, received from the experimenter "objective" test feedback about their degree of social insight, and learned both how their partner presumably rated them on this dimension and how much he liked them. Potter operationalized personalism as agreement between the subject's "true" characteristics (the scores on the tests of social insight) and the other's evaluation of these characteristics, arguing that it is primarily under these conditions that the other's liking or disliking will be seen as "contingent on [the receiver's] own characteristics," and under these conditions that evaluative reactions to the other will be most extreme.

The highly significant ($p < .001$) triple interaction between the subject's "objective" characteristics, the other's estimate of those characteristics, and the other's liking gave strong support to the hypothesis. The subject tended strongly to reciprocate the other's liking or disliking when the other's assessment of his social insight matched the objective feedback. But in the nonpersonalistic cells, other's liking or disliking did not affect the subject's attraction at all. These results suggest that another's hedonically relevant behavior toward us will be particularly impactful in determining attraction when we have reason to believe that the behavior is uniquely directed at us or is based on an assessment of our true characteristics.

But was the resulting attraction mediated by attributional processes? Potter asked his subjects how discerning the partner was. The results showed only that he was seen as more discerning or perceptive in general when he liked rather than disliked the subject. This finding is reminiscent of Lowe and Goldstein's (1970) involved subjects, who rated their evaluator as more perceptive when he gave positive rather than negative evaluations of them. Another question also showed a divergence of attributional vs. attraction results in the Potter study. Subjects indicated "how .... your partner generally responds to the people he meets." Although there was a general tendency to assume that the partner's reactions to others match his liking for the subject, a significant interaction indicated that this was true only in the *non*personalistic conditions. That is, if the other's reaction was *not* based on an assessment of the subject's "true" qualities, this reaction was seen as typical of his responses to others; but if it was personalistic in the sense discussed, it was not seen as informative about his tendencies when dealing with others. Thus, the subject's liking for the other in this experiment seems *not* mediated by an assessment of his general, dispositional qualities, but affected instead by his particular, individual reaction to (and perhaps intentions toward) the subject.

This concludes our review of the relationship between attributions about the cause of another's behavior and liking for that other. Although the review has been selective, considerable stress having been placed on a relatively small number of studies, it accurately reflects the current status of both theory and research on this problem. When there are relatively strong manipulations of factors affecting the actor's presumed degree of choice vs. constraint, attraction seems (though not always) to follow the pattern expected for attributions. In most of this research, attraction has been expected to increase when the perceiver should (or does) attribute positive, desirable dispositional characteristics to the actor. This is particularly likely when positive, helpful, or "good" actions are engaged in with a minimum of apparent environmental constraint.

Additional theory and research has moved somewhat beyond this rather elementary, common-sense derivation. It has been suggested that actions that are hedonically relevant for the perceiver are particularly likely to affect attraction, and that involved participants may be differentially sensitive (compared with observers) to the direction of the actor's behavior and to aspects of the surrounding circumstances. This may be the case regarding both the attribution process and attraction. Finally, evidence has been presented suggesting that actions that appear uniquely directed at the perceiver are particularly impactful in affecting attraction, although the attributional mediation of such effects is not clear. For none of these generalizations, however, is the evidence very systematic or strong. The literature documents many possible determinants of attraction, including frequency of interaction, similarity of attitudes, mere exposure to others, and physical attractiveness, to mention just a few. Although the research thus far demonstrates the plausibility of assuming that sometimes, under some conditions, attraction may be causally mediated by attributions about the other's behavior, the strength of the attributional contribution to interpersonal attraction remains conjectural and may be weak.

EFFECTS OF SELF-ATTRIBUTIONS ON ATTRACTION

Self-Perception

Although it may appear to common sense plausible that another's behavior toward us can have a variety of explanations or be attributable to a variety of factors, the situation has typically seemed radically different when it comes to explaining our own behavior toward others. We may carefully gather evidence on and analyze the environmental conditions that might have caused another's behavior before concluding that the behavior was produced by relatively internal causes (and thus is informative about intentions or motives). But we feel that we need not and do not go through a similar process in understanding our own behavior. If there were internal reasons for it — if, for example, we help another person because we like him — we simply *know* that, directly and noninferentially. We have, it seems, "privileged access" to our own internal states and can discriminate among them reliably and assess their causal role in directing our own behavior without difficulty and without inference.

In his theory of self-perception, originally devised as an alternative explanation for experimental findings consistent with cognitive dissonance theory, Daryl Bem (1967, 1972) has presented a radically different analysis of how we obtain knowledge of our own internal states.[2] Bem's theory, simply put, argues that when internal cues are weak or difficult to discriminate (and this is allegedly often the case), the individual is placed in the same functional position as an outside observer attempting to determine the cause of his behavior. The observer typically assesses the behavior itself and the circumstances in which it occurred in order to infer what, if any, internal state of the actor was responsible for it. The actor himself, Bem argues, does the same thing, following the same rules.

Attitudes are the "internal state" to which the bulk of the research on self-perception theory has been directed. Just as we know what another person's attitude is by observing his behavior and the environmental circumstances in which it occurs, we infer our own attitudes in the same way. Liking for another may be conceptualized as an interpersonal attitude, and self-perception theory rather directly yields the prediction that liking will be affected by our perception of our own behavior toward the person and the constraints (or justifications) existing for that behavior.

Most of the relevant studies were originally conducted to test hypotheses derived from cognitive dissonance theory (Festinger, 1957), but they can be recast so as to indicate the relevance of self-attribution for interpersonal attraction. Davis and Jones (1960) induced subjects to read a very harsh personality evaluation to a target who had previously seemed pleasant and likeable. As predicted from dissonance theory, subjects changed their attitudes toward disliking only when they had choice about reading the evaluation and would not later have an opportunity to meet with the target and explain their behavior. Self-perception theory could make the same prediction. An outside observer would presumably be most likely to conclude that the subject actually disliked the target when he chose to read the harsh evaluation and would not later be able to take it back. Like virtually all experiments where dissonance theory and self-perception theory make the same prediction, there is no convincing way in this case to choose between the alternative formulations.

---

[2]For a more recent and elaborated argument that we typically do not have "direct access" to our own mental states, see the excellent paper by Nisbett and Wilson (1977b).

David Glass (1964) demonstrated the same crucial role of choice in mediating re-evaluation of the recipient of harsh treatment. In his experiment, subjects who had indicated that they were opposed to the use of electrical shock on humans for scientific purposes were led to believe (falsely) that they were delivering a painful set of shocks to a confederate—learner in a concept-formation task. The subjects' self-esteem had been manipulated by giving them very favorable or unfavorable feedback on an elaborate battery of personality scales, and they were then given either free choice or no choice about administering the shocks. Attitudes toward the victim changed significantly toward disliking only when the subjects had free choice about shock administration *and* their self-esteem had been experimentally elevated. The results were interpreted as showing the crucial importance of both choice and high self-esteem for causing dissonance from engaging in harmful, counterattitudinal behavior. Only for subjects high in self-esteem, it was reasoned, would such an action be inconsistent with the self-concept and likely to produce dissonance. Self-perception theory could perhaps handle the results by arguing that, given choice, an outside observer would more confidently infer that the subject truly disliked the victim if his conception of his personality was such that he could act confidently on his principles and beliefs. In any event, for present purposes the experiment demonstrates again that our own behavior and the circumstances in which it occurs can significantly affect liking for another person. The very same behavior toward another has different effects depending on the circumstances (particularly the presence or absence of choice) in which it occurs.

The evidence that one's own *positive* behavior, where self-attributed, will increase liking for another is slender, although the same attributional argument should apply. Jecker and Landy (1969) found that subjects who voluntarily did a favor for a harsh experimenter came to like him more than subjects who did the same favor for somebody else. And Schopler and Compere (1971) found that being kind as opposed to harsh toward another subject increased attraction. But in each of these cases, problems with the experimental design and/or results reduce confidence that the experiments truly show an effect of positive behavior on liking. Such an effect may be difficult to demonstrate in research contexts. Common politeness may be seen as a sufficient reason for positive behavior unless the behavior involves considerable cost to the subject.

Perhaps the most intriguing demonstration that "positive" behavior toward another can increase attraction is the recent experiment by Blanchard and Cook (1976). In their experiment, a naive subject (who was an Air Force recruit) and two confederates ran an involving simulated business in which each was assigned various tasks. One of the confederates pretended to be relatively incompetent, and the situation was rigged such that it was appropriate (although not demanded) for either the subject or the other confederate to help the less competent member complete his tasks. The subject's subsequent liking for the less competent member was assessed, and he was better liked when the subject had helped him than when the other confederate had done the helping. Although the attributional mediation of the liking effect is conjectural (there was, for example, no manipulation of choice about helping), the results do suggest that under appropriate conditions positive behavior toward another can increase helping.

The results presented thus far suggest that our own behavior can significantly affect liking for another person. Although the attributional mediator of the liking effect is never measured directly and though the self-perception explanation of certain of the studies is not completely convincing, in each case the effects on liking are at least roughly consistent with attributional mediation. Where the circumstances are likely to foster an internal self-attribution for the behavior, the effect on attraction is most pronounced.

Overjustification Effects

Another line of investigation, generally interpreted within a self-attributional framework, should be mentioned in this context. This is the burgeoning litera-ture on the "overjustification effect." This literature suggests that when a person is promised overly sufficient justification for engaging in an attractive activity, subsequent interest in the activity is diminished (e.g., Amabile, DeJong, & Lepper, 1976; Garbarino, 1975; Lepper, Greene, & Nisbett, 1973). The ex-istence of the effect itself is somewhat controversial, as is its interpretation (e.g., Kruglanski, 1975; Lepper & Greene, 1976; Reiss & Sushinsky, 1975). One leading interpretation has focused on a distinction between actions that are performed as a means to some other end, versus actions that are performed as an end in themselves. The former actions are said to be "extrinsically motivated," whereas the latter are called "intrinsically motivated." The distinction closely parallels that between external vs. internal locus of causality for behavior. If there are plausible external justifications or pressures for engaging in an action, such as the presence of salient rewards, external surveillance, externally imposed dead-lines and the like, the individual will come to infer that these external factors are the reason he engages in the behavior. This attributional consequence, it is argued, undermines prior internal reasons (such as task interest or enjoyment). The individual comes to see the behavior as supported primarily by external constraints and is less likely to engage in it in their absence.

The vast bulk of the studies on the overjustification effect investigate task performance directly. In a few studies, attitudes toward the task are instead assessed, and these generally parallel the behavioral findings: task attractiveness decreases under the same conditions where task engagement decreases. In all cases, conditions designed to produce the overjustification effect should plausibly lead to the attribution that the behavior was engaged in for external reasons, although this attribution itself is not directly measured.

What might this suggest for interpersonal attraction? The above analysis implies that *if* we perceive our positive behavior toward another person as directed toward obtaining extrinsic rewards, our liking for that person should be less than if we see the same behavior as engaged in for its own sake. There is a stock predicament in certain romantic literature where the wealthy heiress fears that she is being pursued for her wealth and not for her more "intrinsic" characteristics. This analysis suggests that her suitor himself will share this attributional problem and will actually like or love her less *if* he perceives his behavior as directed toward or caused by her money. That, however, is a big "if." We know very little about the factors likely to produce such an attribution. Surely the mere presence of external rewards does not demand the conclusion that behavior is directed toward obtaining them, although this conclusion is more likely than if the rewards were not there in the first place.

There is no published literature investigating the implications of the over-justification effect for interpersonal attraction. But in a recent paper, Seligman, Fazio, and Zanna (1976) present two studies suggesting that increasing or making more salient the extrinsic rewards provided by a romantic partner may make observers infer less liking for the partner and make the participants themselves infer less attraction. In the observer study, subjects heard a tape recording in which a male (Tom) responded to a set of 21 questions about his relationship with his girl friend (Mary). Many of the questions were essentially filler items, which served to indicate that Tom was highly intrinsically interested in Mary. Three of the questions manipulated degree of extrinsic interest. Tom either indicated that Mary was wealthy, physically attractive, and had an influential father (high extrinsic interest) or that she was not wealthy, was not particularly good-looking, and did not have an influential father. Observers then rated how much they thought Tom liked and loved Mary. On both measures Tom was perceived as less attracted to Mary when extrinsic interest was high — when she was wealthy, beautiful, and had an influential father. Presumably, it was because observers perceived Tom's attentions as directed in part toward obtaining external goals (the possibility of money and influence and the glow of being seen with a beautiful woman) that they thought intrinsic reasons (liking and loving) were a less powerful factor in his association with Mary.

Unfortunately, interpretation of the results is somewhat clouded, because Mary not only possessed these external characteristics but Tom portrayed himself as interested in benefiting from them. Of her wealth, he said it would "be really nice to have that kind of money." Of her beauty: "I feel good being seen with her because I get the feeling everybody's looking." Of course, the self-perception hypothesis suggests that external rewards will undermine attraction (and attributed attraction) only if the behavior is seen as a *means* toward obtaining them. A woman's wealth or beauty should not per se undermine attraction to her. But in fastening the "in order to" link in the attributional chain, the authors may have portrayed Tom as a grasping, materialistic, insecure person — a person perhaps incapable of "true" liking or loving for Mary. Nevertheless, the results are suggestive and in accordance with the attributional framework hypothesized; where there are strong external reasons for behaving positively toward another, less liking will be inferred.

The second experiment by Seligman et al. attempted to alter subjects' actual liking and loving for their romantic partners by making salient to them external considerations that might be important in their relationship. Extrinsic reasons for the relationship were made salient in two ways. First, subjects completed a variety of sentences about why they engage in several activities with their romantic partners. The sentences were constructed to encourage extrinsic reasons by including the prepositional phrase "in order to" (e.g., I seek out my girl friend in order to . . .). Then, subjects rank-ordered in importance seven reasons for their romantic involvement, each of which was clearly extrinsic. Intrinsic subjects completed the sentences about reasons for activities following the words "because I," and rank-ordered in importance seven clearly intrinsic reasons for association with their partner. Then all subjects indicated how much they liked and loved their partner. Overall, the results indicated less expressed liking and loving for the partner in the "extrinsic set" condition where external reasons for the relationship had been made salient than in the intrinsic set condition. Although these data are at best suggestive, and further systematic investigation is clearly needed to assess the reliability of the effect and its boundary conditions, the results do support the self-attributional model of liking and shed further light on the ways in which our attributions for our own behavior may influence interpersonal attraction.

Emotional Attribution

Although most of this essay deals with the effects of attributions for behavior — our own and others' — on attraction, there is one further line of attribution research that is clearly relevant to attraction. There is an impressive body of literature showing that we "explain" or understand our own emotional states in part on the basis of cognitions about what has caused them. Recent evidence invites speculation that these attributions about emotional experience may have consequences for interpersonal attraction.

Within social psychology, the dominant theoretical statement on emotional experience has been provided by Stanley Schachter. Originally put forth in his 1962 *Psychological Review* paper with Jerome Singer, Schachter's theory states that emotional experience is a product of two factors: the presence of a general state of physiological arousal, and a set of cognitions about the cause of that arousal. The arousal affords the possibility of an emotional experience; the cognitions determine whether an emotion will be experienced and what label will be applied to the emotion.

There is a convincing set of demonstrations in the literature that the reported quality of an emotional experience, and associated emotional behaviors, can be profoundly affected by altering the "plausible" cognitions for explaining a state of perceived arousal (e.g., Dienstbier & Munter, 1971; Nisbett & Schachter, 1966; Ross, Rodin, & Zimbardo, 1969). In these studies, a person who is given a "nonemotional" explanation for arousal (typically through instructions that a pill or some other "emotionally neutral" device may be causing the arousal) reports feeling less emotion and behaves as if he were experiencing less emotion.

What has all this to do with interpersonal attraction? Most research on the topic by experimental social psychologists has dealt with liking, which is for the most part reflected in positive attitudes toward another indicated by favorable trait ratings and the hope or expectation of future positive interaction. The research typically takes as its unit of analysis a person holding minimal information about another, often without even face—to—face contact. Seldom is there any interactive relationship among the research participants, seldom any real investment of time, effort, or commitment. Strong emotions toward others do not occur in this research. But of course they do in real life. Very strong attraction or repulsion — passionate love and intense hatred — may form a statistically very small proportion of our interactive histories; but their fascination for people is reflected vividly in literature, drama, history, and even the writings of a few psychologists. Romantic love has long been an attractive topic of discussion and speculation but has also been resistant to systematic study. Definitional problems plague researchers, but more importantly, neither our theories nor our relatively restricted methods of investigation have seemed to offer much hope of substantial understanding.

But if we assume that a major difference between romantic attraction and "garden variety liking" is that the former is typically accompanied by intense emotion, the foregoing summary of the attributional aspects of emotion may be relevant. Indeed, Berscheid and Walster (1974) have directly suggested that people will experience "passionate love" when: (1) they are intensely aroused physiologically; and (2) situational cues indicate that "passionate love" is the appropriate label for their intense feelings. It is important to note that the arousal itself may actually derive from any number of sources; but this arousal may lead to the "experience" of love if situational cues are congenial for that label to be applied.

A requirement for the application of that label would be the presence of an appropriate "love object" and a "plausible" connection between that person and the arousal. There does indeed exist some suggestive evidence that "irrelevant" sources of arousal can lead to sexual attraction, provided that there is an appropriate sexual target to whom the arousal can be attributed. In a field experiment, Dutton and Aron (1974) had an attractive female experimenter approach males who were crossing a rather frightening suspension bridge; they then compared their sexual imagery in response to a TAT item and the frequency with which they later tried to contact the experimenter with the behavior of control males who were presumably unaroused. On both measures, aroused subjects indicated more attraction to the female experimenter. Sexual imagery was also higher than that in response to a male experimenter, suggesting that both arousal and an "appropriate" target were necessary to produce the attraction effect. In a laboratory experiment the same investigators found that subjects expecting to receive strong electric shock were more attracted to an attractive female than those expecting weak shock, again suggesting that "irrelevant" sources of arousal can be effective in producing attraction to an appropriate target. Cantor, Zillman, and Bryant (1975) have shown that the effect is strongest when the "true" source of the arousal is unclear. In their experiment, subjects reported most erotic attraction when they viewed a suggestive movie while actually physiologically aroused from a strenuous burst of exercise but believing that the exercise effects had worn off.

The line of reasoning offered above, linking self-attribution of emotion and romantic or passionate love, is admittedly conjectural in the extreme. Although some of the literature seems consistent with the conjecture, none was generated by it; and the laboratory studies cited clearly do not engage passionate love, however one defines it. Neither answering questions of an attractive experimenter, for example, nor viewing an erotic movie can be considered an instance of "love." But the literature does clearly suggest that available cues are utilized to "explain" emotional experience and that the presence of an attractive other can be seen as a relevant basis for interpreting arousal. When so interpreted, reported attraction to the apparent source of the arousal is enhanced.

Three separate lines of research, then, converge in support of the notion that causal attributions about one's own behavior or emotional states may affect liking for another person. For the most part, furthermore, the variables seen as crucial in affecting self-attributions are the same as those implicated in other-attributions, in particular variations in the direction and extremity of the behavior and the degree to which the action seems constrained by environmental forces. Yet in one respect the evidence linking self-attributions to liking is stronger than that demonstrating an effect of other-attributions on attraction. This is because for self-attributions of the sort discussed, the attributional dependent variable *is* liking for the other. Attraction may be plausibly seen as involving a positive attitude toward another person, coupled with positive affect or emotion. In the self-attributional research discussed, attitudes or emotions toward the other are directly inferred by the subject – from a consideration of his behavior (or arousal) and the circumstances in which it occurs. In the case of other-attributions, on the other hand, the subject is expected somehow to move – in accordance with rough plausibility rules – from a dispositional or internal attribution for another's behavior to his own liking for that other. This additional link in the attribution/liking chain sometimes appears to be forged as expected, in a way consistent with the attributions, but by no means always.

But in another respect, the link between self-attributions and liking is arguably weaker. This is because we have no convincing empirical *or* theoretical reason to assume that such attributions play much of a role in generating attraction in natural situations. The evidence suggests that if, for example, a person engages in highly negative behavior toward another, acting under no apparent constraint and holding no prior negative attitude toward the target, he may come to infer that he dislikes the target. Or if a person is inexplicably physiologically aroused in the presence of an attractive other, he may come to infer even greater attraction to that other. But how often do such conditions obtain in life? It has taken considerable experimental ingenuity, subtlety, and stagecraft to rig such conditions in the laboratory. We now have evidence that self-perception processes *can* affect attraction. We would like convincing evidence that they *do*.

EFFECTS OF LIKING FOR ANOTHER
ON ATTRIBUTIONS FOR HIS BEHAVIOR

Throughout the preceding discussion, we have considered attributions as causal mediators of resulting attraction toward another person. Now we consider briefly the other side of the causal coin — the effects of attraction toward another on explanations for his behavior. There are several reasons to expect such effects, particularly when we are reminded that there are a variety of potential causal explanations for any given action, no single one of which is usually demanded by the immediate information about the behavior and its context. People will use whatever relevant information they have to interpret another's behavior, and sentiment toward the other can be seen as highly relevant. If we like another, for example, we tend to describe that person as having a variety of positive characteristics. There is a pervasive "halo effect" that results in the assimilation of a wide variety of characteristics to the predominant affective sentiment (e.g., Nisbett & Wilson, 1977a). One manifestation of this is the tendency for the "cognitive" component of an attitude, measured by the person's beliefs about the attitude object, to correlate highly with the "emotional" or "evaluative" component, typically measured by ratings on the semantic differential (Ostrom, 1969). Overall, a "balanced" state of affairs tends to operate, by which we feel little or no tension between our sentiment toward other people and our beliefs about their characteristics and actions.

Thus, as Heider (1958) suggested, we expect good (or liked) people to perform good actions and bad (or disliked) people to perform bad actions. But what of the cases where things apparently don't work out this way — when, for example, a close friend does something unpleasant, offensive, or immoral? There has been suggested a variety of "modes of belief resolution" to handle such cases of apparent imbalance (Abelson, 1959). Perhaps the clearest methods to restore psychological balance (or to bring belief and sentiment in line again) are to reinterpret the action, seeing it as not "really" a bad one, or to change sentiment toward the actor. Dion (1972), for example, found that observers judge a transgression by an attractive (and presumably favorably evaluated) child as less undesirable than the same transgression by an unattractive child. Much earlier, Zillig (1928) found that the gymnastic performances of sociometrically chosen children were evaluated by their peers more favorably than those of relatively disliked children, even though the latter group had been given special training and their performances were "objectively" superior.

Here we wish to focus on another possibility: that in cases of apparent inconsistency between sentiment toward an actor and the quality of his action, the perceiver may systematically alter his causal attribution for the action so as to provide a "reasonable" explanation for the behavior. This will generally be done by providing a relatively external, situational explanation for behavior that does not seem to "fit" the sentiment toward the actor. Conversely, behavior that is consistent with affect — good or positive behavior by a liked other, bad or negative behavior by a disliked other — will tend to be attributed internally, to the characteristics and intentions of the actor. Such attributional plasticity and responsiveness to sentiment, if it occurs, would have two consequences of note. In the first place, "balance" between the sentiment toward (and presumed characteristics of) the actor on the one hand, and his behavior on the other, is maintained. But we do not argue that it is *in order* to achieve a subjectively pleasing or harmonious state of balance that the attributional outcome is achieved. We see the perceiver, instead, as using the evidence available to him in arriving at an attribution for behavior that may be puzzling or unexpected. If the perceiver believes that the actor has generally positive or admirable qualities, then his apparently "bad" behavior cannot be produced by or reflective of those qualities. That is, the perceiver utilizes what he already knows (or thinks he knows) about the actor's qualities to decide whether a particular instance of behavior was "caused by" those qualities. In the second place, as a consequence of differential attribution depending on liking for the actor and the characteristics of the action itself, there is no need to re-evaluate sentiment toward the actor on the basis of isolated bits of apparently disconfirming behavioral evidence.

Employing this reasoning, Regan, Straus, and Fazio (1974) conducted two experiments intended to demonstrate the responsiveness of attributions for another's behavior to sentiment toward that other. In the first experiment, liking for another was manipulated, and attributions for that other's skilled or unskilled performance on a task were assessed. Subjects were undergraduate women, and their liking for the confederate–target was varied through a double-barreled manipulation. Subjects learned that the other person held attitudes either very similar or dissimilar from their own; they furthermore observed the target (on a video tape disguised as a "live" television transmission) behave either pleasantly or obnoxiously toward the experimenter. They then observed the target attempt a game of skill; the confederate's performance was crosscut with liking toward her, and she behaved either very skillfully or ineptly. Subjects then answered a variety of attributional questions, the most direct of which was to assess the degree to which the experimental task was a good indicator of underlying ability.

The predicted interaction was highly significant. Subjects thought the game a good measure of ability if the liked actor performed well or if the disliked actor performed poorly. They thought it a very poor measure of ability if the liked actor performed poorly or the disliked actor performed well. It should be noted that the behavior observed was by no means "morally" significant; it involved a physical skill. And the liking manipulation contained no direct evidence about physical ability. Nevertheless, subjects assigned credit or blame internally if the level of performance was "consistent" with affect and attributed it externally if it was not.

A second experiment by the same authors, perhaps closer to "real-life" attributional situations, investigated the relationship between liking or disliking for actual friends and acquaintances and attribution for behavior of potential moral import. Subjects nominated someone they liked and someone they disliked for an experimental session, having been told that they would later be asked to explain this person's behavior. The friend or enemy was never actually contacted, but subjects subsequently took part in an experimental session in which they were told that one of their nominees had been put through a standard scene and his behavior had been observed. They were told that their friend or enemy had been asked by an experimental accomplice to do a favor. The accomplice had asked the acquaintance to go to the infirmary and get some pills for him, since he had recently been in an automobile accident and was experiencing discomfort. In all cases, subjects were told that their acquaintance, whom they had said they liked or disliked, had agreed to do it. They were asked why he had behaved as he did, both in an open-ended question and in a structured, forced choice among alternatives reflecting either internal (dispositional) or external (situational) reasons for the behavior. The results were very strong, indicating that this prosocial behavior was attributed to internal causes if done by a friend and to external causes if done by an enemy.

Typical open-ended explanations for a friend's behavior were: "He's just like that, always wanting to help people." "She has such a warm and generous nature, I'm not surprised." But when the behavior was supposedly engaged in by someone nominated as disliked, subjects provided explanations like, "I suppose he felt he had to; after all, he had the time and the guy was hurt."

The author knows of no other experiments directly implicating liking for another in determining causal explanations for the other's behavior. But it seems likely that the simple internal—external dichotomy suggested above does not do justice to the variety of attributional possibilities affected by sentiment. An apparently positive action by a disliked other, for example, could plausibly be attributed to an internal but *negative* characteristic. In the experiment described above, one subject explained the behavior of someone he disliked by saying that the person had few friends and thus was possibly trying to make one by doing a favor. Furthermore, there are doubtless limits on the hypothesized attributional plasticity. We would suggest that the stronger the affect toward the other, the more likely a person is to attribute the other's behavior in a way consistent with that affect. This should be particularly likely when the behavior has occurred only once and when the conditions surrounding it do not themselves strongly indicate a particular causal attribution. Overall, we suggest that there is typically more "slippage" in the attributional process than in already-formed sentiment toward another and that such sentiment is therefore likely to exert a considerable influence on attribution.

CONCLUDING REMARKS

From the preceding review of the relationship between attribution and attraction, no simple, overall perspective emerges. Attraction and attribution seem mutually interrelated in complex ways. We have seen that in some settings, causal attributions about another's behavior seem to be affected by some of the same variables as attraction to the other. In particular, the apparent presence of choice vs. constraint in another's behavior typically affects interpretations of the causal origin of that behavior. And attraction to others sometimes develops in a way consistent with the common-sense notion that we like others, in part, when we see their positive behavior as expressive of their internal, dispositional characteristics.

But the evidence also indicates that it does not always work out that way. Furthermore, there is the suggestion in some studies that as a person becomes involved with another whose actions are hedonically and personalistically relevant for the perceiver, sensitivity to "contextual" factors (as opposed to the positive vs. negative direction of the other's behavior) may be reduced. It may be the case that the highly rational processing of contextual information that forms such a central aspect of attributional theories is not likely to dominate evaluative reactions to another when the perceiver is significantly involved in the interaction.

We have also seen that self-attributions may be implicated in attraction to others and that attraction may itself exert a strong effect on resultant attributions for another's behavior. From all of this, it should be evident that we are far from developing any clear and general attributional theory of liking. Some time ago Kelley (1972) stated that attribution theory "should be able eventually to predict causal interpretations, attributed characteristics and intentions, perceived trustworthiness or sincerity, and such, but will never constitute more than a partial basis for predicting such other phenomena as liking and evaluations [p. 17]." The preceding review suggests several reasons why this is so. In the first place, the line of evidence concerning whether attributions about another's behavior are consistently associated with "resultant" attraction is arguably the weakest and least consistent that we have discussed. Furthermore, the mutual interplay of attraction and attribution that emerges in the material reviewed is resistant to being placed in a linear conceptual sequence. The literature does not suggest that stimulus characteristics produce attributions or interpretations about the characteristics of others, which directly and reliably cause liking or disliking. Judgments of the characteristics of others may normally be associated with attraction (as indicated, for example, by work on the halo effect), but it does not seem that they typically are formed in advance of, and constitute the primary determinant of, attraction. Moreover, there are several variables known to affect attraction (such as proximity, physical attractiveness, frequency of exposure, expectation of future interaction) that can be cast in attributional terms — if at all — only by stretching and straining. And finally, attributions may themselves be a rarer phenomenon than attraction. We may frequently not bother to form them in the first place, unless there is an interpersonal problem to be explained (Orvis, Kelley, & Butler, 1976) or unless considerations about the future make it seem important to come to an understanding of the dispositional characteristics of another (Berscheid, Graziano, Monson, & Dermer, 1976).

Regarding this last point, it is worrisome that the bulk of the literature reviewed (and most of the other literature on attribution) may be inherently insensitive to the functions that attributions serve in the context of meaningful, ongoing social interactions. Consider the studies cited on the effects of attributions about another's behavior on attraction. For the most part, these experiments present a perceiver with some information about another's behavior and the surrounding circumstances. The subject then reports his causal interpretation of the other's behavior and typically indicates his degree of attraction to the other. In no case is there any extended interaction among the participants; in particular, the perceiver does not himself act on the attributions he may have formed.

This focus on the "mechanics" of the attribution process (and the concomitant truncation of interaction in research contexts) may be historically understandable, because the attributional perspective in social psychology has paid particular attention to uncovering the rules in accordance which people supposedly make sense of their social environments. But to what use are such attributions put? What functions do they serve? What actual difference does it make for the future of a relationship if one, rather than another, causal attribution is formed?

Two recent lines of research offer encouragement that an enriched view of the role of attribution in interpersonal attraction and interaction may be achieved if the research enterprise begins to focus on less static and restricted situations. The work on "behavioral confirmation" being done by Mark Snyder and his colleagues (Snyder & Swann, in press; Snyder, Tanke, & Berscheid, in press) begins at the point where the typical attribution study ends. Given that an attribution about some dispositional characteristic of another has been formed (for example, by the perceiver's having stereotyped notions of the characteristics of a group to which the target belongs or by the perceiver's being directly given information about the target's characteristics), what use is made of this information in subsequent interaction? Snyder's work suggests that individuals act on their attributions and that their behavior then shapes the behavior of the target in a way likely to produce "behavioral confirmation" for the attribution. Thus, for example, if a male interacts with a female whom he expects to have certain qualities because of her physical attractiveness, he will behave in such a way as to encourage her to demonstrate these qualities (Snyder, Tanke, & Bersheid, in press). Attributions are then seen not simply as cognitive epiphenomena, helpful largely in achieving the illusion of understanding for the perceiver, but as working hypotheses exerting a strong effect on the future course of interaction.

The correlational work by Kelley and his students (Orvis, Kelley, & Butler, 1976) on attributional conflicts in heterosexual couples also offers an enriched view of the development and functioning of attributions in meaningful interpersonal contexts. These authors have tentatively suggested that attributions are formed, not primarily to achieve cognitive understanding and mastery of the social environment, but in order for a person to explain and justify his own actions and criticize those of others in cases of interpersonal conflict. Attributions serve an important function in *communication* with others with whom significant disagreements arise. According to Orvis et al. (1976): "Overt explanations serve to define, emphasize, and acknowledge the basic understanding within the relationship about the conditions under which various behaviors may and may not occur [p. 378]." Previous concentration on the mechanics of the attribution process may have hindered understanding their actual functioning within relationships. It is to be hoped that continuation of research of these types will move us closer to an appreciation of the complex relationships between our causal explanations for behavior and our attraction to other individuals.

REFERENCES

Abelson, R. P. Modes of resolution of belief dilemmas. *Journal of Conflict Resolution,* 1959, *3,* 343–352.

Amabile, T. M., DeJong, W., & Lepper, M. R. Effects of externally imposed deadlines on subsequent intrinsic motivation. *Journal of Personality and Social Psychology,* 1976, *34,* 92–98.

Bem, D. J. Self-perception: An alternative interpretation of cognitive dissonance phenomena. *Psychological Review,* 1967, *74,* 183–200.

Bem, D. J. Self-perception theory. In L. Berkowitz (Ed.), *Advances in experimental social psychology* (Vol. 6). New York: Academic Press, 1972.

Berscheid, E., Graziano, W., Monson, T., & Dermer, M. Outcome dependency: Attention, attribution and attraction. *Journal of Personality and Social Psychology,* 1976, *34,* 978–989.

Berscheid, E., & Walster, E. A little bit about love. In T. L. Huston (Ed.), *Foundations of interpersonal attraction.* New York: Academic Press, 1974.

Blanchard, F. A., & Cook, S. W. Effects of helping a less competent member of a cooperating interracial group on the development of interpersonal attraction. *Journal of Personality and Social Psychology,* 1976, *34,* 1245–1255.

Cantor, J. R., Zillman, D., & Bryant, J. Enhancement of experienced sexual arousal in response to erotic stimuli through misattribution of unrelated sexual excitation. *Journal of Personality and Social Psychology,* 1975, *32,* 69–75.

Davis, K. E., & Jones, E. E. Changes in interpersonal perception as a means of reducing cognitive dissonance. *Journal of Abnormal and Social Psychology*, 1960, *61*, 402–410.

Dickoff, H. *Reactions to evaluations by others as a function of self-evaluation and the interaction context.* Doctoral dissertation, Duke University, 1961.

Dienstbier, R. A., & Munter, P. O. Cheating as a function of the labeling of natural arousal. *Journal of Personality and Social Psychology*, 1971, *17*, 208–213.

Dion, K. Physical attractiveness and evaluation of children's transgressions. *Journal of Personality and Social Psychology*, 1972, *24*, 207–213.

Dutton, D. G., & Aron, A. P. Some evidence for heightened sexual attraction under conditions of high anxiety. *Journal of Personality and Social Psychology*, 1974, *30*, 510–517.

Festinger, L. *A theory of cognitive dissonance.* New York: Row, Peterson, 1957.

Garbarino, J. The impact of anticipated reward upon cross-age tutoring. *Journal of Personality and Social Psychology*, 1975, *32*, 421–428.

Glass, D. C. Changes in liking as a means of reducing cognitive discrepancies between self-esteem and aggression. *Journal of Personality*, 1964, *32*, 530–549.

Heider, F. *The psychology of interpersonal relations.* New York: Wiley, 1958.

Homans, G. C. *Social behavior: Its elementary forms.* New York: Harcourt, Brace & World, 1961.

Jecker, J., & Landy, D. Liking a person as a function of doing him a favor. *Human Relations*, 1969, *22*, 371–378.

Jones, E. E. *Ingratiation.* New York: Appleton–Century–Crofts, 1964.

Jones, E. E., & Davis, K. E. From acts to dispositions: The attribution process in person perception. In L. Berkowitz (Ed.), *Advances in experimental social psychology* (Vol. 2). New York: Academic Press, 1965.

Jones, E. E., Davis, K. E., & Gergen, K. J. Role playing variations and their informational value for person perception. *Journal of Abnormal and Social Psychology*, 1961, *63*, 302–310.

Jones, E. E., & Harris, V. A. The attribution of attitudes. *Journal of Experimental Social Psychology*, 1967, *3*, 1–24.

Kelley, H. H. Attribution theory in social psychology. In D. Levine (Ed.), *Nebraska Symposium on Motivation.* Lincoln, Neb.: University of Nebraska Press, 1967.

Kelley, H. H. Attribution in social interaction. In E. Jones, D. Kanouse, H. Kelley, R. Nisbett, S. Valins, & B. Weiner (Eds.), *Attribution: Perceiving the causes of behavior.* Morristown, N.J.: General Learning Press, 1972.

Kruglanski, A. W. The endogenous–exogenous partition in attribution theory. *Psychological Review*, 1975, *82*, 387–406.

Lepper, M. R., & Greene, D. On understanding "overjustification": A reply to Reiss and Sushinsky. *Journal of Personality and Social Psychology*, 1976, *33*, 25–35.

Lepper, M. R., Greene, D., & Nisbett, R. E. Undermining children's intrinsic interest with extrinsic rewards: A test of the "overjustification" hypothesis. *Journal of Personality and Social Psychology*, 1973, *28*, 129–137.

Lott, A. J., & Lott, B. E. The role of reward in the formation of positive interpersonal attitudes. In T. L. Huston (Ed.), *Foundations of interpersonal attraction.* New York: Academic Press, 1974.

Lowe, C. A., & Goldstein, J. W. Reciprocal liking and attributions of ability: Mediating effects of perceived intent and personal involvement. *Journal of Personality and Social Psychology*, 1970, *16*, 291–297.

Mettee, D. R., & Aronson, E. Affective reactions to appraisal from others. In T. L. Huston (Ed.), *Foundations of interpersonal attraction.* New York: Academic Press, 1974.

Miller, A. G. Constraint and target effects in the attribution of attitudes. *Journal of Experimental Social Psychology*, 1976, *12*, 325–339.

Nemeth, C. Effects of free versus constrained behavior on attraction between people. *Journal of Personality and Social Psychology*, 1970, *15*, 302–311.

Nisbett, R. E., & Schachter, S. Cognitive manipulation of pain. *Journal of Experimental Social Psychology*, 1966, *2*, 227–236.

Nisbett, R. E., & Wilson, T. D. The halo effect: Evidence for unconscious alternation of judgments. *Journal of Personality and Social Psychology*, 1977, *35*, 250–256. (a)

Nisbett, R. E., & Wilson, T. D. Telling more than we can know: Verbal reports on mental processes. *Psychological Review*, 1977, *84*, 231–259. (b)

Orvis, B. R., Kelley, H. H., & Butler, D. Attributional conflict in young couples. In J. Harvey, W. Ickes, & R. Kidd (Eds.), *New directions in attribution research* (Vol. 1). Hillsdale, N.J.: Lawrence Erlbaum Associates, 1976.

Ostrom, T. M. The relationship between the affective, behavioral, and cognitive components of attitude. *Journal of Experimental Social Psychology*, 1969, *5*, 12–30.

Pilkonis, P. A. *The relationships among status, causal attributions, and interpersonal attraction.* Unpublished manuscript, University of Pittsburgh, 1976.

Potter, D. A. Personalism and interpersonal attraction. *Journal of Personality and Social Psychology*, 1973, *28*, 192–198.

Regan, D. T., Straus, E., & Fazio, R. Liking and the attribution process. *Journal of Experimental Social Psychology*, 1974, *10*, 385–397.

Reiss, S., & Sushinsky, L. W. Overjustification, competing responses, and the acquisition of intrinsic interest. *Journal of Personality and Social Psychology*, 1975, *31*, 1116–1125.

Ross, L., Rodin, J., & Zimbardo, P. G. Toward an attribution therapy: The reduction of fear through induced cognitive–emotional misattribution. *Journal of Personality and Social Psychology*, 1969, *12*, 279–288.

Schachter, S., & Singer, J. E. Cognitive, social, and physiological determinants of emotional state. *Psychological Review*, 1962, *69*, 379–399.

Schopler, J., & Compere, J. S. Effects of being kind or harsh to another on liking. *Journal of Personality and Social Psychology*, 1971, *20*, 155–159.

Seligman, C., Fazio, R. H., & Zanna, M. P. *Consequences of extrinsic rewards for impressions of liking and loving.* Unpublished manuscript, Princeton University, 1976.

Snyder, M., & Jones, E. E. Attitude attribution when behavior is constrained. *Journal of Experimental Social Psychology*, 1974, *10*, 585–600.

Snyder, M., & Swann, W. B. Behavioral confirmation in social interaction: From social perception to social reality. *Journal of Experimental Social Psychology*, in press.

Snyder, M., Tanke, E. D., & Berscheid, E. Social perception and interpersonal behavior: On the self-fulfilling nature of social stereotypes. *Journal of Personality and Social Psychology*, in press.

Steiner, I. D., & Field, W. L. Role assignment and interpersonal influence. *Journal of Abnormal and Social Psychology*, 1960, *61*, 239–246.

Thibaut, J. W., & Kelley, H. H. *The social psychology of groups.* New York: Wiley, 1959.

Thibaut, J. W., & Riecken, H. W. Some determinants and consequences of the perception of social causality. *Journal of Personality*, 1955, *24*, 113–133.

Zillig, M. Einstellung und aussage. *Zeitschrift Psychologie*, 1928, *106*, 58–106.

9

Attribution in the Context of Conflict and Separation in Close Relationships

John H. Harvey
Vanderbilt University

Gary L. Wells
University of Alberta

Marlene D. Alvarez
Vanderbilt University

> *Canst thou not ...*
> *Pluck from the memory a rooted sorrow,*
> *Raze out the written troubles of the brain;*
> *And with some sweet oblivious antidote*
> *Cleanse the stuff'd bosom of that perilous matter*
> *Which weighs upon the heart?*
> — Shakespeare (*Macbeth,* V; iv)

In this chapter, we describe some embryonic work on people's attributions and perceptions in situations involving conflict and separation in heterosexual, close relationships. A basic assumption behind this line of investigation is that people try to explain why they are encountering problems in their close relations or why these relations have ended. As we describe, their attempts to understand begin at least at the inception of conflict and seem to become pronounced after separation. Whether this striving for understanding is based on a need for future control (Kelley, 1971), simple curiosity (Heider, 1976), or some other mechanism, causal attribution in this context often manifests itself in elaborate, interpretive rationales filled with such feelings as anger, hate, despair, failure, and self-deprecation (Weiss, 1975; Orvis, Kelley, & Butler, 1976). Attributional analysis by individuals in these situations does not conform often to the amotivational, emotionless flavor so often ascribed to it (e.g., Bem, 1972).

It should be clear that our work described in this chapter is not so much concerned with probing pure attributional processes (cf. the discussion between Jones and Kelley in Chapter 13, this volume) as it is with employing attributional notions to understand important, quite complex, little understood phenomena.

We have employed these notions in this early work, simply because of the great breadth and common-sense heuristic value an attributional framework seems to provide. Cognitive-social research in the area of close relationships, and in particular with relevance to conflict and separation, is at an early stage. In the following review, we discuss some of the early major contributions in the development of this research domain.

A REVIEW OF SOCIAL PSYCHOLOGICAL WORK
ON CONFLICT AND SEPARATION

The work of Levinger and his colleagues (e.g., Levinger, 1964, 1965, 1966; Levinger & Snoek, 1972; Levinger, 1976; Huesmann & Levinger, 1977) represents the most concerted, long-term program of research on social psychological aspects of close relationships. In its focus on long-term relationships, this work stands in distinction to the substantial literature on liking and physical attraction (e.g., Byrne, 1971; Berscheid & Walster, 1974), which is very important to the field because it has elucidated basic principles. Levinger and his associates have directed much of their attention upon the marital dyad and the attractions and barriers that seem most critical to the continuation or dissolution of that dyad. They have derived some of their conclusions about these attractions and barriers from a study by Levinger (1966). The procedure in this investigation involved examining interview records of a sample of 600 couples in the Cleveland area who had filed for divorce and, in the process, had participated in mandatory interviews with marital counselors. Levinger compiled a set of 12 categories that reflected each spouse's complaints about his/her partner. Among these categories were: neglect of home or children, financial problems, physical abuse, verbal abuse, infidelity, sexual incompatibility, drinking, in-law trouble, mental cruelty (i.e., jealousy, suspicion, deceit), lack of love, and excessive demands.

As we see throughout this review, sex differences are quite evident in the dynamics of separation. In Levinger's (1966) study, it was found that wives' complaints exceeded husbands' two to one. In particular, wives complained more than husbands about physical and verbal abuse, financial problems, drinking problems, and various related issues (e.g., mental cruelty and lack of love). Husbands' complaints exceeded those of wives on only two counts — in-law trouble and sexual incompatibility.

Importantly, these sex differences in complaints reported by Levinger may be translated into perceived causality terminology, with complaints viewed as reflecting the perceived causes of problems. We may wonder whether (and if so, why) women tend to perceive a greater number and different types of causes as related to their separation than do men. In part, the answer to this question may be found in the typical patterns of behavior followed in the divorce procedure and may not be so fascinating for attribution investigators. Regardless of the contemporaneous perceptions about why conflict is occurring in a relationship, explanations presented later at the time of separation or divorce may reflect the obligation of the participants to develop rationales that are socially appropriate and facilitative of the legal and/or psychological dissolution (Weiss, 1975). Typically, the female files for divorce and is obligated to develop a more extensive causal analysis. An important issue in this area is that if we as social psychologists are to be successful in examining the adequacy of our analyses of conflict and separation, we must be able to rule out alternative and relatively artifactual explanations, such as the foregoing one, and base our conclusions upon an indepth examination of these phenomena as they are occurring.

In his recent theoretical statement, Levinger (1976) presents a Lewinian-flavored conceptual framework emphasizing individuals' perceptions of attractions and barriers in a relationship. Included among the attractions are: material rewards such as family income and home ownership, and symbolic rewards such as companionship, esteem, and sexual enjoyment. Included among the barriers are: material costs such as financial expenses; symbolic costs such as obligation toward marital bond, religious constraints, and pressures from primary and community groups; and affectional costs such as feelings toward dependent children. Levinger also discusses alternative attractions such as independence, self-actualization, and alternate sex partners. Levinger (1976) summarizes his theoretical position as follows:

> People stay in relationships because they are attracted to them and/or they are barred from leaving them; . . . consciously or not, people compare their current relationships with alternative ones. If internal attraction and barrier forces become distinctly lower than those from a viable alternative, the consequence is breakup [p. 32].

Rubin and his colleagues (Hill, Rubin, & Peplau, 1976) conducted a 2-year, longitudinal study of dating relationships among college students in the Boston area. They used a survey questionnaire technique to investigate perceptions and events associated with over 200 dating relationships among college students (mostly sophomores and juniors). The questionnaires were administered at intervals of 6 months, 1 year, and 2 years after the initial questionnaire session. At the time of the first questionnaire, one-third had been dating 5 months or less; one-third, 5 to 10 months; and one-third, more than 10 months. In terms of the external validity of Hill et al.'s data for understanding conflict and separation in long-term relationships, it should be emphasized that their couples were not consistently living together and that there were few if any definitive commitments between members of the dyads.

Because of the longitudinal nature of their work, Hill et al. obtained valuable data relevant to factors associated with whether or not their couples' relationships endured for the 2-year period. They found that factors related to breakups included unequal involvement in the relationship and dissimilarity with respect to age, educational aspirations, intelligence, and physical attractiveness.

Hill et al. report some data that are relevant to attributional questions. Former partners exhibited little attributional agreement on the contribution of factors internal to the relationship (e.g., differences in interests and backgrounds) and moderate to high agreement on external factors (e.g., interest in other persons). Further, partners showed a systematic bias with respect to the question of who wanted to break up. There was a general tendency for respondents to say that they themselves were the ones who wanted to break up. The perceived causality data showed that women tended to be more sensitive than men to problem areas in the relationships and that women were more likely than men to compare the relationship to alternatives, whether potential or actual. Hill et al. found that women more readily terminated relationships than did men. Presumably, this last datum reflected the women's greater sensitivity (as revealed in their causal attributions) to relationship problems and their greater need to move quickly in deciding that the relationship should be terminated. Hill et al. also interpreted this effect in terms of women's greater sense of practicality in romance (i.e., that they are pressured by the mores of society to recognize that it is not wise or beneficial for women to dally too long in romance with the wrong person). Finally, Hill et al. found that adverse emotional effects (e.g., depression, loneliness) were much less pronounced for persons who instigated the breakup.

Hill et al. offer the following recommendation, which is central to the work reported later in this paper:

> Given the fundamental asymmetries that often characterize a weakening relationship, it may be almost inevitable for each party to see events somewhat differently. . . . This difference in perspective leads to the recommendation that students of marital separation make every effort to obtain reports of both partners [p. 165].

Weiss (1975), a sociologist, conducted a very informative study of marital separation that has strong attributional implications. He conducted in-depth interviews with separated persons living in the Boston area. Most of the respondents had been separated for a relatively brief period of time (less than 10 months was the limit in most cases). The data are essentially case-history reports. Both males and females were interviewed, but no attempt was made to compare responses of ex-partners; participants came mainly from the organization called Parents Without Partners.

Weiss (1975) suggested that people involved in the act of separation develop an account of what happened and why:

> The account is of major psychological importance to the separated, not only because it settles the issue of who was responsible for what, but also because it imposes on the confused marital events that preceded the separation a plot structure with a beginning, middle, and end and so organizes the events into a conceptually manageable unity. Once understood in this way, the events can be dealt with. They can be seen as outcomes of identifiable causes and, eventually, can be seen as past, over, and external to the individual's present self. Those who cannot construct accounts sometimes feel that their perplexity keeps them from detaching themselves from the distressing experiences. They may say, "If only I knew what happened, if only I could understand why [pp. 14–15]."

In further discussion of the accounts that people often develop in the act of separation, Weiss implies that divergent explanations are quite pervasive: "None of the events significant to him appeared in her account, nor were any of the events significant for her included in his account [p. 15]."

Weiss noted that these divergent accounts usually are selected by individuals from a bewilderingly complex range of events preceding and surrounding a separation, and he noted that accounts, as well as the perceived and actual events upon which they are based, may constantly be reviewed in an obsessive-like ritual performed by the individual in trying to make sense out of what has happened. The accounts represent themes centering on such matters as infidelity and betrayal, a desire for new things in life, perceived overreactions to activities (such as various degrees of extramarital intimacy), and freedom from constraints imposed by the partner. Apparently his respondents did not attribute as much causality to factors such as financial problems and physical and verbal abuse as did the persons whose interview responses were analyzed by Levinger (1966). It seems possible that this difference may stem from the fact that Levinger examined a wider data base along the socioeconomic dimension.

Also, in contrast to the findings by Hill et al. (1976) regarding who was rejected initially as compared to who did the rejecting, Weiss (1975) suggested that a rather different view may emerge from analysis of long-term intimate relationships: "and sometimes a husband or wife who had insisted on separation later had a change of heart and wanted to become reconciled, but the other spouse now refused. In all these circumstances, the identification of one spouse as leaver and the other as left oversimplifies a complex interactive process [p. 63]."

Weiss' work provides provocative suggestions for an attributional analysis of separation. However, it seems essential that a broader methodology be brought to bear on the problem. Systematic probing via questionnaires, asking respondents to react to simulated conflict situations, and again, collection of perceived causality data from both members of a former dyad may provide additional important evidence.

A final investigation to be reviewed here was reported by Orvis, Kelley, and Butler (1976) in the first volume of *New Directions in Attribution Research*. In this study, attributional conflict surrounding problematic behavior among young couples was examined. In all, 41 couples in the Los Angeles area comprised the sample; 11 were living together, 5 were married, and 21 were dating. Approximately one-half of the sample had been together for 2 years or more, and most of the respondents were students between the ages of 18 and 26. They were paid for their participation. Orvis et al. did not obtain data from couples who were separated or who necessarily were contemplating separation. In their procedure, couples initially were asked to list examples of behavior for self and partner for which each had a different explanation. They were requested to avoid listing behavior that might be sensitive to the dyad (it was hoped that with this procedure, possible irritation of troublesome areas of interaction would be avoided). In the same questionnaire in which the couples listed examples of their behavior, they also presented their own explanation and their perception of their partner's explanation for the behavior.

Orvis et al. reported that respondents' examples of behavior for which they thought different explanations obtained fell into a large number of categories, including these predominant categories: actor criticizes or places demands upon the partner; actor is too involved in outside relationships and activities (e.g., with his/her own family or even in other close relationships); actors' behavior inconveniences others (e.g., inconsiderate, late for appointments); actor has distorted view of self or others (too elevated view of self or poor attitude toward others); actor behaves emotionally or aggressively; actor has undesirable practices (e.g., drinks alcohol or smokes tobacco); actor engages in or wants to engage in activity (this activity often was one in which the partner was not involved; e.g., educational or personal recreation); actor avoids activity (e.g., will not try new things or avoids serious discussions).

Orvis et al. found a few significant sex differences in attributions. For example, in line with Deaux's (1976) review and analysis of attributional sex stereotyping, they found that the behavior of the female was more often attributed to the environment and to inability to cope with tasks than was the behavior of the male. Further, these stereotypes were found in the reports of both males and females. Also, and importantly, their actor–partner attributional data offer some support for Jones and Nisbett's (1971) divergent perspectives hypothesis. It was found that actors quite often explained their behavior in terms of external causes (e.g., state of the environment and external objects or events). But they also frequently ascribed their behavior to their concern for their partner. In contrast, partners quite often pinpointed features of the actor (his/her characteristics or attitudes) as causal factors; also, the tone of these explanations was mainly negative. Orvis et al. (1976) articulate their view regarding the relevance of their results for Jones and Nisbett's hypothesis in the following quote:

> Although our actor–partner differences conform in large part to the Jones and Nisbett hypothesis, we doubt that they are based on the process that Jones and Nisbett emphasizes as underlying the actor–observer discrepancy. This process involves *informational differences* that lead actors and observers to infer different causes of behavior. Our actors and partners are clearly much more than simple "actors" and "observers", and differ in more drastic ways than merely in the information they possess. Our actor is a person who, in a close relationship with another person, has . . . behaved in an unfavorable or at least questionable manner. The relationship requires that he/she be concerned about justification and exoneration. Similarly, the partner is no mere observer. Having been affected negatively by the behavior, the partner is concerned about its meaning, about redress or retribution, and about preventing its recurrence [p. 364].

Orvis, Kelley, and Butler (1976) concluded their presentation with some intriguing suggestions regarding attribution in close relations. They suggested that in close relationships, divergent causal interpretations of behavior represent a common and important part of the interaction process. They propose that in these *continuing* relationships, attributions of causality serve a communicative role in defining, emphasizing, and acknowledging the basic understanding within the relationship about the conditions under which various behaviors may and may not occur. They assume that attributional analyses in close relationships are evoked primarily in situations of conflict of interest and that occur largely in the context of justification of self and criticism of other.

STUDIES OF ATTRIBUTIONS IN SITUATIONS INVOLVING CONFLICT AND SEPARATION

With the foregoing early work as a foundation, we have completed two exploratory studies designed to examine attributions in problematic relationships. The first study (conducted by Harvey and Wells) investigated important areas of divergence and lack of knowledge of divergence on the part of young adults living together and experiencing conflict in their relationships. The second study (conducted by Harvey and Alvarez) investigated the explanations and perceptions of persons who had separated recently after generally long-lasting marriages.

Harvey and Wells

This work is methodologically similar to Orvis et al.'s (1976) study, though it differs in certain key respects. Compared to Orvis et al.'s investigation, the present work was designed to probe more comprehensively different types of possible divergence relating to areas of conflict over *sensitive issues* for couples who had been *living together* and who perceived their relationship as involving *considerable conflict*. The Harvey and Wells work involved asking participants to rate the importance of predetermined areas of possible conflict (the areas were chosen after consideration of findings deriving from the work reviewed above).

What are some issues about which divergence in understanding is likely to exist in close relationships? Common yet quite significant issues may include dissimilarities in terms of attitudes, beliefs, educational and career aspirations, sexual incompatibility, and the influence of alternative possible partners and acts of perceived disloyalty. It seems likely that in relationships that have existed for some time and in which serious conflict has emerged, individual members will maintain divergent views on the role of at least some of these and other important issues in fostering their conflict. Not only should divergent views exist, but also individuals may not understand very well the extent to which they do differ in their attributions.

Our questionnaire procedure involved asking individual members of dyads for their own attributions and to predict what attributions they thought their partner would make. The data obtained, therefore, pertained to sex differences in attributions and to how well each partner understood the other's perspective. We also obtained data concerning individuals' satisfaction with their relationship and feelings of freedom to pursue their own personal interests and careers.

Method

Respondents.   A total of 72 individuals (36 unmarried couples) were solicited through an advertisement in the Ohio State University newspaper. When potential respondents called concerning the study, they were screened to assess their suitability for the study; it was made clear that the study was a research enterprise, that no therapy would be offered, and that their interview would be held in the strictest confidence. Couples invited to participate met the following criteria: (a) unmarried but defined one another as the principal close relation in their heterosexual love life; (b) had lived with one another at least 4 nights out of each week for a minimum of 6 months; (c) indicated having experienced conflict in the relationship (e.g., perceived difficulty to the point that they had at some time considered terminating their relationship). Each couple was paid $5 for their participation.

Participants ranged in age from 19 to 30 years, with an average age of 24. All participants had attended college for at least 6 months; both members of four couples had college degrees. The average length of time that couples had been engaged in their close relationship was 11 months. Approximately 95% of the sample indicated that currently or in the past they had discussed marriage with their present partner.

Procedure and Description of Close Relationship Questionnaire.   Participants were seen individually in sessions lasting approximately 1 hour. Each session involved the participant's response to a lengthy close relationship questionnaire and an in-depth interview of the participant concerning his/her close relationship. Generally, the interview covered the same topics included in the questionnaire but also matters about which participants considered the questionnaire to be inadequate or about which they had difficulty in indicating answers via the questionnaire.

The close relationship questionnaire consisted of 65 items. An initial set of items pertained to autobiographical information and information concerning the participant's present close relationship. A subsequent set of items covered participant's *own* and the predictions of *partner's* perceptions about the extent to which the following factors were causes of the conflict they were experiencing in their relationship: dissimilarity in social and political attitudes, dissimilarity in religious—philosophical—ethical beliefs and values, dissimilarity in educational aspirations and career—occupational goals, incompatibility in sexual relations (i.e., at least one person's dissatisfaction), influence of friends, influence of parents, influence of one or more possible alternative partners (e.g., lovers), influence of important events in their relationship (e.g., act of disloyalty), less practical and convenient than it was initially, financial problems, and stress associated with one or both partners' work or educational activities. Finally, individuals were asked to give their own perceptions of and their predictions of their partner's satisfaction with the relationship and freedom in the relationship to pursue personal interests and career. For all items concerned with conflict, satisfaction, and freedom, questions were answered on an 11-point scale, with high numbers indicating large amounts.

After the interview, participants were given general information about the project, and again it was emphasized that the participants' answers to the questionnaire would be held in the strictest confidence. Participants were asked to be careful in assessing whether to discuss with their partners at a later time their responses to the various items on the questionnaire and in the interview. The investigator indicated that the research team was concerned about the possibility that such discussion could exacerbate conflict in relationships.

Results

Analysis of Data.   Our primary interest was with patterns of sex and perspective differences within each of the 11 conflict areas, the satisfaction measure, and the freedom measure. The statistical design for each of these measures was a 2 (male, female) × 2 (self-attribution, prediction of partner's attribution) mixed factorial, with the latter being the within-subjects factor. Though the nature of this design does not yield readily interpretable main effects, it does provide mean square error terms that were pooled (Winer, 1971, p. 378), allowing pair-wise comparisons within each 2 x 2 using Duncan's Multiple Range Procedure (Winer, 1971, p. 196).

Attributions of Conflict.   Means for attributions of importance of sources of conflict as a function of sex and perspective of respondent are shown in Table 9.1. It can be seen from this table that in terms of absolute magnitude, the perceived conflict responses generally are near the lower end of the 11-point conflict measures. Crosscutting sex and perspective, stress associated with work or educational activities, financial problems, the influence of alternative possible partners, and dissimilarity in religious–philosophical–ethical beliefs and values had the highest ratings as sources of conflict.

Table 9.1 shows that for both own and partner's perspective, females and males showed similar responses for many of the measures. However, there were measures that did reveal differences. On the sexual incompatibility measure:

1. Males rated incompatability in sexual relations as a more potent source of conflict than did females.
2. Females underestimates males' attribution of conflict to sexual incompatibility.
3. Males overestimated females' attribution of conflict to sexual incompatibility.

For the measure concerning the influence of important events (e.g., disloyalty), a similar pattern emerges. As can be seen in Table 9.1, males inputed greater importance to this influence as a source of conflict than did females; also, females underestimated males' attribution to the influence of important events. However, unlike the factor of sexual incompatibility, males appear to have understood that females held divergent views. Males rather accurately predicted females' attributions about the influence of important events.

Regarding the role of financial problems as a source of conflict, Table 9.1 shows that:

1. Females attributed more importance to this source of conflict than did males.
2. Females overestimated males' attribution to this source.
3. Males underestimated females' attributions of conflict to financial problems.

For the measure regarding stress associated with work and educational activities, Table 9.1 reveals that females attributed more importance to such stress than did males. Further, females overestimated males' attributions to this source, and males underestimated females' attributions.

TABLE 9.1
Means for Attributions of Importance of Sources of Conflict

Conflict Measure	Answering from Own Perspective		Answering from Partner's Perspective	
	Females' Attributions	*Males' Attributions*	*Females' Prediction of Males' Attributions*	*Males' Prediction of Females' Attributions*
1. Dissimilarity in social and political attitudes	3.2_a	3.2_a	3.0_a	3.2_a
2. Dissimilarity in religious—philosophical—ethical beliefs and values	4.2_a	4.1_a	3.6_a	4.1_a
3. Dissimilarity in educational aspirations and career—occupational goals	3.7_a	3.9_a	3.6_a	3.2_a
4. Incompatibility in sexual relations	2.6_a	3.9_b	2.7_a	3.7_b
5. Influence of friends	3.0_a	2.6_a	2.4_a	2.2_a
6. Influence of parents	3.6_a	3.6_a	3.7_a	3.8_a
7. Influence of one or more alternative possible partners	3.7_a	3.8_a	3.4_a	3.8_a
8. Influence of important events in relationship	3.9_a	4.6_b	3.4_a	3.8_{ac}
9. Less practical and convenient than it was initially	2.0_a	2.0_a	2.0_a	2.2_a
10. Financial problems	4.9_a	3.6_b	4.6_{ac}	4.1_c
11. Stress associated with one or both partners' work or educational activities	5.6_a	4.9_b	5.8_a	5.0_b

Note: The higher the number, the greater the perceived role of the factor as a determinant of conflict (1- to 11-point scale). Means not sharing a common subscript, within a conflict dimension, differ at the .05 level by Duncan's multiple-comparison procedure.

It should be noted from Table 9.1 that across conflict measures, both females and males predicted that partners would make attributions very similar to their own. Except for the perception of the influence of important events, there were no significant differences between males' predictions of the partner's attributions and males' own attributions. For females, there were no significant differences between their own attributions and their predictions of their partner's attributions for any of the measures. This lack of significant data was true, despite the fact that males' and females' own attributions significantly differed in four major conflict areas.

The foregoing data may be related meaningfully to evidence regarding an index of average total discrepancy across all items for the attributions of conflict. The overall discrepancy average for one partner's attributions versus the other's was 11.27, which was significantly different from zero $[t\ (71) = 2.7, p < .01]$. This evidence is in contrast to evidence showing that the index of discrepancy between own attributions and predictions of partner's attributions across all conflict items did not significantly differ from zero. In others words, across conflict categories, there is significant discrepancy between partners but *no overall perceived* discrepancy. It should be noted that this discrepancy index is independent of sex of the partner since it is based on the *absolute* direction or the discrepancy between partners.

Satisfaction and Freedom.   An analysis of the satisfaction measure revealed that females indicated significantly greater satisfaction with their close relationship than did males (mean for females = 10.2; for males = 9.1); males significantly underestimated females' satisfaction (mean for females' own ratings = 10.2; for males' prediction = 9.1). The analysis of the freedom measure did not reveal any significant effects, although males indicated that they felt slightly more free to pursue their own interests and career than did females (mean for males = 8.7; for females = 8.3).

Interview Data.   Participants' responses during the interview were coded and analyzed for relevance to the questions of interest in the study. These responses were mainly relevant to attributions of conflict in the relationship. First, it should be noted that regardless of the magnitude of problems pinpointed by participants, they indicated that they expected their relationship to continue "indefinitely" or "forever."

With respect to perceived causes of conflict in the relationship, the coded responses showed two sex differences of interest. Approximately one-third of the males (11) mentioned sexual matters (e.g., differential interest) as a source of conflict, whereas only 3 females explicitly mentioned this concern. On the other hand, more females (9) than males (5) mentioned the presence of other possible lovers in the life of their partner as a source of great conflict in their relationship.

Finally, there was a tendency for both males and females to emphasize strongly their partner's activities in areas such as involvement with other possible lovers or interest in outside sexual behavior as determinants of conflict in their relationship. A great majority of the respondents centered their discussion on their partner's activities (often indicated to be perceived as misdeeds) to a much greater extent than they focused on their own behavior (cf. Orvis et al., 1976).

Discussion

Although the data in the Harvey and Wells investigation do show attributional divergence, we should point out that the results reveal considerable convergence in perceptions. On the conflict dimensions, participants exhibited similar responses on these items: attitudes–beliefs–values; educational–career goals; the influence of friends, parents, and alternative possible partners; and practicality–convenience.

In terms of attributional divergence, sex differences were found showing that males imputed more importance to sexual incompatibility and the influence of important events (disloyalty) than did females. On the other hand, females attributed more importance to financial problems and stress associated with work or educational activities than did males. Males showed inaccuracy in: (a) overestimating females' attribution of importance to sexual incompatibility; (b) underestimating females' attribution of importance to financial factors; and (c) underestimating females' attribution of importance to stress associated with work or educational activities. Further, males underestimated females' self-ratings of satisfaction. Females showed inaccuracy in underestimating males' attribution of importance to sexual incompatibility and the influence of important events; also, females overestimated males' attribution of importance to financial problems and stress associated with work or educational activities.

The evidence showing both overestimation and underestimation for both males and females in trying to predict one another's attributions is quite noteworthy. Even in these relatively long-term close relationships, individuals tend to be inaccurate in their perceptions of how their partners see the world in terms of areas of conflict. Also, there is strong support for the contention that couples tend to be egocentric in their attributions. Both females and males very clearly tended to indicate that their partners' perceptions would be similar to their own. This inaccurate and egocentric posture may constitute a major determinant of conflict in close relations. Obviously couples will differ in their views about the importance of some sources of conflict in their lives. But do they know that they differ? The comparison of results showing a significant, overall discrepancy across conflict dimensions between own and partner's attributions compared to the nonsignificant overall discrepancy between own attributions and predictions of partner's attributions suggests that they do not know that they differ. It seems reasonable to speculate that a *lack of understanding of divergence* is as critical or more critical to the viability of close relationships than is actual divergence.

Were females and males differentially inaccurate or egocentric in the sense of being insensitive in their attributions of their partner's weighting of the conflict areas? The data do not suggest clear sex differences. There were several instances of inaccuracy on the part of both sexes. Perhaps of more importance are the particular sex differences revealed in the data. Males' relatively great ascription of importance to sexual incompatibility and to the related issue of the influence of important events (in the interview, most participants indicated that they understood this item to refer specifically to an act of disloyalty typically involving sexual activity) reflects their emphasis upon sexuality as a point of conflict in their close relationship. The interview data showing males' greater mentioning of differential sexual interest as a source of conflict also attests to this emphasis. These data are consistent with results recently reported by Peplau, Rubin, and Hill (in press) in research concerned with sexual attitudes in the development of dating relationships.

Why did females impute more importance to finances and stress associated with educational or work activities than did males? The answer to this question is not clear, but it could relate to an insight developed by Hill et al. (1976). These investigators suggested that presumably because of greater insecurity, women may be more pragmatically oriented than men in their perceptions and decisions about continuation of close relationships. This point seems more tenable for the finances dimension than for the stress dimension. Interestingly, both of these dimensions relate to problems mostly *outside* the relationship. In further support of this point, females in the interviews spontaneously mentioned other possible partners (perhaps the most important outside pressure for them) as a source of conflict more often than did men. However, we should emphasize that in these nonmarital relationships, both males and females imputed a relatively high amount of potency to other partners as a source of conflict.

In conclusion, Harvey and Well's data provide some insight about attributional aspects of conflict in close relationships. They suggest that divergence in attribution, lack of understanding of divergence, and sex differences in attributional perspective are important factors in the dynamics of conflict.

Harvey and Alvarez

Since only single individuals could be solicited for interviewing in the following study involving recently separated persons, the divergency patterns found by Harvey and Wells could not be re-examined. However, Harvey and Alvarez's report suggests that these attributional divergencies may be enmeshed in the complex accounts developed by individuals involved in the dilemma and frequent trauma of separation.

The main foundation for this investigation was the provocative case history-type analysis of separation by Weiss (1975). We, too, collected a small sample of case histories of persons who recently had separated; we interviewed each person on two to four different occasions for a period of about 6 months. Our objective was to employ relatively stringent criteria for participation and in sampling, but the great difficulty of finding research participants to share these often-embarrassing intimacies about their lives and the fact that many more females than males expressed interest led to a small sample, composed mainly of females, In this regard, our sample appears to be similar to Weiss's sample (although the details of his procedure are scantily reported in his 1975 book).

Important a priori questions in this study were: What kinds of themes will individuals develop in their attributional analyses of their separation? Would there be discernible sex differences in these analyses? Would the analyses center on blame regarding infidelity (an important, perceived source of conflict in the Harvey and Wells study), lack of affection, intolerable personal habits, reduced involvement, etc.? Or, in another domain of possibilities, would the analyses emphasize a desire for independence, freedom from the constraints of marriage, and movement into different lines of work and professional activity? How much work would individuals indicate that they had exhibited in trying to understand their separation? How would this search for understanding vary as the separation grew longer?

Method

Participants.   A total of 10 individuals, 8 females and 2 males, were recruited through an advertisement in a Nashville, Tennessee, newspaper. They were solicited to participate in a study of approximately 6 months' duration concerned with their explanations of and feelings about their recent separations. Each participant was paid $10 for the initial interview and $5 for each follow-up interview.

Participants had been separated from 1 month to 10 months. Separation was operationally defined in terms of a respondent's perception of emotional severance of ties and actually living in a different physical location than his/her ex-spouse. At the time of the initiation of the research, none of the participants had gone through divorce proceedings. However, two participants had finalized divorces by the end of the project 6 months later, and only one participant had been reunited with her husband; the rest planned to file or complete divorce proceedings some time in the future.

The length of marriage varied from 1 year to 31 years, with a median of 9 years. Participants ranged in age from 21 to 49; most of them were in their mid- to late 30s. All participants had finished high school, and three had obtained bachelor's degrees. Of the two men in the sample, one worked as a part-time clerk, and the other was working on an advanced degree in engineering. Six of the eight women held jobs, all of which were clerical or secretarial in nature; the other two women were not working outside of their homes. All but one of the women had children; all of these seven women received some degree of financial assistance from their ex-spouses.

Procedure.   Participants were screened for eligibility when they called concerning the newspaper advertisement. They were questioned carefully to determine if they met the criteria of: (1) being separated, in line with the operational definition outlined above; (2) having had a recent separation — no longer than 1 year and; (3) having been married for at least 1 year. As in the Harvey and Wells study, it was made clear to callers that the study did not involve therapy; individuals who were or had been involved, in therapy were not permitted to participate. The delicate nature of the questions to be asked and the follow-up facet of the work were described and consented to by the individuals in the telephone interview.[1]

The interviews were tape-recorded and later typed into transcripts. The initial interview lasted an average of 1 hour and 30 minutes. Follow-up interviews lasted an average of 30 minutes. During the initial interview, participants were asked a standard set of questions pertaining to: (1) biographical factors; (2) the nature of their separation; (3) their explanations (in their own words) of why the separation occurred; (4) what they thought their ex-spouses perceived regarding the grounds for the separation; (5) the possibility of reconciliation with their ex-spouses or of the development of new close relationships; (6) their practical and emotional experiences in being separated; and (7) their most candid self-perceptions regarding their own present identity and perceptions of their ex-spouse. The interviewer probed in a detailed fashion perceived causes of the separation, which were touched upon but not initially elaborated.

The follow-up interviews involved questions about how matters had changed in the interim since the last interview. In particular, these interviews focused on changes in explanations, feelings, and experiences with regard to the separation or ex-spouse. Each person was interviewed for a follow-up report at least twice (with 1 to 2 months of intervening time); some were interviewed three or four additional times.

Throughout each of the interviews, the investigator was highly sensitive to the possibility that the interview experience might be too demanding or stressful

---

[1]The difficulty of securing participants is illustrated by the fact that although about 50% of the callers (90% of whom were women) were eligible, most of those eligible decided not to participate when told of the in-depth, delicate nature of the interviews or simply failed to appear when scheduled — which also may reflect anxiety about the questioning. Thus, as is true in most studies in this area, the sample of participants may differ from other separated people along many dimensions including willingness and ability to talk to strangers about this area of their lives and perhaps perceived need to talk to someone (although as noted, none of the participants were or had been involved in therapy).

for the participant. Whenever the participant seemed to become upset, a relaxation period was begun — or in some instance the interview was terminated, and a period of relaxed conversation followed. Prior to participation, respondents were assured that the interview responses would be held in the strictest confidence and that no names would be kept with data after the completion of the entire project. Further, they were fully informed about the purposes of the project and were warned to consider carefully whether discussion of their separation might lead to a further worsening of their relationship with their ex-spouse.

During the course of each person's participation in the project, the participant was asked to keep a written record of his/her feelings about the separation and especially any new perceptions regarding why it had happened, as well as any other experiences occurring in his/her life. All but two participants cooperated and wrote about many of their feelings and perceptions duing this period.

Results

All of the transcribed and written material were coded into very general categories by advanced, undergraduate students who were unfamiliar with the ideas behind the project. The categories pertained only to attributions of causality. The authors of this chapter further analyzed the evidence for additional, important information. We should emphasize that even for the small number of participants it would be impossible to provide all of the meaningful data collected. Our presentation will be selective and will focus on gross percentage-type data across the sample and information derived from individual cases that illustrates particularly well certain points.

Causal Attribution Categories:

1. *Romantic involvement outside of marriage.* All but one of the participants (who was a woman who had been married 31 years and who had a rather distinctive marital history, as will be described in part below) reported that they perceived this factor as an integral aspect of their separation. Interestingly, one of the men and four of the women indicated that they believed their ex-spouses would not have designated this outside involvement factor as a major cause of the separation. All of these five persons indicated that while their ex-partners had been engaged in such activities, they had not had "affairs" themselves.

2. *Insensitivity—lack of affection and warm sexual intimacy.* A rather multi-faceted category that emerged for 80% of the participants concerned their perception of insensitivity on the part of their ex-partners as well as themselves (though the emphasis was on the ex-partner's liability in this regard). This insensitivity was reported to be manifest especially in a lack of affection (hugs and kisses, pleasant conversation on an enduring basis, remembrances on special occasions, availability in times when emotional support or mere presence was desired) and warm sexual intimacy. During their preseparation conflict periods, participants reported having had sexual experiences with their partners. However, a great majority of the sample reported that these experiences represented no

more than "mechanical acts" on most occasions and that they lacked love and emotional giving; two reported an almost total lack of sexual activity during their final 2 years of marriage.

3. *Quest for freedom from constraints of marriage and desire for new lifestyle.* Six of the eight women and one of the two men ascribed considerably significance to this factor as contributing to their separation. Again, they generally saw the locus of the quest for freedom and desire for new lifestyle in their ex-spouses, although three of the women believed their ex-partners would attribute importance to their own desire to no longer be housewives and to have jobs of their own. Often the characteristics of this category were defined by the participant in terms of the ex-spouse's desire to be more of a "playboy" or "playgirl," as the case may be. This activity involved expression of the desire to have dates outside of marriage or to travel on one's own, unencumbered by family. An interesting aspect of this category is that most who emphasized this behavioral and perceptual pattern believed that it either *emerged* after several years of marriage or that it became much more salient after such other major problems arose; hence, the category relates to the issue of "growing apart," which has received attention in popularized accounts of separation.

4. *Different personal values and habits.* Five of the eight women and both of the men indicated that different values regarding thriftiness in financial matters and home care, personal cleanliness, and physical fitness were important sources of conflict in their former relationships. None of these participants felt that their ex-partner would pinpoint these issues as causes of their problems. The general feeling expressed was that their ex-partners were not frugal enough in spending money and tended to be "slobs" in areas of personal or home upkeep.

5. *Differential religious orientation.* Four of the women and one of the men emphasized a floundering orientation toward religion on the part of their ex-partner as a cause for their separation. They viewed this neglect as an indication of their ex-partner's failure to carry our important family responsibilities.

6. *Alcoholism and physical abuse.* Three of the women inputed major importance to extreme alcohol consumption on the part of their ex-partner as a cause of the separation. This consumption was reported to be accompanied sometimes by physical abuse toward the women. The woman who had been married 31 years had an alcoholic husband who had frequently become violent and required hospitalization. She indicated that she stayed with him so long in the hope that he would quit drinking, but to no avail. Parenthetically, of all the participants this woman exhibited the most positive adjustment to separation and the most vigor in seeking a new, satisfying life. The husbands of the other two women were not alcoholics in the fullest sense, but they did drink heavily on occasion. Both women indicated that the "clincher" in their decision to separate occurred when their husbands physically assaulted them.

7. *Escalation of conflict and commitment to separation.* A somewhat amorphous temporal category that embodies elements of all of the foregoing was

implied to be of great importance by all of the participants who had been married for as many as 4 years (8 of the 10). This category involved these patterns of reported behavior and feeling: much verbal fighting (often screaming matches), threats, feelings of discontent and general misery, being lonely even in the presence of the ex-partner, maneuvers to exercise more control in the relationship, and finally periods of "trial separation." These behaviors and feelings were reported to have escalated to a point that continuation of the relationship was viewed as intolerable piror to separation. Females, in particular, reported having become too exhausted to continue the relationship. In her written record that described events prior to separation, one woman said: "Together all day but miserable. Everything was pleasant except for the end of evening. We got into a terrible fight."

In all of these relationships, separation had been discussed for *more than 2 years* before it was carried out in earnest. All participants who by implication pinpointed the various facets of this category felt that both members of the former dyad had been active contributors to the conflict and escalation of negativity. They believed their ex-partners also would feel that both had contributed.

Attribution Over the Course of Separation.   The records kept by eight of the participants provided information about how their understanding of their separation changed over the 6-month period. Most individuals indicated in their records that they had engaged in considerable thinking (often focusing on a few thoughts repeatedly) about why their relationship had ended. The intensity and persistence of this cognitive work seemed to be related to the individual's psychological state and new life experiences.

The more lonely and depressed the individual, the greater the concern with rehashing the issues (often participants indicated that they had done this review with friends as well as privately). Furthermore, the less involved the individual in new experiences, the greater the continued review of the separation and its bases. These "why" questions and answers occurred regularly in the participants' records throughout the 6-month period. Essentially, the answers recorded did not change much and paralleled those presented to the interviewer in the initial session. On occasion, respondents would write about an issue that they gave the appearance of representing a "new insight" into why the separation occurred or why it was necessary. However, most often these insights represented little more than slight changes on general themes that previously had been elaborated. Although the need to explain seemed to remain high, the already great amount of explaining that had been done, no doubt, in countless conversations and private ruminations seemed to preclude much in the way of truly new explanation.

Finally, we should note that the causal analysis reflections were reported to have occurred especially at night — often when the person could not go to sleep because of continued thinking about the former relationship — or when the person was unoccupied during the day. Also, for the few people who did not write much about their separation, their records contained many entries relating to such matters as new jobs, shopping trips with friends, activities with relatives, and community activities.

Other Important Data: Initiation of separation. Similar to Hill et al.'s (1976) data deriving from a study with dating couples, we found that for 80% of our sample, the female actually took the actions finally leading to separation (which usually involved demanding that the man move out of the house).

Effects of separation. However bad the marriage may have been, our data suggest that individual's reactions to separation probably were worse and generally debilitating. Parkes' (1972) and Weiss' (1976) concept of separation distress syndrome seems to fit our evidence rather well. Insecurity, confusion, loneliness, depression, despair, and even infrequent thoughts of suicide were common themes in the written and oral records for a great majority of the sample. Some illustrative entries in the written records are:

> I need somebody *now*. I'm falling apart inside. My insecurity is showing. I feel as though I have very little to hang on to. . . . I'm terribly confused. I need help! Somebody please tell me what I'm doing.
>
> Oh, I know all the reasons, I guess, but he shouldn't have left. We loved each other — oh, I know we did — I'm so lonely.

Most of the participants indicated that they still missed and were attached to some degree to their ex-partners. This self-attribution persisted even in two cases in which the spouse had made plans to remarry after divorce. Thus, though they reported disaffection in their behavior, their feelings of attachment and caring persisted (see Weiss, 1976). Despite the expressed feelings of attachment, not one of our participants indicated a high, positive general impression of their ex-partner. Further, 70% of the participants revelaed an unfavorable view of themselves. Their disdain for their ex-partner was rivaled by their negative self-attributions. One said, "I'm nobody now." Another said, "I've learned the hard way and am broken because of it." Still another wrote in her record, "To sum up, my life has gone to pot." Some had begun to cope by the time the project was concluded, but only two indicated feeling much happiness.

Plans for the future and the development of new close relationships. Consistent with their present feelings of low self-esteem, most of the participants indicated that they perceived bleak prospects for future, close relations. The one person who had become reunited with her husband by the time the project ended indicated that she was very unhappy still and did not know how long the relationship would continue. All seemed to want to have other close relations but not the problems they had encountered in their previous relationships. They indicated much fear of anything approaching a complex, romantic relationship in the near future. Some comments taken from the written records of different participants are illustrative:

> I can't have a relationship with anyone now — not until I can get myself straightened out. I'm really screwed up.
> Someone called me and asked me to go out last night. I said no. They called back tonight and asked for tomorrow. No, again. I think I feel I will never have that romantic feeling again — that I felt for _____.
> I'm feeling that there is no hope for my ever finding anyone with whom to spend what is left of my sometimes miserable life.

Discussion

Representativeness of Data.   Do the data collected by Harvey and Alvarez have strong relevance for typical attributional aspects and emotional effects of separation? Despite the caution that must be exercised because of the limited sample and possibly special characteristics of the sample, we believe that there are good reasons to place confidence in the representativeness of the data in reflecting the separation process.

In many ways, the data are consistent with the evidence reported by Weiss (1975) in his work with participants in the organization, Parents Without Partners. The elaborate accounts with their central themes appear in our data just as they did in Weiss' data. The participants' reports of attributional confusion and emotional agony appear to be similar in the two investigations. Perhaps because Weiss apparently interacted personally with his respondents more often (as part of a therapeutically oriented seminar) and over a longer period of time than was true in the present work, his report contains evidence about a greater incidence of adjustment and happiness than does our report. Also, the fact that therapy and research apparently were being carried out intentionally at the same time in Weiss' work may help explain the greater adjustment he reports. In the Harvey and Alvarez study, the project objective was strictly collection of information.

A further argument in support of the representativeness of our data derives from the diverse characteristics of the sample. The participants differed along many dimensions including age, socioeconomic position, and nature of their marriage. Yet, the basic attributional and other findings generalized to a considerable degree across the set of individuals. The one exception to this point is that the emotional effects for those who had been married a relatively short time or a very long time, in the one instance, and for those who had been married an intermediate length of time (between 5 and 15 years in this study) differed in intensity, with the intermediate-range persons indicating a more difficult adjustment.

Finally, the data in our separation study make sense on intuitive grounds. That is, in light of the times, they have common-sense face validity. They reveal a grim picture of separation in this society. However, we do not have to be reminded of the role separation plays in so many tragic events and instances of personal misery to have some recognition of the pathos involved in conflict and separation in contemporary close relations.

ATTRIBUTIONAL SIGNIFICANCE OF DATA

What general attributional conclusions can be drawn from our study of separation? We would argue that an attributional theoretical approach in this area has merit, because it is directly attuned to individuals' intensive and apparently long-term efforts to understand their situations. In general, the separation situation is one in which people do not have well-learned repertoires of responses. It is not "well-scripted," to use Abelson's (1976) concept. And hence, it hardly appears to represent the social setting that elicits the type of "mindless" behavior that Ellen Langer described in Chapter 2, this volume. In fact, Langer does suggest that one situation in which people engage in thought, including attribution, is when they encounter a novel situation, for which, by definition they have no well-learned script. Our interview and written record data suggest that, if anything, individuals are overly "mindful" during separation. Weiss (1975, 1976) reports that separated persons show heightened vigilance and great restlessness. Our data suggest that these vigilant, restless periods may be filled often with incessant causal analyses. Further, the written records over 6 months in the separation study imply that extended, continued causal analysis may make the difficulty of separation somewhat more palatable.

The results of these studies may be interpreted to suggest that there is a complex interplay between behavior and attribution in the development of conflict and separation in close relationships outside of the research setting. But do the results suggest any particular sequence involving attribution and behavior? Possibly so. Although couples seem to be very much interested in understanding the bases of their conflict prior to separation, attributional analysis seems to be of more import and to be carried out in a more intense fashion *after* separation. Data from the separation study are consistent with this notion that in-depth attributional analysis *lags* somewhat behind critical behavior in the sequence of events from conflict to separation. There is a pronounced emphasis upon specific behavior and behavioral patterns in the participants' interview and written responses regarding events leading to separation. For example, they emphasize the importance of romantic behavior outside of marriage, too few signs of affection such as kisses and hugs, verbal fights and screaming matches, physical abuse, unacceptable personal hygenic practices, frequent discussions of the possibility of separation, and small-scale trial separations. Attributional analysis is reported to occur during this preseparation period, but the frequently exhaustive nature of conflict at this time seems to preclude the amount and intensity of attributional analysis that occurs when the relationship has reached the point of a major separation. When such a separation occurs, there is not only the time but also apparently a strong psychological need for frequent, intensive attributional probes.

A final, general point about attribution was implied in the introduction to this work. The type of attribution displayed by these individuals in their stressful situations is only remotely related by comparison to the bland type of attribution displayed by college students on experimental tasks. The type of attributions that we have studied is similar to that type studied by Orvis, Kelley, and Butler (1976), who emphasized the communicative, justificatory and defensive nature of the attributions displayed by their respondents.

In terms of more specific attributional conclusions, a strong parallel between the data in the conflict study and the data in the separation study was the participants' emphasis upon the role of outside romantic involvements as a source of difficulty. This factor received a relatively high attribution of importance by both males and females in the conflict study and was seen as representing one of the foremost problem areas in their marriages by persons in the separation study.

Although we could not directly investigate ideas about divergent causal perspectives in the separation study, it is clear that the participants felt that important areas of divergence obtained. They often thought that their ex-partner did not give enough weight to his/her negative behavior or characteristics in appraising their problems. Though the ex-partners indeed may have been negligent in their causal analyses, we must remember that the general tone of the participants' perception of their ex-partner was highly negative. All data considered, the ex-partner essentailly was imputed the greatest percentage of blame for the marital difficulties. Again, these data are similar to those reported by Orvis et al. (1976) for couples living together in their attributions about troubling behaviors in the relationship. Given the complexity of the accounts reported in the separation study, it seems unlikely that either partner had a good understanding of the extent of divergence in the relationship. To feel confident that they knew their own minds was difficult for many of our participants. And often they expressed even more uncertainty about their ex-partner's perspective and attributions.

THE IMPORTANCE OF SEX DIFFERENCES

We cannot speak with much confidence about sex differences in separation because of the female-dominated nature of our sample.[2] However, the overall pattern of sex differences in the first study and the individuals' reports of sex differences in the second study suggest that sex differences represent a factor of great importance in conflict and separation. Our separation data are consistent with those of other investigators (e.g., Hill et al., 1976) in indicating that women generally are more attuned to and perceptive of relational difficulties than men. Further, women appear to be quicker to conclude that relationships should be terminated. This conclusion may be quite difficult for the woman, because she may have much more practical security and a better standard of living if she stays with her partner. Nevertheless, women in their youth and middle-aged years seem increasingly capable of arriving at this conclusion.

Because of the continued existence of a double standard in extramarital sexual activity as well as other domains, men who are in troubled relationships generally may be insensitive to the seriousness of the problems, may not be willing to engage in much cognitive work to find out what is wrong while the relationship is still viable, then may be reluctant to terminate when discussion of this possibility becomes serious, and consequently may get "dumped" when the woman can no longer tolerate the difficulties in the relationship. Granted this line of reasoning is highly speculative, but it maps onto the *reports* of the participants in our separation study quite well. Hill et al. suggest that the person who is left in a broken relationship suffers the most depression and loneliness. We could not examine this possibility in our separation study. But if it is true and if the foregoing reasoning contains some truth, it is certainly ironic that males who typically possess more opportunity and power in this society may be more psychologically harmed by separation than are females. In this same speculative vein, we may predict difficulty in close relationships for males who have been given much exposure to so-called macho patterns of socialization that emphasize double standards, unbending masculine strength, and sexual superiority. Not only may these individuals be trained to hold incorrect attributions about their roles in close relations, but they also may not develop enough skill at working hard to analyze and understand the dynamics of their close relationships.

---

[2]This datum is itself intriguing and may relate in part to the discussion of socialization differences that follows. Generally speaking, men may feel that is is a sign of weakness to perceive the need to talk to others (even strangers) about matters that may be as embarrassing as separation. As noted previously, the majority of inquiries by potential participants about this research came from females.

CONCLUSION

Regardless of one's sex, people do not easily adjust to conflict and separation. And relational problems, including most vividly separation and divorce, are increasing greatly in scope and magnitude in this society. Social psychologists who are trained to theorize about and investigate peoples' cognitions, attributions, emotions, and motivations, as well as their interaction patterns, may be particularly well equipped to contribute to understanding close relationships. But at present our contributions to the separation and divorce area does not begin to match those of family sociologists (e.g., Goode, 1956), who have been working in the area for decades now. Unfortunately, there has been little direct emphasis on people's thinking, emotions, and motivations in this long history of work by family sociologists. It seems likely that problems in the area will be addressed most adequately by approaches that integrate ideas from family sociology, clinical psychology, and rich cognitive domains of social psychology such as attribution theory.

ACKNOWLEDGMENTS

This work was supported by grants from the Vanderbilt University Research Council and the Spencer Foundation to the first author. We are indebted to various scholars who have commented upon both the general plan of this work and its specifics, including: Charles Hill, Bill Ickes, Harold Kelley, Bob Kidd, George Levinger, Anne Peplau, and Zick Rubin. Also, we are grateful to the individuals who participated in this research and who shared with us some of the intimate and often painful details of their close relationships.

REFERENCES

Abelson, A. P. A script theory of understanding, attitude, and behavior. In J. Carroll & J. Payne (Eds.), *Cognition and social behavior.* Hillsdale, N.J.: Lawrence Erlbaum Associates, 1976.

Bem, D. J. Self-perception theory. In L. Berkowitz (Ed.), *Advance in experimental social psychology* (Vol. 6). New York: Academic Press, 1972.

Berscheid, E., & Walster, E. H. Physical attractiveness. In L. Berkowitz (Ed.), *Advances in experimental social psychology* (Vol. 7). New York: Academic Press, 1974.

Byrne, D. *The attraction paradigm.* New York: Academic Press, 1971.

Deaux, K. Sex: A perspective on the attribution process. In J. H. Harvey, W. J. Ickes, & R. F. Kidd (Eds.), *New directions in attribution research* (Vol. 1). Hillsdale, N.J.: Lawrence Erlbaum Associates, 1976.

Goode, W. J. *After divorce.* New York: Free Press, 1956.

Heider, F. A conversation with Fritz Heider. In J. H. Harvey, W. J. Ickes, & R. F. Kidd (Eds), *New directions in attribution research* (Vol. 1). Hillsdale, N.J.: Lawrence Erlbaum Associates, 1976.

Hill, C. T., Rubin, Z., & Peplau, L. A. Breakups before marriage: The end of 103 affairs. *Journal of Social Issues, 1976, 32,* 147–168.

Huesmann, L. R., & Levinger, G. Incremental exchange theory: A formal model for progression in dyadic social interaction. In L. Berkowitz (Ed.), *Advances in experimental social psychology* (Vol. 9). New York: Academic Press, 1977.

Jones, E. E., & Nisbett, R. E. *The actor and the observer: Divergent perceptions of the causes of behavior.* Morristown, N.J.: General Learning Press, 1971.

Kelley, H. H. *Attribution in social interaction.* Morristown, N.J.: General Learning Press, 1971.

Levinger, G. Note on need complementarity in marriage. *Psychological Bulletin, 1964, 61,* 153–157.

Levinger, G. Marital cohesiveness and dissolution: An integrative review. *Journal of Marriage and the Family, 1965, 27,* 19–28.

Levinger, G. Sources of marital dissatisfaction among applicants for divorce. *American Journal for Orthopsychiatry, 1966, 36,* 803–807.

Levinger, G. A social psychological perspective on marital dissolution. *Journal of Social Issues, 1976, 32,* 21–47.

Levinger, G., & Snoek, J. D. *Attraction in relationship: A new look at interpersonal attraction.* Morristown, N.J.: General Learning Press, 1972.

Orvis, B. R., Kelley, H. H., & Butler, D. Attributional conflict in young couples. In J. H. Harvey, W. J. Ickes, & R. G. Kidd (Eds.), *New directions in attribution research* (Vol. 1). Hillsdale, N.J. Lawrence Erlbaum Associates, 1976.

Parkes, C. M. *Bereavement.* New York: International Universities Press, 1972.

Peplau, L. A., Rubin, Z., & Hill, C. T. Sexual behavior in dating relationships. *Journal of Social Issues,* in press.

Weiss, R. S. *Marital separation.* New York: Basic Books, 1975.

Weiss, R. S. The emotional impact of marital separation. *Journal of Social Issues, 1976, 32,* 135–145.

Winer, B. J. *Statistical principles in experimental design.* New York: McGraw–Hill, 1971.

10 Attributional Strategies of Social Influence

Avi Gottlieb
William Ickes
University of Wisconsin

> *Since Eve and the Serpent began their fateful talk, the transmission of selected information has remained the most important means by which people have manipulated each other. It probably always will be. Education, prayer, rhetoric, propaganda, demagoguery, romantic seduction, and advertising are all typical efforts to this end, though they are only a fraction of the whole. All of them are forerunners of the technology of control by information, and some have grown sophisticated enough even to be considered bona fide parts of this technology.*
>
> (London, 1969, p. 35)

Although we would hardly dispute London's (1969) conclusion that the technology of information control and dissemination has become quite developed, we have some doubts about whether a parallel science has developed concomitantly, at least on the social psychological level. It is a matter of some concern to us that although manipulations of information are used constantly as the primary means of influence in social interaction, the factors involved in the selection, strategic use, and subsequent processsing of such information remain relatively obscure. We know very little about the relationships among the provision of information, the processing of this information, and the ultimate effects of this information on attitude and behavioral change — the very issues that the study of influence is presumably all about. Even the recently developing convergence of cognitive and social psychology (Taylor, 1976) has culminated merely in a restricted examination of the mechanics of information processing and has disregarded the questions of why this information was provided in the first place or what behavioral consequences may result from its reception.

261

This, of course, is not to say that social psychologists and sociologists have not recognized the centrality of influence processes in social interaction. Influence has been examined in such diverse areas as mass communication (Hovland, Lumsdaine, & Sheffield, 1949), interactions with bureaucratic and public institutions (Katz, Gurevitch, Danet, & Peled, 1969), transactions with professionals such as physicians (Svarstadt, 1976) or psychiatrists (Lewis, 1972), and day-to-day interactions involving ingratiation (Jones, 1964), impression management (Goffman, 1959), and other forms of "strategic interaction" (Goffman, 1969). In empirical, and particularly in experimental, investigations, however, disproportionate attention has been paid to those means of influence that involve manipulations of environmental contingencies such as the use of rewards, punishments, and coercion (Blau, 1964; Gergen, 1969; Homans, 1961; Thibaut & Kelley, 1959). The disproportionate emphasis given to these more "heavy-handed" influence strategies appears to have persisted in spite of the recognition by some authors that there may be quite severe limitations on the use of such strategies in social interaction. Not only do they require control over practically unlimited resources and continuous surveillance in order to assure compliance (French & Raven, 1959), but they are often considered illegitimate or proscribed by law, as is the case for bribery, assault, and extortion. Similar limitations apply to the verbal equivalent of these strategies, such as promises and threats (Tedeschi, 1970).

When rewards and punishments are inapplicable, inappropriate, or illegitimate, the most viable means of interpersonal influence is the use of verbal communication.[1] As McGuire (1969) has pointed out, persuasive communications might be categorized most conveniently on the basis of the distinctions made by Aristotle in his *Rhetoric*. Aristotle distinguished among appeals relying primarily upon the characteristics and personality of the persuader (*Ethos*), appeals relying on the emotional arousal of the audience (*Pathos*), and appeals relying on logical argumentation and the presentation of evidence (*Logos*). The concept of Ethos has received attention in the social psychological literature in the form of studies on source characteristics in persuasive communications (Kelman & Hovland, 1953). Closely related to the notion of Pathos is the research on fear communications (Janis, 1967; Leventhal, 1970), guilt arousal (Carlsmith & Gross, 1969), and the invocation of reciprocity norms (Gouldner, 1960).

There is, however, only sparse and indirect evidence that communications have any persuasive impact when they are contingent only upon the presenta-

---

[1]Facial expressions, gestures, body postures, etc. are also potential sources of influence and have been treated as such (Goffman, 1959; Hall, 1969; Laing, Phillipson, & Lee, 1966; Scheflen, 1974). These nonverbal aspects of influence are beyond the scope of this paper, however, and are not discussed here.

tion of information (*Logos*). Most of this indirect evidence comes from studies instigated by Carl Hovland and his colleagues in the Yale Persuasive Communications Research Project (Hovland, 1957; Hovland & Janis, 1959; Hovland, Janis, & Kelley, 1953; Hovland, Lumsdaine, & Sheffield, 1949; see also McGuire's review, 1969), which have demonstrated postcommunicative attitude change even when the communications contained emotionally *uninvolving* information (for example, on the issue of the economic principles involved in currency devaluation; Hovland & Mandell, 1952). Further corroboration for the notion that information, in and of itself, is a sufficient impetus for attitude change comes from Rokeach's studies on value change. Rokeach and McLellan (1972), (for example, have demonstrated that feedback about other person's value hierarchies, which, as we shall see, corresponds to consensus information in Kelley's (1967) attributional terminology, will produce changes in subjects' own values as well as in their behavior.

What are the processes by which such informationally induced changes occur? For Hovland et al. (1953), "The most relevant considerations . . . are those derived from the analysis of *motivational factors*, . . . [and] theoretical issues concerned with *learning factors* [p. 99, italics in the original]." Rokeach (1973) traces cognitive (value hierarchy) changes to the arousal of a state of self-dissatisfaction.

Obviously, in order for a communication to be motivating, induce learning, or cause dissatisfaction with the state of one's own value hierarchy, the information that is contained in the communication must first be processed. Furthermore, since the consequence of influence is typically conceptualized and measured as postcommunicative attitude change, it is important to determine on what basis attitude change may occur. A quite reasonable answer seems to be that an individual's attitude is based upon a particular perception of the situation to which this attitude pertains. The term "situation" is used here to denote what might be referred to more formally as the "stimulus complex," which is comprised of the stimulus proper (the object, event, entity, etc.) and the stimulus-surround (the "field" or "ground" in which the stimulus proper is embedded and phenomenally defined; see Heider, 1958).

If attitudes are based on a particular situational perception, attitude *change* can be seen as precipitated by changes in situational perception; and, in the case of persuasive communications, these changes in perception may well be instigated by information about the situation. We therefore propose that a two-stage causal sequence of cognitive processes is involved in influence: Specific bits of communicated information lead to changes in the attributor's perception of the situation. These changes, in turn, may generalize into a change in the person's attitude and/or behavior with respect to this situation.

What remains elusive in this causal sequence is the amorphous term "perception of the situation." In concurrence with Heider (1958), we think that it may be appropriate to substitute for this term the phrase "a set of causal inferences (or attributions)." When we try to understand the reality that surrounds us and the events we observe (including internal events; Nisbett & Valins, 1971), we tend to do so by causal analysis. Suppose, for example, we observe two people fighting on Main Street. An answer to the question "Why are they fighting? — that is, a causal explanation — will give us the subjective feeling that we "understand" the event. But even more importantly, this explanation may determine our judgment and evaluation of the adversaries and their fight: "Who cares!" if they are drunk; "Nice kids!" if they are friends in a mock fight; "Dirty criminals!" if they are Mafia members starting an underworld war. The particular causal explanation we formulate may also determine how we decide to react to the event — intervene, call the police, walk on by, etc.

The interrelationship of causal attributions, attitudes, and social conduct is fundamental to the social psychologist's interest in the causal inference process. This interrelationship is also fundamental to our understanding of the processes by which interpersonal communication and the dissemination of information can serve as potent strategies of social influence. If causal attributions are typically unstable or changeable in nature — and we will argue that they are — they can be altered by changing the information upon which they were initially based. If attributions are related to attitudes and behaviors, informationally induced change in attributions may lead to subsequent changes in attitudes and behaviors.[2]

We believe, therefore, that attribution theory can provide a parsimonious theoretical framework for the analysis of the processes involved in influence via verbal communications and information dissemination. The present paper attempts to develop such an integration between informational influence and attribution theory and will proceed as follows. First, we will briefly review Kelley's (1967, 1973) model of causal attribution, which will serve as the basis for the present analysis. After examining Kelley's (1967) attempt to apply his model to the influence process and pointing out some major ambiguities in his approach, we will develop a more differentiated model of "attributional strategies of social influence" and discuss some of its major aspects and implications.

---

[2]We will momentarily proceed under the simplifying assumption that the communicated information is perceived to be valid and accurate. The importance of the communicator's perceived intent and credibility as determinants of the magnitude of attitude change following persuasive appeals has been demonstrated repeatedly in the literature (Kelman & Hovland, 1953; Walster & Festinger, 1962; etc.). Other characteristics of the communicator (attractiveness, similarity, power, etc.), as well as the discrepancy between the position advocated in the communication and the "target's" initial attitude position (Hovland, Harvey, & Sherif, 1957) may also modify the impact of persuasive appeals. We shall return to some of these issues in our discussion of a proposed attributional model of social influence.

ATTRIBUTION AND INFLUENCE:
PAST INTEGRATION ATTEMPTS

The most systematic model of the attributional process, and the one we will draw upon most heavily here, was proposed by Kelley (1967, 1973). Kelley's attribution theory is concerned with the process by which the naive observer arrives at causal inferences.[3] Central to Kelley's model is his argument that causality is inferred by applying the principle of covariation. This principle, adapted from J. S. Mill's method of difference, postulates that a given effect is attributed to that one of its possible causes with which it covaries. According to Kelley (1967), such covariations can occur along three mutually exclusive, possibly exhaustive dimensions that comprise the "attributional cube" [p. 195]. (See Fig. 10.1.)

The three dimensions on which any given effect may vary are those of entities, time/modality, and persons. The entities dimension refers to the presence or absence of different entities or objects at the time the perceived effect occurs. The time dimension corresponds to the occurrence of the effect across a series of repeated observations, whereas the modality dimension reflects the constancy or variability of the effect across different modes of interaction with the entity or object. Finally, the persons dimension refers to whether different individuals in the same situation have experienced or observed a similar effect.

Thus, for example, a person may enjoy movie x (the effect to be explained) but also movies y and z (entities). Since the effect (enjoyment) is invariable across different entities (perfect covariation), it is not likely to be attributed to any one of them. Similar analyses can be applied to variations of the effect across the attributional dimensions. If the movie is enjoyed over repeated viewings (time) and/or when viewing it on TV, at drive-ins, and in theaters (modality), enjoyment will not be dismissed as accidental. If others enjoy the movie as well, the person will not attribute his or her favorable attitude toward the movie to his or her own idiosyncrasies.

Any given effect to be explained may or may not vary across any of the three dimensions. In theoretically "ideal" cases, each dimension exhibits either unique covariation or complete constancy of the effect. Such variations have different

---

[3]Kelley's model also applies to the naive analysis of non-causal relationships. Thus, the heuristics used to associate *causes* with *effects* are also applied when associating *attributes* with *entities,* antecedent *conditions* with *consequences, contingencies* with *outcomes,* etc. Although these relationships may be causal in the strict, scientific sense, they frequently reflect correlations or other non-causal associations. The reader should note that although we have chosen to use the cause–effect terminology to analyze the illustrative examples in this chapter, we are not necessarily suggesting that these examples reflect strictly causal associations.

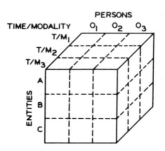

FIG. 10.1. Kelley's (1967) "attribution cube."

significance and meaning depending upon the attributional dimension on which they occur. For the least equivocal, most fundamental, causal inference to be made, the observed effect must be constant over time (movie *x* is always enjoyed) and must occur regardless of modalities of interaction (movie *x* is enjoyed everywhere). This invariability of the effect fulfills what Kelley (1967) has denoted as the *consistency* criterion. Moreover, the attribution of an observed effect to one single cause (object) requires that the cause and the effect be uniquely associated. Contrary to the time/modality dimension, which should ideally exhibit effect consistency, a unique covariation or effect *distinctiveness* is therefore desirable on the entities dimension (the effect of enjoyment occurs only with movie *x,* not with other movies). Finally the corroboration of one's own perception by others requires that they perceive and report a similar cause—effect relationship, that is, establish a *consensus* (everyone thinks movie *x* is enjoyable). Obviously, certain effects may not exhibit ideal patterns of complete constancy or unique covariation, in which case more complex analyses are required (Kelley, 1971).

Kelley's model has been applied primarily to social perception. The model stipulates that effects that are perceived to be nondistinctive, consistent, and nonconsensual will be attributed to a person. Causality for an event perceived to be distinctive, consistent, and consensual, on the other hand, will be attributed to a particular object or entity in the environment. In the first case, one person (but not others) enjoys all movies regardless of time and modality and might therefore be seen as a "film freak." In the second case, one particular movie (but not others) is enjoyed by everyone regardless of time and modality and might therefore be seen as a "good" movie. Any deviation from one of these attributional patterns is postulated to result in decreased attributional stability, as evidenced by reduced confidence in these attributions. Such deviations may also result in "unique interaction" attributions (for example, a *particular* person liking a *particular* movie).

The first extensive test of Kelley's attribution model was carried out by McArthur (1972). McArthur provided subjects with written information about hypothetical stimulus persons who were associated with hypothetical effects (for example, "John laughs at the comedian"). The information provided indicated whether the effect (John's laughter) varied on dimensions of distinctiveness ("John does not laugh at any/laughs at every other comedian"), consistency ("In the past John as almost always/never laughed at the same comedian"), and consensus ("Almost everyone/hardly anyone who hears the comedian laughs at him"). Subsequently, McArthur measured subjects' decisions regarding the cause of the described effect and their confidence in these decisions. The predictions stipulated by the model were largely borne out, with one major exception: A relatively low proportion of variance in attributions was accounted for by consensus information. The reliability of this observation and its theoretical significance are issues we discuss later in the chapter.

The important point about this study and similar others that were carried out subsequently (for example, Orvis, Cunningham, & Kelley, 1975) is the fact that *subjects made causal inferences that varied markedly as a function of the information communicated to them.* Due to the experimental procedure employed in these studies, respondents were called upon to make inferences about issues they had never considered before. The attributions they made in the process of reading the information provided were therefore novel in their content. But if the presentation of attributionally relevant information leads to the *establishment* of novel causal attributions, it is not unreasonable to assert that such information may also effect *change* in existing attributions.

Kelley (1967) in fact addressed this possibility in discussing the relevance of his theoretical model to the influence phenomenon. He proposed that the criteria of distinctiveness, consistency, and consensus can be used to substantiate the veridicality of the attributions made by the individual. Should a person's attributions not be fully supported by all of these criteria (in other words, should covariation not be perfect), causal inferences will be unstable (that is, more susceptible to change) and will be made with less confidence.

A person's "state of information," according to Kelley, can be defined as a ratio of the degree of attributional differentiation to the degree of attributional stability. Differentiation is a function of the distinctiveness of a given effect to a given entity, whereas stability is based upon the consistency of the effect over time and modality, and upon consensus. In an analogy to the F-ratio of an analysis of variance, differentiation can be seen as the between-conditions variance or "effect"; whereas stability across time, modality, and persons (consensus) constitutes the denominator or "error term." If both differentiation and stability are high (that is, distinctiveness, consistency, and consensus are all high), the level of information a person will feel (s)he has about an entity is, by definition, high; and the person will not attempt to acquire further information to increase confidence in his or her attributions. If this is not the case, however, information dependence will be experienced, and the person will be motivated to search for additional information to raise the level of attributional stability and confidence.

In short, if we denote a person's level of information as L, then according to Kelley:

$$L = f \left(\frac{\text{distinctiveness}}{\text{consistency, consensus}} \right) = f \left(\frac{\text{differentiation}}{\text{stability}} \right)$$

The logic of this reasoning suggests that L has to be at a certain indeterminate level that fluctuates depending upon the presence of another person who can provide information that increases it. This is the case because information dependence, like the conceptually similar notion of outcome dependence (Thibaut & Kelley, 1959), is treated as an interpersonal rather than an intrapersonal phenomenon. Thus, a person is informationally dependent upon another only when the latter actually can increase the person's level of information or is thought to be able to do so. Therefore, a person can exert influence only when (s)he can or is believed to be able to increase L, and then only by increasing the attributional stability (consistency and consensus) of the attributor (Kelley, 1967, pp. 199—200).

Although the concept of information dependence seems to emphasize the attributor's vulnerability to being influenced, Kelly encounters this impression by giving a similar emphasis to the influencer's vulnerability to being disbelieved. In his distinction between instruction [providing means to obtain attributional consistency by introducing new evidence, perspectives, and procedures (paraphrased from Kelley, 1967, p. 201] and persuasion [providing consensual information about the influencer's own conclusions], Kelley notes with regard to the latter strategy that "[The communicator's] *message is itself* an effect, and [the target's] problem is to attribute it, either to that part of their common environment under discussion (in which case it is considered valid), to [the communicator] himself (his role, desires, etc.), or to the situation or the target (. . . the particular circumstances) [pp. 201–202, italics in the original]." We suggest, however, that the judgments attributors make about the validity of communications are not limited only to cases in which consensus information is provided. On the contrary, informational validity is probably questioned to some degree whenever the information is provided second-hand from a communicative source rather than obtained from direct interaction with the environment. As Hill (1974) has pointed out, such information is nothing more than a public attribution, and there may be many reasons to doubt its accuracy (sincerity, competence, ulterior motives, etc.) — especially when it diverges from the receiver's private attribution.

Kelley's discussion of influence is very provocative, but, upon reflection, several ambiguities become apparent. First, Kelley chooses to base the initiation of the influence process on the concept of information dependence — that is, on a relative lack of information. Presumably, this dependence is the prerequisite for information seeking, without which influence cannot occur. It is evident, however, that people are also exposed to and influenced by communications that provide information they did not actively seek to elicit (for example, in advertisements, political campaigns, and frequently in the context of interpersonal relations are well). Unsolicited information may often be as influential as solicited information, since attributions are rarely stable and since someone who can provide new information is frequently at hand. Information dependence, then, seems to be the rule rather than the exception. Thus, though we would agree that the level of existing information might be one determinant of susceptibility to influence and active information seeking, information seeking itself does not appear to be a *necessary* condition for the success of persuasive appeals. We might note that one indirect corroboration for this argument comes from the Hovland et al. studies reviewed above, in which subjects changed their opinions after exposure to communications they did not actively seek but which they were experimentally induced to listen to or to read.

Our second divergence from Kelley's analysis relates to the attributional dimensions by which influence can be mediated. Kelley refers only to attributional stability (consistency and consensus) but fails to consider induced change in attributional *differentiation* — that is, change in the attributor's perception of effect distinctiveness. We believe that the provision of distinctiveness information may in fact be an important means of influence. Consider, for example, our perception that John is exceptionally competent. There are basically two ways that we could substantiate this inference. First, we could encounter situations that would provide direct experiential evidence on the dimensions of consensus (other people observing and reporting instances of his competence), consistency over time and modality (our repeated experience that John succeeds in everything he does, every time, and in any setting), and distinctiveness (when we observe others performing the same tasks, they make a mess of them). Second, we could gather evidence through verbal feedback or communication, such as a conversation with a friend. This conversation could provide socially mediated information relevant to any or all of the three dimensions. Thus our friend could say that he thinks John is competent or that John's name came up in a conversation among other people who all thought highly of his competence (consensus); he could report John's performance at certain times and in certain places about which we know nothing (consistency); and he could compare John's performance with that of other people whom we do not know (distinctiveness). There is no reason to believe that provision of the latter type of information — regarding the distinctiveness of John's performance — would be a less viable strategy of influence than any other type of communication.

Third, we see a little reason to believe that only information that *increases* either attributional stability or differentiation will mediate influence. Kelley's (1967) discussion of influence implies that people have to seek information in order to be exposed to influence, a notion that ignores the possibility that they may be exposed to influence by information they have not sought. Once we relinquish the notion that exposure to influence must be voluntary, we can see that it may frequently be in the communicator's interest to *reduce* the level of attributional stability that the attributor has already achieved. This, perhaps, is the implicit mechanism underlying attitude change that involves giving up a positive and adopting a negative position toward an object, or vice versa: previously established attributions are dismantled, and new, alternative attributions are consolidated. This notion is similar to Lewin's (1947a, 1947b) description of the three-stage process of attitude change: unfreezing a given attitude, changing it, and refreezing the newly developed attitude. Returning to our prior example, a hostile communicator might want to raise doubts about our evaluation of John as competent by informing us that (s)he and/or others think John is an incompetent slob (reduction in consensus), that he really blew it that day when he tried to repair Jane's car (reduction in consistency), and that when you see how Andy handles mechanical things, you will know what competence is really all about (reduction in distinctiveness).

We argue, then, that communications can effect the consolidation of previously established causal attributions by providing information that substantiates the person's perception of the situation on any or all of the three attributional dimensions. On the other hand, the person's confidence in pre-established attributions may be reduced, and these attributions may be changed by exposure to information that reduces the stability of prior attributions. Thus, both communications that stabilize and those that destabilize attributions may have a persuasive effect.

A fourth ambiguity in Kelley's (1967) statement should be mentioned, and it may be the most crucial one. Kelley's analysis leaves obscure precisely what the ultimate purpose of the influence process is. If the provided information is intended merely to change prior attributions, we would have little to add to what Kelley has said. However, this is not likely to be the case. Most people who engage in influence attempts probably try to change others' attitudes and behaviors as well as their attributions. Therefore, when we want to evaluate the effectiveness of a given communication, we would want to measure not only increases in attributional stability (by confidence ratings, for example), but also the degree of discrepancy between pre- and postcommunicative attitudes (McGuire, 1969, pp. 173–174), or the degree of postcommunicative compliance.

This last point raises the question of the nature of the relationships among attributions, attitudes, and behaviors, a question we address at some length below. For now, let us note that one might initially expect that information about only one of the attributional dimensions would alter only the perception corresponding to that particular dimension. For example, the communication that John once mounted the car engine in the trunk would probably lead to the inference that "he is not always competent" rather than to the perception that "he is definitely incompetent." These are both attitudinal statements, but they obviously differ in the extent to which they are qualified. In the least equivocal case, the communication would effectively lead to a change in prior perceptions on *all* attributional dimensions and allow no qualifications. The new attitude would then reflect the information that John is less competent than anybody else, at any time and place, and that everyone agrees with this assessment. In short — he is definitely incompetent.[4] Furthermore, these attributional changes may or may not have affective connotations ("I really despise John, that incompetent slob") or behavioral implications ("I'm leaving town tomorrow so I'll never have to encounter that incompetent slob John again"). Neither of these are necessarily inherent in the attributional change process itself, however.

ATTRIBUTIONAL STRATEGIES OF SOCIAL INFLUENCE: A MODEL

The preceding discussion of Kelley's (1967) attempt to apply the "attributional cube" to the influence process, and the criticisms we raised regarding this attempt, lead to what on the surface appears to be a straightforward attributional model of influence by verbal communication. Extending Kelley's reasoning, we argue that three types of information can be conveyed in persuasive communications. This tripartition corresponds to Kelley's attributional dimensions of effect distinctiveness, consistency, and consensus about the effect or about the cause—effect relationship. Correspondingly, information can be provided about:

1. *Distinctiveness* — the extent to which the described effect occurs only when a specific cause is present, or the extent to which a given object or entity is unique and different from other objects or entities;
2. *Consistency across time/modality* — the extent to which the described effect occurs regardless of when and where the observation is made;
3. *Consensus* — the extent to which others agree on the existence of the effect, or the extent to which they concur on the cause—effect relationship;
4. *Any combination of the above.*

Let us illustrate this attributional approach to influence with a relatively straightforward task of persuasion. Suppose that we want to convince the more naive among our departmental colleagues that the report of statistically significant findings (cause) is the basic prerequisite for publication in scientific journals (effect). Given this task, we might proceed by providing the following information.

---

[4]An entirely unqualified attitude is obviously only hypothetically possible. The concept is used here merely for purposes of explication.

1. Articles with statistically significant findings are published, whereas those with insignificant findings are not. (Published articles are *distinct* in that they contain significant findings.)
2. Statistically significant findings are prevalent in the published literature from 1970 to 1977, or in different scientific journals. (Published articles *consistently* report significant findings regardless of year and/or outlet.)
3. (a) We, ourselves, and others agree that statistical significance is a prerequisite for the publication of findings; *or* (b) we and others have experienced acceptance of articles with significant findings and rejection of articles with insignificant ones. In (a), *consensus about a cause–effect relationship* that is unrecognized by the attributor ("Other people agree with me, not with you") is conveyed. In (b), *consensus about a common experience* that diverges from the attributor's ("We have experienced this situation; either you haven't, or you have managed to misconstrue it") is conveyed.

As we have already pointed out, all of these communications may be rejected as invalid if the communicator is perceived to lack expertise or knowledge or if his or her perceptions are thought to be biased. If the attributor does not doubt the credibility of the communication, however, the persuasion process appears to be rather simple: We convey one or more of the delineated types of information, thereby changing our naive colleagues' attributions on the dimensions of effect distinctiveness and/or consistency and/or consensus, and thus establishing a novel perception of the situation. In less time than it takes to do a covariance analysis, our once-naive colleagues, who previously thought that scientific merit → publication, now believe that statistically significant findings → publication.

Things are, of course, not quite as simple as that. Although it may be valid to assert that communicated information about effect distinctiveness, consistency, or consensus will change the "target's" corresponding attribution, we believe that this notion is not adequate for a complete explanation of the influence process. First, and most importantly, we have so far failed to consider that the influence process frequently involves changes not only in causal attributions but in attitudes and behaviors as well. Second, there is reason to believe that the complexity of information processing in attribution cannot be represented by a simple one-to-one relationship between the conveyed information and causal attributions. The remainder of this chapter is devoted to the explication of a theoretical model of attributional strategies of social influence (Fig. 10.2), which attempts to address some of these more complex aspects of the influence process. Though the postulates advanced here have not yet been subjected to empirical test, they can for the most part be supported by indirect evidence that will be discussed when applicable.

The following aspects of the model can be derived from Fig. 10.2 and are discussed at some length later:

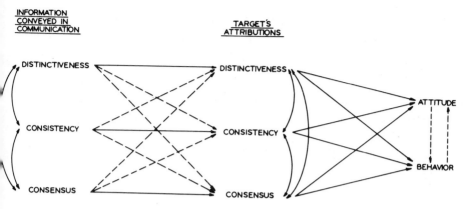

FIG. 10.2.   Attributional strategies of social influence — a model.

1. We will discuss the relationships among attributions, attitudes and behaviors, and the implications of these relationships for the influence process. We will argue that there is no necessary one-to-one correspondence between attributional and attitudinal change and that attributional change may lead to behavioral change both with or without a mediating change in attitude.

2. We will examine the interrelationships among communicated bits of information. We also consider interrelationships among causal attributions and the effects of different types of information on attributional change. As Fig. 10.2 indicates, we hypothesize that both the informational and the attributional dimensions are intercorrelated and interdependent and that any given type of information may affect all attributional dimensions (carry-over). From these notions, predictions regarding the differential impact of different types of information on attributions and attitudes will be derived.

3. Finally, we will discuss two issues that are not evident from the graphic representation in Fig. 10.2 but are crucial in determining whether the communicated information will have any impact in the first place. Both are aspects of the correspondence between the conveyed information and the influence "target's" cognitions: the relevance of the communicated information to the "target's" attributions and attitudes, and the discrepancy of this information from his or her initial "state of knowledge."

Before discussing these aspects of the model, one general caution should be introduced. Although it ideally would be desirable to apply path-analytic methods to the relationships hypothesized in Fig. 10.2, so that the direct and indirect informational effects on both attributions and the global attitude could be estimated, it is in fact doubtful that the model can be fully traced by empirical observation.[5] The major difficulty lies in establishing empirically the validity of the hypothesized temporal order of information presentation → attributional change → attitude change. Specifically, we would have to demonstrate that attitude change occurred following, not prior to, changes in attributions. Although the reversed sequence of causal ordering (attitude change → attributional change) is theoretically unlikely, it cannot be ruled out (it may, for example, occur in the case of emotional appeals). However, given the short timespan during which these processes occur, the problems of reactivity in repeated measurement, and subjects' general inability to report accurately on complex cognitive processes, evidence on the temporal order of these processes would be difficult to collect. These measurement limitations thus allow us only to examine the relationships among these cognitive processes (for example, by means of partial correlations), since they impose constraints on the specification of causal ordering.

Attributions, Attitudes, and Compliance

Attributions and attitudes.   When, in our earlier example, we informed our colleagues about the perfect covariation between report of significant findings and scientific publication, we induced a partial change in their "perception of reality" by providing information that presumably affected their attributions about the causal link between two observable variables. But when we speak about attitudes, we do not refer merely to the perception of causal relationships in our environment. Rather, there is an extensive literature on attitudes that claims that in addition to the "cognitive" component, which most closely resembles such causal perceptions, one should distinguish the affective and behavioral components of attitudes as well (Kahn, 1951; Katz, 1960; Ostrom, 1969; Rosenberg, 1956; Woodmansee & Cook, 1967). These three attitudinal components may be highly interrelated (Kahn, 1951) but should nevertheless be treated separately, particularly when the issue of attitude *change* is addressed.

Cognitive attitudinal change, as in our example, would probably suffice if we simply wanted to engage in an intellectual debate about the validity of certain factual statements. However, most persuasive communications attempt to go farther than that. We may want to persuade our colleagues to accept not only our interpretation of reality but also our value judgments about this reality. Our colleagues' agreement with the statement, "Publication decisions should not be made on the basis of the statistical significance of the findings," implies change in the affective attitudinal component as well as the cognitive one. If we

---

[5]In addition to the methodological problems of empirical traceability, there are also several statistical assumptions that would be violated if this model were subjected to path analysis. For example, the residuals are likely to be intercorrelated, resulting in an underspecified model.

further wish to impose our view that action should be taken to alter the presently applied criteria for publication, it is incumbent on us also to influence the behavioral component of our colleagues' attitude.

A multicomponent view of attitudes does not necessarily imply that, in addition to simple covariation information, persuasive communications must also contain emotion-arousing messages or indicate the desired response — although these characteristics may be conducive to attitude change. This view does imply, however, that attributional change will not necessarily lead to change in the affective or behavioral components of the attitude. We must recognize, for example, that attributional information can influence the *affective* attitudinal component only if the influencer and the attributor make similar value judgments about the effect. In some cases, such as when we convey information about the covariation of smoking and cancer, this requirement is easily met. Even the most suicidal among us are not likely to regard cancer as a desirable effect. But for many issues there may be no implicit consensus regarding the desirability of a given effect; indeed, there may be a variety of conflicting views on the subject. For example, the argument that information about private citizens should be computerized because this process increases their "identifiability" and "accessibility" is not likely to persuade those who do not consider this outcome to be a desirable one.[6]

Thus, the transition from attributions to attitude change may often require a transition from causal statements to valence statements about an entity's "attributes" or its "effects" (good, bad, etc.), and the prerequisite for this transition is a correspondence between the communicator's and the attributor's perception of effect desirability. In the absence of this tacit agreement, the conveyed information is ineffective and will further polarize the attributor's initial attitude ("That makes it even better/worse than I thought") or leave the attributor totally unaffected ("So what?"). The first of these responses can be illustrated by means of the above example. If computer technology, in addition to all of its other dehumanizing aspects, also robs us of our privacy, we should be even *more* opposed to it. The second response, "So what?", implies that the effect is affectively neutral and thus irrelevant to attitude change. The question of informational relevance and attitude change will be treated at some length below.

Compliance-gaining functions. Our discussion has so far been confined — mostly for reasons of parsimony — to the analysis of postcommunicative attitude change and has not addressed the issue of behavior change. This emphasis is shared by the long tradition of research on persuasive communications that

---

[6]However, if the attributor is aware that disagreement about effect desirability exists, (s)he may also be able to argue this issue (effect desirability) in strictly attributional terms.

we have already cited above. It is quite obvious, however, that the purpose of many persuasive appeals is not merely to change attitudes but to elicit certain behaviors or to gain compliance. It may not be sufficient merely to convince chain-smokers that smoking causes cancer, that cancer is bad, and that one should do something to prevent it; we may want them actually to stop smoking.

Although there is little reason to believe that the basic processes of attributional influence differ in behavior as opposed to attitude change, an array of extraneous factors and complex relationships has to be considered when one is predicting compliance. First, as indicated by the direct paths from attributions to behavior in Fig. 10.2, informationally induced changes in attributions may affect behavior directly — that is, behavioral change does not have to be mediated by attitude change. An example of this process can be derived from Tedeschi, Schlenker, and Linskold's (1972) concepts of mendations and warnings. According to Tedeschi et al., mendations stand in the same analogical relationship to promises as warnings do to threats; thus, both concepts refer to the communication of information about anticipated situational contingencies that are *not* under the communicator's control ("If you do *x,* I know that *y* will happen, although I have no control over the occurrence of *y*"). In other words, the communication brings certain *distinctive* features of a given situation to the attributor's attention. Tedeschi et al. (1972) reason that the person is then likely to behave so as to maximize (or minimize, depending on the desirability of *y*) the likelihood that these contingencies materialize. Thus, the warning, "If you shoot your neighbor, you will be tracked down, arrested, and incarcerated," may effectively check a behavioral impulse while having little or no impact on the attitude underlying it.

Whether attitude change will ultimately follow compliance in the case of mendation and warning communications is an open question. However, the notion that behavior may determine attitude, rather than vice versa — and that behavior change therefore leads to attitude change — is certainly not new in the social psychological literature (Bem, 1972). When attributional change leads directly to compliance, a reverse sequence may result in which attitude change is the final stage of the influence process, rather than the stage that mediates the informational impact on behavior. In such cases, attributional information about behavioral consequences over which the communicator has no control can nonetheless be used strategically to effect behavioral change.

However, even in those cases in which the compliance-gaining function of the communication is mediated by a prior change in attitudes, the relationship between attributional influence and behavior may be clouded by the fact that attitudes are not necessarily related to behavior (Schuman & Johnson, 1976; Wicker, 1969). Many factors extraneous to the present model, such as physical and social constraints on behavior, are among the important suppressors of the attitude–behavior relationship. Moreover, the notion that such constraints may override the attitude–behavior relationship suggests an interesting implication for our attributional analysis of the influence process, which can be illustrated with the following example: Suppose you are able to successfully convince your 16-year-old son, by means of attributional information, that smoking is dangerous since it causes cancer. Your son, however, paradoxically continues to smoke despite his nightmares about cancerous lungs; and it is not until you question him about this inconsistency that you discover that he thinks that all his friends regard smoking as an expression of masculinity. This, of course, is the problem: a social constraint is limiting the degree of attitude–behavior correspondence. But once this obstacle has been identified, it can potentially be removed by providing your son with additional attributional information – namely, that his friends actually think that smoking is a lousy habit too but suffer from pluralistic ignorance about each others' views.

Analytically speaking, in cases such as these, we confront attitude A as being unrelated to correspondent behavior B because of some perceived constraint C (a social or physical constraint, or a conflicting attitude). The provision of additional attributional information regarding C may be employed strategically to attenuate or eliminate its impact and thus establish the desired A–B relationship. In principle, of course, the successful operation of this process requires knowledge, if not insight, about the form and nature of such constraints. Social psychologists interested in identifying them have not come very far, and lay persons involved in social influence have probably advanced only slightly more in this respect.

Attributional Information:
Carry-Over, Overlap, and Effectiveness

As we have tried to demonstrate in the preceding section, the influence process involves more than a change in causal perceptions. Once we have recognized that influence ultimately results in attitudinal or behavioral change, it becomes possible to engage in a more extensive analysis of the causal sequence of information provision → attributional change → attitude change. In particular, we may want to examine the ultimate effects of different types of attributional information on attitudes and/or behavior. We believe, however, that such an analysis requires an understanding not only of the processes by which causal attributions are transformed into the various attitude components, but also of the prior information-processing mechanisms involved in changing one's attributions after given bits of information are communicated. We now turn to these information-processing aspects of attributional influence. Following this discussion, we will propose a set of more general hypotheses regarding the effects of different types of information on attitude change.

Carry-over effects. As indicated in Fig. 10.2, the model postulates that any given type of attributional information affects not only the attribution to which it corresponds, but noncorrespondent attributions as well (Fig. 10.2, broken arrows). Some empirical evidence in support of this assumption can be found in a study by Orvis, Cunningham, and Kelley (1975). These authors have shown that attributors who are given only partial information about cause–effect covariations will make fairly predictable inferences with respect to covariations about which they have not obtained any information. For example, subjects who were informed only that a given effect was consensual inferred that it would also be highly consistent and distinctive.

To take a second example more relevant to the social influence context, let us assume that a given communication contains only unidimensional information on effect distinctiveness: "Relative to all alternative programs, medical services under the auspices of a National Health Insurance (NHI) system will reduce existing consumer costs." Note that such distinctiveness information implicitly conveys a comparison: Cost economy is a feature of NHI but is not a feature of the present system or any other system. We suggest that although the communication conveys only comparative information about distinctiveness, attributions about effect consistency and consensus will also be made. Thus, the attributor is likely to infer that consumer costs will remain low and that there is some agreement on the validity of the statement that NHI reduces costs. A similar rationale applies to analogous carry-over effects from consistency and consensus information to their noncorrespondent attributions. For obvious reasons, the correspondent attributional effect should in any event be stronger than the noncorrespondent effects will be. To say that any given attributional information is most likely to affect the particular attributional dimension it addresses is almost a tautological statement.

In considering the possible determinants of such carry-over effects, one could argue that at least part of the change in noncorrespondent attributions is carried not by the information itself but by the interdependence among the three attributional criteria (as denoted by the correlational paths among attributions in Fig. 10.2). If we are informed that NHI entails reduced costs (as compared to other systems), but that costs are likely to increase subsequently and that cost-reduction does not apply to the poor and disadvantaged, we are likely to conclude that there is nothing distinctive about this effect after all, since it is inconsistent across modalities and disappears over time. Similarly, we might infer that in this case there cannot possibly be any widespread consensus about cost-reduction.

If such attributional interdependencies in fact occur, they would be reminiscent of similar notions of internal cognitive consistencies developed elsewhere (Heider, 1958; Osgood, Suci, & Tannenbaum, 1957). This hypothesized tendency for different attributions to be consistent with each other — to form an attributionally "good" gestalt — further suggests that attributors may exhibit informational closure in their processing of communications, implicitly completing patterns of data for which information is not provided on all attributional dimensions

(Orvis et al., 1975). Further exploratory work along these lines is currently being conducted by Gottlieb (in progress) and West (personal communication).

Note that the model specifies intercorrelations not only among the target's attributions, but also among the categories of communicated information (Fig. 10.2). The rationale for this "shading over" is basically that the three types of attributional information are not as distinct as they may appear to be. This argument will be developed more fully in the following section.

Overlap in attributional information. We have just suggested that there may be overlap in the attributional information conveyed by a given communication. That is, information ostensibly correspondent with only one attributional dimension may often be informative regarding other attributional dimensions as well. In order to explain why this may be the case, it is necessary for us to first examine some important differences between information that is obtained first-hand in a direct, experiential manner and information that is obtained indirectly by means of social influence.

There are obviously numerous differences between the information conveyed by direct observation of cause—effect relationships as opposed to information contained in communications about these covariations. One of these differences has already been mentioned in our discussion of Kelley's (1967) model. Any communication, whether it contains information about effect distinctiveness, consistency, or consensus, has to be subjected to secondary or higher-order attributional analyses in order to determine its validity. Direct experience with the effect, on the other hand, does not require such analyses. Thus, if we observe that the chemical product z is produced by a chemical interaction of x and y (a distinctive effect, since z is not produced by any other interaction), we are likely to make a distinctiveness attribution with some confidence. If the same information is conveyed verbally by a poet, however, we had better first find out how (s)he got the information before we attempt to draw any inferences from it.

A second difference between experienced and communicated information, which we will discuss at some length, is this: When cause—effect covariations are observed by the attributor, the distinctions among the three attributional dimensions are relatively clear-cut. However, when the same information is conveyed verbally, these distinctions are likely to be obscured. Having personally observed the chemical reaction of $x + y = z$ and noted its distinctiveness $(a + b \neq z; x + m \neq z;$ etc.), we know perfectly well that we have *not* experienced a consistent covariation (always $x + y = z$), or consensus about the cause—effect relationship (everyone observes $x + y = z$). This is not to say that we will not make such noncorrespondent *attributions;* carry-over effects and the tendency toward informational closure may still obtain. But the dimensionality of the *information* derived from our experience is fairly clear-cut and unequivocal.

Contrast this clear separation among attributional dimensions with the ambiguity that is inherent in the following two communications:

1. "Only the combination of chemicals x and y results in z."
2. "I think that only the combination of chemicals x and y results in z."

Ostensibly, the first statement pertains only to effect distinctiveness; in the second statement this distinctiveness information appears to have been converted into consensus information by indicating that effect distinctiveness is a matter of personal opinion rather than objective truth. But the distinction between these two types of information is obscured in the everyday communicative process. Indeed, the distinctions among *all three* types of information are frequently blurred in everyday discourse, both in the encoding and in the decoding of such messages.

The first source of attributional ambiguity occurs when the message is encoded by the communicator. Ambiguity may result because communicators often transmit messages that are linguistically equivalent to Statement 2, although they are really providing the information contained in Statement 1. In other words, we frequently attach disclaimers ("This is only my personal opinion") to our objective reality statements — in part because absolute truths rarely exist and in part because it is socially more acceptable to convey an impression of modesty. The effect of this linguistic tendency is that many apparently "consensus" communications are not really intended as such. Ambiguity in encoding may also result when the opposite process occurs, for there are many occasions in which communicators express conjectures and opinions as if they were matters of fact (most educators, no doubt, have had this experience frequently). In these cases, the messages will be linguistically equivalent to Statement 1, although they really provide the information contained in 2. Because communicators frequently fail to say what they mean and mean what they say, the attributional information conveyed in their communications will often be ambiguous.

The second source of this ambiguity occurs when the information is processed by the attributor. The results of this decoding process, however, may not be strictly parallel to those above. We suggest that attributors may have a pervasive tendency to construe *any* communication as conveying 2, although it may be linguistically parallel to 1. That is, even if the communicator explicitly argues that the effect is distinctive or consistent, the attributor is liable to take this as a personal opinion. In terms of Kruglanski's (1975) distinction between exogenous and endogenous attributions, one could conjecture that any influence attempt that is recognized as such will be perceived as a means that mediates a further goal rather than as an end in itself; for such "exogenously" attributed actions, "knowledge of their content does not reveal their underlying intentions [p. 392]." Typically, if the communicator's intentions are unknown, the veridicality of his or her statements will be questioned. This is precisely the reason for secondary attributional analyses of *any* communicated information — such information is mediated by a social agent whose perceptions are, for many reasons, questionable. This is not to say that the opposite process will never occur, for what is offered as opinion will at times be taken as fact. But although this latter process may characterize some influence situations and some attributors (for example, the young, the innocent, the gullible), we contend that the former process is represented much more frequently in typical influence situations.

Of course, the observation that attributional dimensions are overlapping and obfuscated in everyday communications should not cause the social scientist to despair, since these difficulties can be circumvented in controlled situations such as the laboratory experiment. Communicator motives, credibility, expertise, etc., can be controlled for, and subjects' perceptions of these factors can be elicited. The distinctions among attributional categories of information can be deliberately augmented (clean manipulations), and the correspondence between the conveyed information and subjects' perceptions of it can be measured (manipulations checks). As in most controlled experiments, the purpose of removing these reality-based ambiguities by control and measurement is not to replicate social reality but to test hypotheses derived from a theoretical model. This brings us to our next topic, for the most central hypotheses in the present framework concern the potential differences among the three categories of attributional information in their impact on attitudes and attitude change. In fact, it is the communicator's implicit or explicit awareness of these differences as (s)he strategically selects the information to be presented that makes attributional strategies of social influence possible.

Informational effectiveness.   The question of the differential effectiveness of persuasive communications in changing attitudes is obviously not new. Research by the above-mentioned Yale Communication Project examined the impact of persuasive communication by considering characteristics of the source (credibility, attractiveness, power), the target (persuasibility, intelligence), the "destination" or time dimension (temporal decay, delayed effects), the channel (mass media, face-to-face interaction), and the message. Of these factors, the target, "destination," and channel are presently of no concern; and though we have addressed source characteristics tangentially above, we will not pursue them further here. Hovland et al.'s analysis of message characteristics seems to come closest to the current considerations, but this research addressed only the effects of certain formal structural characteristics of the communication such as omission or inclusion of conclusions (Hovland & Mandell, 1952), refutation or ignoring of opposing arguments (Hovland, Lumsdaine, & Sheffield, 1949), and order of presentation of pro-and-con arguments (Lund, 1925). In contrast to these message characteristics, we are presently concerned with the effects of the *content* of communications – to be specific, with the attributional information contained in them. If our categorization of attributional information is viable, we may ask whether, for example, we could expect more attitude change after distinctiveness versus consensus information, or vice versa.

As we noted in our discussion of Kelley's (1967) model, the degree of post-communicative attitude change is generally inferred from pre–post differences in attitude. These differences in attitude are typically measured by unidimensional, Likert-type rating scales anchored at definitely-agree–definitely-disagree points. Such differences might be represented more accurately, however, if they were measured in terms of the number of qualifications that respondents impose on their global attitude statements. From this perspective, which would allow us to take a more process-oriented approach to communication effectiveness, the extent to which different types of attributional information affect global attitudes would depend on the extent of informational carry-over to noncorrespondent attributions. The extensiveness of carry-over may be a function of the type of information conveyed. The greater the number of attributions that are altered by unidimensional information, the less qualified the resulting attitude will be – thus leading to attitude change of greater magnitude.

The possibility that distinctiveness, consistency, and consensus information may be differentially effective in changing attitudes is suggested by prior attributional literature. Applications of Kelley's (1967) model to social perception have repeatedly found that consensus information accounts for a relatively small proportion of the explained variance in subjects' attributions. McArthur (1972), for example, reports that consensus information accounted for only 2.9% of the attributional variance, whereas distinctiveness and consistency information explained approximately 10% and 20% of the variance, respectively (Table 2, p. 172). Similar findings are reported by Orvis, Cunningham, and Kelley (1975) and McArthur (1976). Some caution must be exercised in interpreting these results, however, since they all may be due to a procedural artifact. All three studies presented consensus information first or second in the consensus–consistency–distinctiveness sequence. When the order was randomized in a recent study, the differences in the variance explained by the three types of information disappeared (Ruble & Feldman, 1976).

However, although these direct tests of Kelley's model may be criticized on methodological grounds, their results have been independently replicated in other research settings. A number of studies also suggest that consensus information has little effect on attributions, self-perceptions (Nisbett, Borgida, Crandall, & Reed, 1976), and behavioral predictions (Nisbett & Borgida, 1975).

Even more importantly, consensus information seems to be most easily discounted when subjects also have direct information about the stimulus (Feldman, Higgins, Karlovac, & Ruble, 1976). Though this finding is particularly reminiscent of Festinger's (1954) notion that physical reality takes precedence over social reality, it is also quite compatible with Kelley's (1967) argument that consistency criteria (time/modality) may be more important than the consensus criterion, and with Kahneman and Tversky's (1973) finding that base rate (consensus) information has virtually no effect when additional data are given to subjects.

The notion supported by all of these arguments and data is that when both direct (physical) and indirect (communicated) information is available, the former will carry more weight. When applied to the influence context, this seems to be a rather common-sensical notion. Having observed the chemical reaction $x + y = z$ consistently time after time, no one in his right mind would change his causal attributions if we were to inform him that we think that $x + y \neq z$.

Less common-sensical is our suggestion that this notion is applicable even if no direct physical corroboration of the communicated information can occur. In order to understand why this may be true, the nature of the effect itself should be considered. We suggest that it is more important to distinguish between effects that, *in the abstract,* do or do not have a physical referent than it is to distinguish between effects that the attributor, *in situ,* can or cannot physically experience. For example, if we wish to dissuade you from moving to California, it may be quite effective for us to argue that California has experienced more earthquakes than any other state in the United States and that earthquakes have consistently occurred there at average intervals of n years. It may be much *less* effective, however, for us to inform you that Frank Faultline also thinks that California is earthquake-prone. This difference in the effectiveness of our two communications may occur despite that fact that you do not have any direct experience with the absolute or relative frequency of California earthquakes.[7] We suggest that it occurs because the particular issue or effect we are concerned with here can be designated as *physically defined.* We believe that distinctiveness and consistency information will be more persuasive than consensus data for any communicated effect that is defined in strictly physical terms.

In contrast, many other effects can be designated as *socially defined,* since they are not corroborated easily by physical evidence. Many culturally-determined issues or effects (standards of beauty, matters of taste, fashion, etc.)

---

[7]This example does not represent a generic causal statement, since it is obviously not the state of California, but geological processes at its geographical location, that cause earthquakes. This distortion is not crucial for the present purposes, however, since it is unreasonable to assume that naive attributors will always trace a complete chain of causal relationships. For example, sailors attribute the rocking motions of their boat to waves, not to the gravitational forces of the moon that cause the waves.

fall within this category. We argue that in the absence of unequivocal physical referents, the social definition of cause—effect relationships — as conveyed by means of consensus information — will be more potent than the provision of information about consistency and/or distinctiveness. For example, if we wish to convince you that St. Laurent is a marvelous designer, it may be more viable to provide information about all the famous, well-dressed people who think so, rather than to inform you that St. Laurent's designs are better than others (distinctiveness) or that he has repeatedly won awards for his creations (consistency). The latter two items of information are of course not unimportant, but since a designer's reputation is often defined more by social consensus than by objective criteria, they may be less persuasive. A study by Hansen and Lowe (1976) may provide some evidence for this notion. In this study, subjects based their dispositional attributions for other persons' musical preferences (socially defined effects) on consensus rather than on consistency information.

We reason, then, that information about the distinctiveness and consistency of an effect may be more persuasive for physically defined issues, whereas consensus information is more likely to generate attitude change for issues that are socially defined. This argument suggests that the strategic use of attributional information should be based on the distinction between physically versus socially defined issues.[8] It also implies that when more than one category of attributional information is provided — which may frequently be the case — the impact of these multiple-information categories on attitudes will not necessarily be additive. For physically defined issues, distinctiveness and consistency information may well produce additive effects, with combinations of these types of information generating more attitude change than either one of them alone. However, the addition of consensus information to such a message may add nothing at all to its persuasiveness. Thus, the knowledge that frequent earthquakes both are distinct to California and occur there consistently may be more important than knowledge about either one of these covariations, but Frank Faultline's opinions may matter not a whit. The picture for socially defined issues is reversed: Information about the celebrities' opinions may be most conducive to persuasion, whereas distinctiveness or consistency data about St. Laurent's talent may not add much impact.

---

[8]We should note that the present analysis does *not* pertain to contexts in which influence is effected by forces other than the informational value of the communication. For example, consensus information may be more effective regardless of the issues addressed by the communication if the influence "target" is motivated primarily to gain rewards or avoid punishments (for example, Kelman's [1958] "compliance") or to establish a favorable relationship with the other person (for example, Kelman's [1958] "identification," and Deutsch & Gerard's [1955] "normative influence"). When influence is purely "informational" (Deutsch & Gerard, 1955) and the influence target merely perceives a need to restructure his or her knowledge (Katz, 1960), variations in informational effectiveness due to the nature of the underlying issue are more likely.

Two final points about physically versus socially defined issues should be mentioned. First, the physical/social elements should probably be viewed as the endpoints of a continuum and not as the mutually exclusive classes of a dichotomy. In fact, we err frequently in our assumption that much of our reality is factual rather than grounded in social consensus. Our unquestioning acceptance of many social taboos as "facts" having a physical rather than a strictly consensual basis is only one example of this bias. Secondly, issues that are defined exclusively or primarily by physical reality are often of less interest than socially defined issues in the context of influence and attitude change. We are unlikely to have any strong attitudes toward Einstein's relativity theory; but whether our children should be bused to school to promote racial integration or whether marijuana should be legalized has not only cognitive but also affective and behavioral ramifications. Dramatic exceptions to this generalization may occur, however, when a physically defined issue suddenly has major implications for issues that were previously socially defined. Witness the impact of "the pill" and "the bomb."

Information and Cognitions — Relevance and Discrepancy

We have argued so far that the provision of information may be conducive to attitude and behavior change by influencing the "target's" perception of the situation upon which these attitudes and behaviors are based. Further, we have suggested that different types of attributional information may be differentially effective as means of influence, depending on the particular kind of issue to which the communication pertains. It is obvious, however, that information is not necessarily conducive to influence, and not only because of communicator characteristics such as perceptual biases, lack of knowledge or credibility, etc. We now intend to discuss two additional factors that may determine the success of influence by the provision of attributional information. Both are related to the correspondence between the content of the communicated information and the attributor's prior cognitions. The degree of *discrepancy* between the provided information and the attributor's prior knowledge, as we will see in the next section, is conceived as a function of the origin of the communicated information in relation to the attributor's "domain of attributional knowledge." The question of informational relevance, which we will now examine, is somewhat more difficult to trace, which may be one of the reasons that — in contrast to the issues of communicator credibility and communication discrepancy — it has not often been addressed in the literature.

Informational relevance.   Any given bit of information may or may not be relevant to either the attributor's attitudes or his or her attributions. As we will see, the mechanisms involved in attitudinal and attributional relevance are somewhat different, but the end result — the failure of the influence attempt — is basically the same.

Regarding attitudinal relevance, one major analytical difficulty lies in our inability to specify a priori which information will be relevant to the person's attitude and which will not be. An example might help to clarify this point. Suppose that I want to provide a series of distinctiveness statements about Bob in order to persuade you to terminate your friendship with him. After some detective work I have discovered that Bob has been prosecuted for shoplifting, is a compulsive gambler, and has had to repeat 2 years in high school. Although all these statements may affect your attributions about Bob, none of them will necessarily change your attitude toward him. The attributional change activated by my information simply may not have anything to do with your attitude toward Bob, which is determined not by his criminal tendencies or stupidity but by his social skills and his willingness to provide you with manual help when you need him (issues about which I did not provide any information).

Thus, information may change a person's attributions without being necessarily relevant to his or her attitudes. It is important to recognize, however, that information may be irrelevant not only to the person's attitude but also to the very attribution(s) it is supposed to change. We suggest that whereas "attitudinal irrelevance" results from communication errors related to the attributional *domain* (distinctiveness attributions regarding shoplifting, etc., have no function in determining the attitude toward Bob), "attributional irrelevance" results from communication errors related to the *level* of attributional comparisons. Another example might serve to clarify the latter point.

Suppose it is 1979. We have agreed to work for Carter's next presidential campaign. The location: Libtown, featuring a relatively liberal constituency. We decide to inform the voters that Jimmy is really quite a liberal and to exemplify this point by emphasizing his civil rights record. We therefore provide distinctiveness information about Carter's civil rights activities during his first presidential term. The information is probably *attitudinally* relevant, since the described activities are important for this particular constituency's attitudes toward Carter. The information may or may not be *attributionally* relevant, however, because distinctiveness statements require that we provide some comparison(s) (Carter, but not others, promote civil rights); and the nature of these comparisons is crucial to informational relevancy. If we compare Carter's acts to those of Attila the Hun, Adolph Hitler, and Joseph Stalin, our constituency may have some trouble attributing liberalism to our candidate. Obviously, the compared politicians have to exceed a certain level of liberalism to make the comparison (distinctiveness) information meaningful.

A similar logic applies to the attributional relevance of consensus and consistency information. If Goldwater and Nixon agree that Carter is in the vanguard of the radical left, their consensus still may not convince the folks in Libtown, since the invoked consensus must come from persons with perceptions similar to the attributor's own. Conceptual support for this notion has been developed by Wells and Harvey (1977), who demonstrated that individuals do base their behavioral predictions on consensus data — but only when these data were described as taken from a random or representative sample. In other words, in order to draw inferences from consensus data, the persons who provide this consensus must at least be representative of the population about which I am required to draw these inferences. In a similar fashion, consistency data are also irrelevant when they are derived from unordinary circumstances. If the friendly cop on the beat ordinarily will not hurt a fly but on three occasions shoots someone in order to save innocent lives, we still have no evidence for a basic killer instinct.

In sum, attributional information is irrelevant and therefore not persuasive in two instances. The information is *attitudinally irrelevant* when the causal attributions affected by it do not pertain to the attitude that is intended to be changed. The information is *attributionally irrelevant* if the events, objects, or entities it describes lie outside that range of covariation that the target considers acceptable as the basis for formulating his or her attributions.

Information discrepancy and influence tactics.   The communication-attitude discrepancy has ordinarily been viewed in terms of the discrepancy between the influencee's attitudinal position prior to the persuasive communication and the position the communicator desires to establish (for example, Hovland, Harvey, & Sherif, 1957; see McGuire, 1969, pp. 217–224, for a review). There is, however, an additional dimension of discrepancy that is specific to our concern with attributional influence. This aspect is related to the extent to which the communicated indices of covariations — that is, additional entities or instances for which the described effect obtains, or additional persons who agree on the existence of this effect — *have been considered previously by the attributor.* This aspect of informational discrepancy refers to the distinction between attributional information that can be derived from the influencee's subjective "attributional cube" and information that is presently extraneous to it. In contrast to the communicator's choice among *types* of attributional information (distinctiveness, consistency, and consensus), which are relevant to his or her attributional *strategies* of influence, we can refer to his or her choices among origins of information as *tactics* of influence. Attributional strategies of influence, then, refer to the communicator's selection of general types of information based on their relative effectiveness when applied to specific types of attitude issues. Attributional tactics of influence, on the other hand, refer to the communicator's selection of the specific sources of such information relative to the attributor's existing "domain of knowledge."

To illustrate the use of such tactics, let us now take a look at a modification of Kelley's attribution cube in Fig. 10.3. Kelley and we both use the attributional cube primarily as a heuristic device, but it can also be seen for purposes

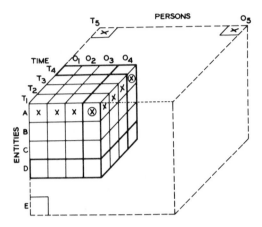

FIG. 10.3. A modification of Kelley's (1967) "attribution cube" illustrating the influence tactics of *interpolation* and *extrapolation*.

of explication as representing an intrapersonal "domain of knowledge." Our hypothetical attributor, according to Fig. 10.3, has the following covariation information regarding a given effect x:

1. x occurs in the presence of entity A, but not in the presence of B or C (distinct).
2. x occurs in the presence of A at times T_1, T_2, and T_3 (consistent).
3. Persons O_1, O_2, and O_3 agree on the occurrence of x in the presence of A (consensual).

The emphasized lines added to the cube surfaces in Fig. 3 indicate that our hypothetical attributor has often wondered, but does not know, whether the following covariation data also obtain for effect x:

1a. x does not occur in the presence of entity D.
2a. x occurs in the presence of A at times T_4.
3a. Person O_4 agrees on the occurrence of x in the presence of A.

Now, if we provide information about D, T_4, O_4, or any combination thereof, we merely add covariation data that the attributor wanted to obtain anyway.[9] We thus offer new covariation data that does not exceed the attributor's attributional "planes" (Fig. 10.3). The derivation of information from within the

---

[9]Consistent with our notion that stabilization as well as destabilization of attributions is persuasive, it is obviously possible to provide information that is *in*consistent with the attributor's covariation data. One could communicate that x *does* covary with D, does *not* covary with A at T_4, and that O_4 has *not* observed x in the presence of A. The argument developed here applies to both attributional stabilization and destabilization.

attributor's inquiry domain, but extraneous to his or her present knowledge, is referred to as the *interpolation* tactic.

A second tactic — differential *salience* — is related to interpolation, since it also makes use of information within the attributor's existing domain of knowledge. It differs from interpolation, however, in that the communicator merely gives a differential emphasis to particular covariation data currently available to the attributor, rather than providing any new information. The fact that differential attention and informational salience are important determinants of the attribution process has already been demonstrated in the literature (for example, Arkin & Duval, 1975; Duval & Hensley, 1976; Duval & Wicklund, 1973; Regan & Totten, 1975; Storms, 1973; Taylor & Fiske, 1975; Wegner & Finstuen, 1977). In the influence context, this implies that the influencer, though still operating within the given domain of knowledge as in interpolation, can selectively emphasize given aspects of this domain, or de-emphasize others. If the attributor infers lack of covariation on the persons dimension (O_1, and O_2 have reported x in the presence of A, but O_3 and O_4 have noted its absence), it may be effective either to increase the perceived salience of the consensus among O_1 and O_2, or to distract the attributor's attention from the lack of perfect covariation, or to do both.

These tactics may be particularly effective when two causes covary with almost exactly the same effects or when two objects or entities covary with almost exactly the same attributes. For example, car buyers who try to decide between two essentially similar vehicles are often easily convinced by their dealers to buy car A, which provides better gas mileage, over car B, which provides more passenger room. Given the rising gas prices and the declining birth rate, it seems tactically most effective to increase the salience of the gas mileage effect (energy crisis) or to dismiss the importance of passenger room (small family). Note that in doing so, the relative perceived relevance of these two attributes is also altered.

It should be clear from our examples that both interpolation and salience tactics operate within the person's existing attributional domain. However, the communicator is not limited to only these tactics, since (s)he can also provide data about covariations that the attributor has not considered previously. Information of this type may contain statements to the effect that x does not occur in the presence of entity E, that x occurs in the presence of A at time T_5, and that person O_5 has also observed x in the presence of A (see Fig. 10.3). The difference between the previously described tactics and what we will label here *extrapolation* is that this last tactic invokes entities, instances, or persons that are completely novel to the attributor. If our attributor has always wondered about Frank Faultline's opinions about Californian earthquakes and we tell him that Faultline in fact thinks that California is earthquake-prone, we apply interpolation tactics. But if our attributor never heard about Frank Faultline in the first place, the consensus data are extrapolated.

Everyday examples of extrapolation tactics are rare, particularly regarding entity and time covariations (that is, distinctiveness and consistency attributions). It may be relatively difficult to find new entities or instances with which a given effect covaries but of which the attributor is completely unaware. This situation may be less esoteric in science, however, where new discoveries and new evidence sometimes challenge established scientific theory and ultimately lead to the introduction of a new paradigm (Kuhn, 1962). Scientific research thus produces evidence that theoretically important effects do or do not covary with novel entities and instances and promotes new professional agreement or disagreement about these covariations. But as Kuhn's analysis demonstrates, "one-shot" information about such covariations is often insufficient to change scientific paradigms and may frequently be unsuccessful in changing lay theories and lay attributions as well.

When is such information likely to change lay attributions? Extending Kuhn's (1962) conceptualization, we would predict that replications and some accumulation of the new evidence are necessary. Once this has taken place, the dimensions of the attributional cube have been extended to include E, T_5 and O_5, and the communicator can return to the use of interpolation tactics. Thus, the discrepancy between the knowledge domain from which attributional information is taken and the attributor's present knowledge domain typically has to be reduced and bridged in order to increase the effectiveness of persuasive communications.

A final influence tactic, *informational contrast,* is based on the relationship between extrapolation tactics and attributional relevance. As we have pointed out in our previous discussion of relevance, consensus information pertaining to persons who are dissimilar from oneself is unlikely to be persuasive (for example, Nixon's opinion that Carter is an extreme liberal). However, attributional irrelevance may sometimes be turned to the influencer's advantage. This may be especially true in the case of "persons" covariations (consensus) and particularly when the attributor features him- or herself as one of the persons on the consensus dimension. Suppose you regard yourself as a liberal but are opposed to busing since you think it promotes violence instead of racial integration. You have consensus data (agreement) from other liberals similar to yourself. One bright morning you discover that members of the American Social–Nationalist Party, the majority of whose opinions you actively detest, are also in complete agreement with you. We suggest that the provision of this consensus information may actually lead to a destabilization of the existing attitude, simply because the attributor is reluctant to be on the same "consensus plane" with others who are perceived to be so dissimilar. This prediction seemingly contradicts Goethals' (1976) notion that dissimilar others' agreement is persuasive. In fact, the contradiction lies in the degree of dissimilarity and in the implications that the dissimilar others' agreement have for one's attributions about oneself. Both of these factors were not extreme in Goethals' studies, in contrast to the present example.

In terms of Heider's balance theory, we have created an imbalanced situation in which the interrelationships among the party, yourself, and busing are all negative; so that change is likely. In terms of our own model, we have extrapolated consensus information to irrelevant (dissimilar to self) points on the persons dimension. These points should, by the attributor's preconceptions, show lack of covariation. By *contrasting* this preconception with information that covariation does obtain, attitude change becomes likely.[10] The *informational contrast* tactic is thus similar to extrapolation in that it extends the covariation matrix to new data points, but it differs from extrapolation in that the irrelevance of these data points is utilized to demonstrate that unexpected covariations occur (or expected covariations do not), a process that should produce at least some destabilization of the attributor's existing attitude.

INFLUENCE – A POSTSCRIPT

Let us now summarize the theoretical arguments developed here. We have argued that Kelley's tripartition of attributional dimensions can be applied to social influence processes by using it to categorize information as addressing either distinctiveness, consistency, or consensus attributions. The choice among these three types of information, either singly or in combination, provides the basis for attributional strategies of social influence. We have pointed out to what extent these strategies may be interrelated, under what conditions they can be expected to be differentially effective in changing attributions, and under what conditions attributional influence may affect not only causal inferences but also attitudes and/or behaviors. Finally, we have developed the notion of tactics of social influence to refer to the origin of the communicated information in relation to the influence target's a priori covariation data, and we have speculated on the relationship between the degree of informational discrepancy from these data and the likelihood of successful influence.

We now wish to add one final comment that pertains to any form of social influence, but that is particularly relevant to influence that involves relatively subtle interpersonal dynamics. This comment pertains to a question that is rarely asked in the study of social influence processes, namely: What aspect of the process should be studied, or alternatively, what conceptual perspective should be taken in approaching social influence phenomena?

There are three perspectives from which such processes may be studied. The first, *reactive* perspective, predominates in the empirical literature on influence. It is limited, however, to examining only the effectiveness of certain strategies

---

[10]Obviously, one could also think of cases in which the attributor conceives of himself or herself as similar to others, but the information indicates that covariation does *not* occur. The implications are essentially the same.

of influence in accomplishing certain desired interpersonal goals (attitude change, compliance, etc.).

A second, *proactive* perspective, in contrast, concentrates on the influencer's selective utilization of these influence strategies rather than on their effectiveness. Numerous attempts have been made to develop an exhaustive classification of influence strategies (Blau, 1964; Etzioni, 1961; Homans, 1961; Marwell & Schmitt, 1967; Parsons, 1963; Skinner, 1952), and some empirical research has addressed individuals' actual preferences among different influence strategies — primarily with regard to the differential use of rewards and punishments (Goodstadt & Kipnis, 1970; Leventhal & Michaels, 1971), ingratiation (Jones, 1964), self-presentation and impression management (Goffman, 1959), and altercasting (Weinstein & Deutschberger, 1963). In the present theoretical framework, a proactive view would require, among other things, an elaboration of the mechanisms by which such strategies are acquired (for example, in childhood socialization) and an analysis of both cognitive and motivational biases that may impede the rational use of attributional strategies.

The third, *interactive* perspective, combines both the reactive and proactive aspects of the influence process and views influence as a reciprocal process in which two or more persons attempt to influence each other or to resist each other's influence attempts. This approach is the most pertinent to influence processes in natural settings, because it views influence as a reciprocal relationship that is embedded in social interaction. It is also the most pertinent to attributional influence, since the provision of information is more easily resisted and opposed than, for example, coercion. This view requires somewhat more complex analyses, because we also have to consider problems related to the interactants' awareness that an influence attempt is taking place, their mutual attributions of intent, etc. Further, issues such as the likelihood of resistance (Prus, 1975) and strategy modifications in response to the other person's behavior (opposition, indifference, counterinfluence, etc.) would have to be considered.

An elaboration on these three perspectives on influence is beyond the scope of this paper, but we do wish to emphasize one final point. In our view, it is insufficient to conceptualize interpersonal influence as a unidirectional process, since such a perspective cannot possibly capture the complexity that is actually entailed in this process. Influence, as any other interpersonal and social phenomenon, is a reciprocal process; and in order to understand it, we do not have any choice but to approach it as such. In the words of Harré and Secord (1972):

> When we try to persuade the person to "see" his situation differently, to attend to other aspects of a situation than those he is considering . . ., we are trying to change the meaning of the situation for him. We try to get him to ascribe different meanings to things and situations from those he ascribed before which influenced his past choice of mentalistic predicates. If this is true, *it must be possible for [the person] to counter-persuade us to see the matter his way too* [p. 113, italics added].

ACKNOWLEDGMENTS

We would like to acknowledge our thanks to Jay Y. Brodbar, Philip Brickman, John Harvey, Gerald Marwell, and Shalom Schartz for their comments on earlier drafts of this chapter.

REFERENCES

Arkin, R. M., & Duval, S. Focus of attention and causal attributions of actors and observers. *Journal of Experimental Social Psychology*, 1975, *11*, 427–438.

Bem, D. Self-perception theory. In L. Berkowitz (Ed.), *Advances in experimental social psychology*, Vol. 6. New York: Academic Press, 1972.

Blau, P. M. *Exchange and power in social life*. New York: Wiley, 1964.

Carlsmith, J. M., & Gross, A. E. Some effects of guilt on compliance. *Journal of Personality and Social Psychology*, 1969, *11*, 232–239.

Deutsch, M., & Gerard, H. B. A study of normative and informational social influences upon individual judgment. *Journal of Abnormal and Social Psychology*, 1955, *51*, 629–636.

Duval, S., & Hensley, V. Extensions of objective self-awareness theory: The focus of attention–causal attribution hypothesis. In J. Harvey, W. Ickes, & R. Kidd (Eds.), *New directions in attribution research*. Vol. 1. Hillsdale, N.J.: Lawrence Erlbaum Associates, 1976.

Duval, S., & Wicklund, R. A. Effects of objective self-awareness on attribution of causality. *Journal of Experimental Social Psychology*, 1973, *9*, 17–31.

Etzioni, A. *A comparative analysis of complex organizations*. New York: The Free Press of Glencoe, 1961.

Feldman, N. S., Higgins, E. J., Karlovac, M., & Ruble, D. M. Use of consensus information in causal attributions as a function of temporal presentation and availability of direct information. *Journal of Personality and Social Psychology*, 1976, *34*, 694–698.

Festinger, L. A theory of social comparison processes. *Human relations*, 1954, *7*, 117–140.

French, T. R., & Raven, B. The bases of social power. In D. Cartwright (Ed.), *Studies in social power*. Ann Arbor, Mich.: Institute for Social Research, 1959.

Gergen, K. J. *The psychology of behavior exchange*. Reading, Mass.: Addison–Wesley, 1969.

Goethals, G. R. An attributional analysis of some influence phenomena. In J. Harvey, W. Ickes, & R. Kidd (Eds.), *New directions in attribution research*. Vol. 1. Hillsdale, N.J.: Lawrence Erlbaum Associates, 1976.

Goffman, E. *The presentation of self in everyday life*. New York: Doubleday Anchor, 1959.

Goffman, E. *Strategic interaction*. Philadelphia, Pa.: University of Pennsylvania Press, 1969.

Goodstadt, B. E., & Kipnis, D. Situational influences on the use of power. *Journal of Applied Psychology*, 1970, *54*, 201–207.

Gouldner, A. The norm of reciprocity: A preliminary statement. *American Sociological Review*, 1960, *25*, 161–178.

Hall, E. T. *The hidden dimension*. New York: Doubleday Anchor, 1969.

Hansen, R. D., & Lowe, C. A. Distinctiveness and consensus: The influence of behavioral information on actors' and observers' attributions. *Journal of Personality and Social Psychology*, 1976, *34*, 425–433.

Harré, R., & Secord, P. F. *The explanation of social behavior.* Totowa, N.J.: Littlefield, Adams and Co., 1972.

Heider, F. *The psychology of interpersonal relations.* New York: Wiley, 1958.

Hill, C. T. *When do people make causal attributions?* Unpublished manuscript. Harvard University, 1974.

Homans, G. C. *Social behavior: Its elementary forms.* New York: Harcourt, Brace and World, 1961.

Hovland, C. I. (Ed.) *Order of presentation in persuasion.* New Haven, Conn.: Yale University Press, 1957.

Hovland, C. I., Harvey, O. J., & Sherif, M. Assimilation and contrast effects in communication and attitude change. *Journal of Abnormal and Social Psychology,* 1957, *55,* 242–252.

Hovland, C. I., & Janis, I. L. (Eds.) *Personality and persuasibility.* New Haven, Conn.: Yale University Press, 1959.

Hovland, C. I., Janis, I. L., & Kelley, H. H. *Communication and persuasion.* New Haven, Conn.: Yale University Press, 1953.

Hovland, C. I., Lumsdaine, A. A., & Sheffield, F. D. *Experiments on mass communication.* Princeton, N.J.: Princeton University Press, 1949.

Hovland, C. I., & Mandell, W. An experimental comparison of conclusion-drawing by the communicator and by the audience. *Journal of Abnormal and Social Psychology,* 1952, *47,* 581-588.

Janis, I. L. Effects of fear arousal on attitude change: Recent developments in theory and experimental research. In L. Berkowitz (Ed.), *Advances in experimental social psychology* (Vol. 3). New York: Academic Press, 1967.

Jones, E. E. *Ingratiation.* New York: Appleton–Century–Crofts, 1964.

Kahn, L. A. The organization of attitudes toward the Negro as a function of education. *Psychological monographs,* 1951, *65,* (Whole No. 330).

Kahneman, D., & Tversky, A. On the psychology of prediction. *Psychological Review,* 1973, *80,* 237–251.

Katz, D. A functional approach to the study of attitude. *Public Opinion Quarterly,* 1960, *24,* 163–204.

Katz, E., Gurevitch, M., Danet, B., & Peled, T. Petitions and prayers: A method for the content analysis of persuasive appeals. *Social Forces,* 1969, *47,* 447–462.

Kelley, H. H. Attribution theory in social psychology. In D. Levine (Ed.), *Nebraska Symposium on Motivation.* Lincoln, Neb.: University of Nebraska Press, 1967.

Kelley, H. H. Causal schemata and the attribution process. In E. E. Jones, D. E. Kanouse, H. H. Kelley, R. E. Nisbett, S. Valins, & B. Weiner (Eds.), *Attribution: Perceiving the causes of behavior.* Morristown, N.J.: General Learning Press, 1971.

Kelley, H. H. The process of causal attribution. *American Psychologist,* 1973, *28,* 107–128.

Kelman, H. C. Compliance identification and internalization: Three processes of attitude change. *Journal of Conflict Resolution,* 1958, *2,* 51–60.

Kelman, H. C., & Hovland, C. I. "Reinstatement" of the communicator in delayed measurement of attitude change. *Journal of Abnormal and Social Psychology,* 1953, *48,* 327–335.

Kruglanski, A. W. The endogenous–exogenous partition in attribution theory. *Psychological Review,* 1975, *82,* 387–406.

Kuhn, T. S. *The structure of scientific revolutions.* Chicago: University of Chicago Press, 1962.

Laing, R. D., Phillipson, H., & Lee, A. R. *Interpersonal perception.* New York: Harper and Row, 1966.

Leventhal, G. S., & Michaels, J. W. Locus of cause and equity motivation as determinants of reward allocation. *Journal of Personality and Social Psychology*, 1971, *17*, 229–235.

Leventhal, H. Findings and theory in the study of fear communications. In L. Berkowitz (Ed.), *Advances in experimental social psychology*. Vol. 5. New York: Academic Press, 1970.

Lewin, K. Frontiers in group dynamics I. *Human Relations*, 1947, *1*, 2–38. (a)

Lewin, K. Frontiers in group dynamics II. *Human Relations*, 1947, *1*, 143–153. (b)

Lewis, W. C. *Why people change*. New York: Holt, Rinehart & Winston, 1972.

London, P. *Behavior control*. New York: Harper & Row, 1969.

Lund, F. H. The psychology of belief: The law of primacy in persuasion. *Journal of Abnormal and Social Psychology*, 1925, *29*, 183–191.

Marwell, G., & Schmitt, D. R. Compliance-gaining behaviors: A synthesis and model. *Sociological Quarterly*, 1967, *7*, 317–328.

McArthur, L. The how and what of why: Some determinants and consequences of causal attribution. *Journal of Personality and Social Psychology*, 1972, *22*, 171–192.

McArthur, L. The lesser influence of consensus than distinctiveness information on causal attributions: A test of the person–thing hypothesis. *Journal of Personality and Social Psychology*, 1976, *33*, 733–742.

McGuire, W. J. The nature of attitudes and attitude change. In G. Lindzey & E. Aronson (Eds.), *Handbook of social psychology*. Vol. 2. Reading, Mass.: Addison–Wesley, 1969.

Nisbett, R., & Borgida, E. Attribution and the psychology of prediction. *Journal of Personality and Social Psychology*, 1975, *32*, 932–943.

Nisbett, R., Borgida, E., Crandall, R., & Reed, H. Popular induction: Information is not necessarily informative. In J. S. Carroll & J. W. Payne (Eds.), *Cognition and social behavior*. Hillsdale, N.J.: Lawrence Erlbaum Associates, 1976.

Nisbett, R., & Valins, S. Perceiving the causes of one's own behavior. In E. E. Jones, D. E. Kanouse, H. H. Kelley, R. E. Nisbett, S. Valins, & B. Weiner (Eds.), *Attribution: Perceiving the causes of behavior*. Morristown, N.J.: General Learning Press, 1971.

Orvis, H. R., Cunningham, J. D., & Kelley, H. H. A closer examination of causal inference: The roles of consensus, distinctiveness and consistency information. *Journal of Personality and Social Psychology*, 1975, *32*, 605–616.

Osgood, C. E., Suci, G. J., & Tannenbaum, P. H. *The measurement of meaning*. Urbana, Ill.: University of Illinois Press, 1957.

Ostrom, T. M. The relationship between the affective, behavioral and cognitive components of attitudes. *Journal of Experimental Social Psychology*, 1969, *5*, 12–30.

Parsons, T. On the concept of influence. *Public Opinion Quarterly*, 1963, *27*, 37–62.

Prus, R. C. Resisting designations: An extension of attribution theory into a negotiated context. *Sociological Inquiry*, 1975, *45*, 3–14.

Raven, B. H., & French, J. R. Legitimate power, coercive power, and the observability of social influence. *Sociometry*, 1958, *21*, 83–97.

Regan, D. T., & Totten, J. Empathy and attribution: Turning observers into actors. *Journal of Personality and Social Psychology*, 1975, *32*, 850–856.

Rokeach, M. *The nature of human values*. New York: Free Press, 1973.

Rokeach, M., & McLellan, D. D. Feedback information about the values and attitudes of self and others as determinants of long-term cognitive and behavioral change. *Journal of Applied Social Psychology*, 1972, *2*, 236–251.

Rosenberg, M. J. Cognitive structure and attitudinal effect. *Journal of Abnormal and Social Psychology*, 1956, *53*, 367–372.

Ruble, D. N., & Feldman, N. S. Order of consensus, distinctiveness and consistency information and causal attributions. *Journal of Personality and Social Psychology*, 1976, *34*, 930–937.

Scheflen, A. E. *How behavior means.* New York: Anchor Press, 1974.

Schuman, H., & Johnson, M. P. Attitudes and behavior. *Annual Review of Sociology,* 1976, *2,* 161–207.

Skinner, B. F. *Science and human behavior.* New York: Macmillan, 1953.

Storms, M. D. Videotape and the attribution process: Reversing actors' and observers' point of view. *Journal of Personality and Social Psychology,* 1973, *27,* 165–175.

Svarstadt, B. L. Physician–patient communication and patient conformity to medical advice. In D. Mechanic (Ed.), *The growth of bureaucratic medicine.* New York: Wiley, 1976.

Taylor, S. E. Developing a cognitive social psychology. In J. S. Carroll & J. W. Payne (Eds.), *Cognition and social behavior.* Hillsdale, N.J.: Lawrence Erlbaum Associates, 1976.

Taylor, S. E., & Fiske, S. T. Point of view and perceptions of causality. *Journal of Personality and Social Psychology,* 1975, *32,* 439–445.

Tedeschi, J. T. Threats and promises. In P. Swingle (Ed.), *The structure of conflict.* New York: Academic Press, 1970.

Tedeschi, J. T., Schlenker, B. R., & Linskold, S. The exercise of power and influence: The source of influence. In J. T. Tedeschi (Ed.), *The social influence processes.* Chicago: Aldine and Atherton, 1972.

Thibaut, T. W., & Kelley, H. H. *The social psychology of groups.* New York: Wiley, 1959.

Walster, E., & Festinger, L. The effectiveness of "overheard" persuasive communications. *Journal of Abnormal and Social Psychology,* 1962, *65,* 395–402.

Wegner, D. M., & Finstuen, K. Observers' focus of attention in the simulation of self-perception. *Journal of Personality and Social Psychology,* 1977, *35,* 56–62.

Weinstein, E. A., & Deutschberger, P. Some dimensions of altercasting. *Sociometry,* 1963, *26,* 454–466.

Wells, G. L., & Harvey, J. H. Do people use consensus information in making causal attributions? *Journal of Personality and Social Psychology,* 1977, *35,* 279–293.

Wicker, A. W. Attitudes vs. actions: The relationship of verbal and overt behavioral responses to attitude objects. *Journal of Social Issues,* 1969, *25,* 41–78.

Woodmansee, J. J., & Cook, S. W. Dimensions of verbal racial attitudes: Their identification and measurement. *Journal of Personality and Social Psychology,* 1967, *7,* 240–250.

III THEORETICAL INTEGRATIONS

The two papers that appear in this section primarily are concerned with theoretical integrations and critiques. Investigators in the attribution area sometimes have assumed that the scientist and the naive attributor engage in similar conceptual processes when acquiring knowledge about the world. This fundamental epistemic position is adopted by Arie Kruglanski, Irit Hamel, Shirley Maides, and Joseph Schwartz in their chapter, "Attribution Theory as a Special Case of Lay Epistemology." These authors review Kelley's ANOVA formulation of attribution, propose their own theory of lay epistemology, and finally, review evidence pertinent to this new conception. Kruglanski et al.'s theory involves a threefold distinction among: (1) the contents of lay knowledge; (2) the logic whereby such knowledge is validated; and (3) the course of the epistemic episode. The content of the naive epistemology is based on the naive attributor's set of concepts pertaining to his/her experiences with the world. Logic pertains to the assessment criteria in terms of which the naive attributor judges knowledge as valid. The course of the epistemic episode refers to the sequence of cognitive operations initiated by the naive attributor's desire to assess the possibility of a significant increment in knowledge and terminating with the experience that such possibility has been adequately assessed. This epistemological theory broadly speaks to the attainment of knowledge about all possible identity categories and all relations, not just causal relations.

Since its inception, the Schachterian two-factor model of emotion that emphasizes the role of nonspecific physiological arousal and cognitive interpretation in the experience of emotion has been criticized heavily. In his chapter, "Attribution and Misattribution of Excitatory Reactions," Dolf Zillmann reviews work relevant to the two-factor model and concludes that research on the model has demonstrated convincingly the great plasticity of emotional behavior but that it has not generated evidence that would support the assumption of a conscious, epistemic search that ultimately produces highly specific causal attributions or misattributions. Zillmann proposes his own three-factor model, which involves the central proposition that emotional experiences and behavior result from the dispositional, excitatory, and experiential components of the emotional state. Finally, Zillmann reviews concepts and evidence pertaining to his excitation-transfer theory. This theory applies the three-factor approach to the experience of emotional states in sequence and specifies the conditions under which residues of sympathetic excitation from a previous state will intensify a subsequent emotional state.

11

Attribution Theory as a Special Case of Lay Epistemology

Arie W. Kruglanski
Tel-Aviv University

Irit Z. Hamel
Shirley A. Maides
Joseph M. Schwartz
Vanderbilt University

INTRODUCTION

The topic of causal attribution has recently become a dominant research concern in social psychology (see e.g., the bibliographical survey by Nelson & Hendrick [1974] [1] and the books edited by Harvey, Ickes, & Kidd, 1976, and the present volume). The study of attribution has germinated in the writings of Fritz Heider (e.g., 1958) but has derived particular momentum from Jones and Davis's (1965) insightful attributional analysis of social perception and Kelley's (1967) seminal conceptualization of attribution in terms of the analysis of variance model.

A fundamental assumption of attribution theory is that naive epistemology is similar in crucial respects to scientific epistemology (cf. Heider, 1958, p. 297; Kelley, 1967, p. 194). This means that the lay person's epistemic encounters with the world are basically rational, although psychological biases may exist and introduce distortions into the process. The relevance of attribution theory to social psychology derives from the fact that (other) people are both signifi-cant *means,* whereby knowledge about the world is acquired (they provide

---

[1] Nelson, C. A., & Hendrick, C. *Bibliography of journal articles in social psychology.* Kent State University, 1974. (Mimeo).

social reality, consensual validation of beliefs, etc.), and significant *objects* of most persons' interest. This has been amply documented in Kelley's (1967) application of attributional analyses to major social psychological topics such as attitude formation and change, social influence, and the perception of self and others.

The present article shares with previous attributional formulations the assumed parallelism between lay and scientific epistemology, and it accepts Kelley's arguments (1967, 1971, 1972, 1973) regarding the relevance of attribution theory to social psychology. At the same time, the subsequent discussion introduces a novel general perspective for viewing the topic of causal attribution. Specifically, it is argued that the ANOVA model of attribution (Kelley, 1967) is actually a special case of a broader theory of naive epistemology, the rudiments of which will be adumbrated presently.

The outline of the paper is as follows: First, the ANOVA formulation of attribution is reviewed in some detail. Second, a theory of lay epistemology is developed, along the lines of a threefold distinction between: (1) the *contents* of lay knowledge; (2) the *logic* whereby such knowledge is validated; and (3) the *course* of the epistemic episode. Third, experimental evidence is reviewed pertinent to the present characterization of knowledge and attribution.

The ANOVA Model of Attribution

In his classic paper presented at the *Nebraska Symposium on Motivation,* Kelley (1967) proposed that causal attribution may be understood by analogy to an *analysis of variance.* In Kelley's cubic representation of this ANOVA "... along the vertical axis are listed *entities* which correspond to things in the environment (e.g., movies). Along one horizontal axis are various *persons* who interact with the entities and along the other the various *modalities* of interaction and the *times* at which these interactions may occur ... [p. 195] ."

In subsequent discussion we shall refer to the above dimensions of the cube as the *attributional categories.* These should not be confused with the related notion of *attributional criteria* that are the layman's means of assessing (or validating) his attributions and of arriving at confident (subjectively valid) judgments about the external world.

The four criteria for external validity are ... (1) *Distinctiveness:* the impression is attributed to the thing of it uniquely occurs when the thing is present and does not occur in its absence. (2) *Consistency over time:* each time the thing is present, the individual's reactions must be the same or nearly so. (3) *Consistency over modality:* his reaction must be consistent even though his mode of interaction with the thing varies. (4) *Consensus:*

attributes of external origin are experienced the same way by all observers [p. 197].

According to Kelley "To the degree a person's attributions fulfill these criteria he feels confident that he has a true picture of his external world, he makes judgments quickly and with subjective confidence and he takes action with speed and vigor. When his attributions do not satisfy the criteria, he is uncertain in his views and hesitant in action [p. 197]."

An immediate question with respect to the ANOVA model of attribution is the issue of its *universal application*. Common sense and experimental evidence alike suggest that people may make confident and reasonable attributions without laboriously filling out the matrix, as the model implies. Kelley (1973) articulated this clearly:

The attribution process implied by the analysis of variance model is undoubtedly somewhat on the idealized side. It would be foolish to suggest that anything like a large data matrix is filled out with effect observations before a causal inference is made. The framework should be regarded as simply the context within which some limited and small sample of observations is interpreted. Beyond that, it is obvious that the individual is often lacking the time and the motivation necessary to make multiple observations . . . . In these circumstances he may make a causal inference on the basis of a *single observation* of the effect [p. 113].

Kelley went on to suggest that in circumstances where the individual would, for some reason, fail to apply the full-fledged ANOVA, he might draw on his repertory of causal schemata to arrive at a stable attribution. A causal schema "can be viewed as an *assumed pattern of data* in a complete analysis of variance framework" (Kelley, 1973, p. 115). Thus, the ANOVA cube is assumed to provide the reference-structure within which a single observation is placed and interpreted. For example, "Given information about a certain effect and two or more possible causes, the individual tends to assimilate it to a specific assumed analysis of variance pattern and from that to make a causal attribution" (Kelley, 1973, p. 115).

To conclude, then, the ANOVA model just reviewed is portrayed as an ideally rational framework that is universally *applicable* across all types of attributional problems, even though it may not be universally applied (i.e., when it is expedient to supplant the total ANOVA structure by the inferential shortcuts of causal schemata). The foregoing assumption of *universal applicability* will be closely examined in subsequent sections of this paper. But first we will lay out the concepts and distinctions of the present theory of "lay epistemology," of which the ANOVA model is assumed to represent a special case.

A THEORY OF LAY EPISTEMOLOGY:
ON THE CONTENT, LOGIC, AND
COURSE OF NAIVE INQUIRY

A central organizing theme of the present analysis is a threefold distinction between the content, logic, and course of epistemic behavior.[2] The content of naive epistemology is the layman's total set of concepts pertaining to the world of experience. The epistemic logic is the assessment criterion, the fulfillment of which yields a sense of valid knowledge. The course of lay inquiry is the sequence of cognitive operations intended to assess the possibility of significant new knowledge. It is assumed that the contents involved in any instance of knowledge-seeking behavior are specific to the knower's cultural and personal background and to this person's circumstantial interests. By contrast, the assessment logic as well as the epistemic course are assumed universal and invariant (across situations, contents, and people). We now move to consider more closely the content, logic, and course aspects of lay epistemology.

The Content of Lay Epistemology

The layman's concepts about the world of experience may be conveniently classified into those that fulfill the function of *identification* and those that fulfill the function of *interrelation.* Identificational concepts are categories that classify portions of experience under general rubrics. Examples are concepts like "table," "mammalian," or "psychologist." Relational concepts connect two or more identificational categories. Examples are concepts like "love," "ownership," or "causality." The examples just listed hint at the immensely diverse character of identificational and relational concepts that may constitute the content of any layman's knowledge. Furthermore, a person's knowledge is continually evolving under the unrelenting barrage of new experiences. Thus, new identificational concepts may be invented and applied (e.g., culturally invented catagories like "sexist," "psychopath," or "midi-dress") as may new relational concepts (like "divorce," "alienation," etc.). In addition, existing relational concepts may be freshly applied to existing identificational categories (as in discovering that punishment *causes* counteraggression, or that John *loves* Mary).

The lay knower's total conceptual repertory is clearly distinct from the small subset of concepts invoked by him during any given instance of epistemic (knowledge-seeking) behavior, as in John's wondering "whether Paul is engaged to Mary," "whether whales are mammals" or "who stole the cookies from the

---

[2]In keeping with the underlying assumption of attribution theory, the above division has a direct, scientific parallel in the distinction between topic (or subject matter), method, and course of investigation.

cookie jar?" It is the restricted epistemic instance rather than the total store of personal knowledge that is of central concern in the present analysis. But for now, let us disregard the problem of content and focus instead on the logical aspect of lay epistemology; that is, on the rationale employed by the layman to validate his various conjectures about the world.

Logical Consistency Among Beliefs as the Validation Criterion of Knowledge

Some Notes on the Reconceptualization of Scientific Method.   In line with views of major attribution theorists (e.g., Heider, 1958; Kelley, 1967, 1973; Jones & Davis, 1965; Jones & McGillis, 1976), it is presently posited that the logic whereby the layman validates his conceptions and hypotheses essentially resembles the scientific method. It is particularly relevant to note that in recent years there have been exciting new developments in the conception of science that departed strikingly from traditional formulations. Detailed review of these developments is well outside the scope of the present essay (for an excellent summary the interested reader is referred to Weimer[3]), but a few highlights will be mentioned in passing. First, compelling arguments have been advanced against the principle of *induction* by philosophers branded "nonjustificationists " (cf. Bartley, 1962; Weimer[3]). They have challenged the idea that scientific inferences are induced from facts following their appropriately complete examination (cf. Bacon, 1625 [1962 edition]; Mill, 1873 [1963 edition]), or that scientific hypotheses are verified (or are rendered more probably) by large numbers of pertinent factual observations (cf. Reichenbach, 1961). The recognized originator of the nonjustificationist viewpoint was Karl R. Popper (1959, 1963, 1972), whose leading ideas were assimilated and developed by several contemporary philosophers (e.g., Kuhn, 1962, 1970; Feyerabend, 1970; Lakatos, 1968, 1970; Watkins, 1970; and others). In his early works, Popper (e.g., 1959) outlined a deductive *falsificationist* model of science, whereby the role of factual observations is to serve as potential falsifiers of scientific hypotheses. But the influential falsificationist model was later criticized by Lakatos (1968) for its *monotheoretic emphasis,* which accorded a differential status to the *theory* that is the *object* of assessment and the *facts* that are the criterion of assessment. This seems inconsistent with the widely accepted view that facts and theories are equally conceptual or conjectural (cf. Popper, 1959; Hempel, 1959). Furthermore, the falsificationist model is at odds with Duhem's (1906) early argument that no factual observation can clearly falsify a theoretical hypothesis, because of the numerous unarticulated (auxiliary) assumptions involved in conducting any

---

[3]Weimer, W. B. *Notes on the methodology of scientific research.* Unpublished manuscript. University of Minnesota, 1976.

observation. Such auxiliary assumptions rather than the theoretical hypothesis under test could actually be falsified by an inconsistent observation.

The above considerations led Lakatos (1968) to replace the verificationist (inductive) and falsificationist (deductive) criteria of scientific assessment by a *logical consistency* criterion of assessment. In the following famous passage Lakatos (1968) is clear on this point:

> The problem is *not* what to do when "theories" clash with "facts". Such a "clash" is only suggested by the monotheoretical "deductive model". Whether a proposition is a "fact" or a "theory" depends on your methodological decision. "Empirical basis" is a monotheoretical notion, it is *relative* to some monotheoretical deductive structure. In the pluralistic model the clash is between two high-level theories: an *interpretative theory* to provide the facts and an *explanatory theory* to explain them . . . The problem is not whether a refutation is real or not. The problem is how to repair an inconsistency between the "explanatory theory" under test and the . . . explicit or hidden . . . "interpretative" theories, or if you wish, *the problem is which theory to consider as the interpretative one which provides the "hard" facts and which the explanatory one which "tentatively" explains them.* Thus experiments do not overthrow theories as (early) Popper has it, but only increase the problem-fever of the body of science. *No theory forbids some state of affairs specifiable in advance: it is not that we propose a theory and Nature may shout NO.* Rather, we propose a maze of theories and Nature may shout INCONSISTENT [pp. 161–162].

To summarize, recent developments in the philosophy of science implicate the principle of *logical consistency* as the sole assessment criterion of scientific knowledge. To conform with the equivalence assumption between scientific and lay epistemology, it is presently posited that consistency also constitutes the assessment criterion of lay knowledge. In the next section, we discuss explicitly the consistency principle and its workings in the validation of beliefs.

The Consistency Principle in Lay Epistemology

Relevance.   *Two cognitive elements are mutually relevant if and only if they are related by material implication such that one of the elements either materially implies the other, or the negation of the other.* Let R denote the symmetrical relation of relevance and $\sim R$ the symmetrical relation of irrelevance. For any pair p, q of cognitive elements we have:

(1) pRq iff $(p \supset q) \lor (q \supset p) \lor (p \supset \sim q)^4$

(2) $p \sim Rq$ iff $\sim [(p \supset q) \lor (q \supset p) \lor (p \supset \sim q)]$.

---

[4] iff = if and only if; $p \supset q$ = if p then q; $p \lor q$ = p or q; $\sim p$ = not p.

Thus, for a given layman the cognition "being a good student" may imply "doing one's homework in time"; this would render mutually *relevant* the two cognitions just considered. On the other hand, "being a good student" may have no implications for the person's gastronomic preferences, so that this cognition would be *irrelevant* to cognitions about a "strong like" or "dislike" for spicy dishes.

Consistency. *Two mutually relevant elements are consistent if their conjunction is compatible with (i.e., is not negated by) their relevance relation.* Let *K* denote the symmetrical relation of consistency. For any pair, *p, q* of cognitive elements entertained by the layman we have:

(3) pKq iff $(p \supset q) \lor (q \supset p)$.

Thus, *if* for our layman "being a good student" *implies* "doing one's homework," the cognitions "Johnny is a good student" and "Johnny is doing his homework" are consistent.

Inconsistency. *Two mutually relevant cognitions are inconsistent if and only if the negation of their specific conjunction is implied by their relevance relation.* Let $\sim K$ denote the symmetrical relation of inconsistency. For any pair, *p, q*, of cognitive elements we have:

(4) $p \sim Kq$ iff $(p \supset -q)$.

For instance, if some person "being a good student" implies "preparing one's homework," the cognitions "Johnny is a good student" and "Johnny seldom prepares his homework" are inconsistent.

The similarity of the preceding formulations to Festinger's (1957) terminology must be quickly acknowledged. Thus, the present notion of relevance is near identical to (albeit more explicitly defined than) Festinger's "relevance" (Festinger, 1957, pp. 11–13). The present notion of inconsistency is similar to Festinger's dissonant relation (p. 13, save for the present use of the logically precise "negation", instead of Festinger's "obverse" relation), and the present notion of consistency is identical to Festinger's notion of a consonant relation (p. 15). The decision to introduce novel terminology rather than referring to Festingerian concepts is motivated by the wish to avoid importing from dissonance theory the specific psychological meanings associated with logical relations among cognitions. Dissonance is a motivational formulation in which logical relations among cognitive elements are coordinated to the individual's motivational states. By contrast, the present theory is epistemological; it coordinates logical relations among cognitions to the individual's epistemological states (i.e., to states of belief or disbelief with respect to a given matter). Specifically, it is postulated that *the lay person's experience or belief in some state of affairs is a positive function of the degree to which this person's relevant cognitions are consistent, and the experience of doubt and uncertainty is a positive function of the degree to which the person's relevant cognitions are inconsistent.*

Let us see now how the layman might apply the consistency principle toward the validation of specific hypotheses about the world. Consider, for example, how an individual knows whether a given *identify* category, say "table," aptly describes an object observed at a given moment. According to the present view, any hypothesis is (subjectively) validated by checking whether implications of the hypothesis are *consistent* with other beliefs about the situation. Thus, the concept "table" may imply among other things "flat surface," "perpendicular support," "inanimate nature," etc. In order to conclude whether X, observed at a given moment, is a table, our lay investigator may check whether some of the "tabular" implications can be reasonably entertained. If so, the entire cognitive structure that includes the hypothesis, "X is a table," and the observations, "X has four legs, a flat surface, and is inanimate," is internally consistent (precisely in the sense of Definition 2, above) and therefore may be held with some confidence. However, should inconsistency among the relevant cognitions become apparent (e.g., should it suddenly seem that X's inanimate nature is doubtful), the hypothesis, "X is a table," would be rendered uncertain. Under these circumstances, additional cognitive work might be required to be confident of X's identity.

The consistency logic just illustrated with respect to identification of an inanimate object is assumed applicable to person-perception. For example, consider the comment by Weiner, Frieze, Kukla, Reed, Rest and Rosenbaum (1972) regarding conditions under which an attributor might reach the judgment that an actor's success on some task has been due to considerable *effort*. ". . . Covariation of performance with incentive value or . . . with cues as perceived muscular tension or task persistence will lead to the inference that effort was a dominant behavioral determinant . . . [p. 5]." The above analysis by Weiner et al. is readily transformable into the present terms whereby the category "effort" implies things like muscular tension, task persistence, or appearance under high incentive. Evidence that the latter implications may be true should lend credence to the "effort" hypothesis, etc.

In addition to being applicable to knowledge about "external" referents (other persons and objects), the consistency logic is also assumed applicable to "internal," or "self-" knowledge. For instance, this same logic is assumed to mediate the labeling of one's emotions, as in Schachter's and Singer's (1962) classic experiment. In one portion of this research, subjects (in the *ignorant* condition) who were (1) inexplicably aroused, and (2) exposed to situational cues like jokes, laughter, and playful behavior by a peer, showed a stronger tendency to identify their experiences as "euphoria" than did subjects who were not physiologically aroused (*placebo* condition) and than subjects led to believe that their arousal was the result of injection (*informed* condition). In present terms, the "euphoria" concept implies among other things "bodily arousal" and "situational reasons for being euphoric." Thus, for subjects with good reasons to believe the latter two matters (as in the *ignorant* condition, above), it would have been more consistent to also believe the euphoria hypothesis than it would have been for subjects without such consistent support for the euphoria hypothesis (the *placebo* condition) or for subjects with a compelling alternative hypothesis for the arousal (the *informed* condition).

So far, applicability of the consistency logic has been illustrated with examples from the identificational domain. But it is easily demonstrated that the same principle also holds for the layman's inferences of relational concepts. For instance, one might assess whether "John is *in love* with Sara" by considering whether implications of the "in love" relation are consistent with the remaining cognitions that one has about the situation. Specifically, the "in love" concept may imply to our knower that someone involved in such relation should "spend sleepless nights," make every effort to meet with the love object," "write passionate letters," and "propose marriage to this person." To the extent, then, that the knower may have corroborating evidence for these implications, the hypothesis of the "in love" relation might be accepted, in a conformity with the consistency logic.

Course of the Epistemic Episode

The course of the epistemic episode is the sequence of cognitive operations initiated by the layman's desire to assess the possibility of a significant increment in knowledge and terminating with the experience that such possibility has been adequately assessed. Earlier it was implied that the layman might invoke different cognitive contents in different circumstances. In the following outline of the epistemic episode, we consider the circumstantial factors that may guide the invocation of contents.

Background knowledge, epistemic intention, underlying motive and question generation. We begin with the assumptions that epistemic (i.e., knowledge-seeking) behavior; (1) runs its course within the context of *background knowledge*[5] (the so-called given, or the facts) that the individual has about the situation; and (2) is *motivated,* i.e., is guided by a specific intention. This is the intention to assess the possibility of an increment in knowledge relevant to the person's *underlying motive.* By *relevant* we mean that the knowledge increment either

---

[5]The background knowledge is never static but rather is in constant flux; it is endlessly reshaped under the impact of new observations, the retrieval of old memories, and/or the chainlike emergence of associations.

directly satisfies or frustrates the motive in question or suggests action with such consequences for this motive.

Against the background of pre-existing knowledge, the layman proceeds to formulate a question such that at least some potential answers thereto would constitute relevant knowledge increments in the sense just defined. For instance, given that Paul had proposed to Ann and that Paul was seen with an attractive blonde at a cafe, Ann's epistemic intention might be to find out whether this indicates anything relevant about Paul's merit as a potential husband. In these circumstances Ann might formulate the question, "Who is the attractive blonde with Paul at the cafe?" Some answers to this question — say, that the blonde is Paul's secret lover — might have important bearing on Ann's *underlying motive:* evaluating Paul's suitability for marriage.

It is noteworthy that in some cases implications of the very same question may be pertinent to several disparate motives; obversely, the very same motive may sometimes be affected *via* answers to disparate questions. For example, a person wondering *"whether Paul is in love with Ann"* might wish to know the answer in order to decide whether or not to: (1) adivse Ann to accept Paul's marriage proposal; (2) advise Paul's wife to divorce him; (3) seek Ann's assistance in order to influence Paul's business decision; etc. Similarly, someone whose underlying motive is to counsel Ann with respect to Paul's proposal (as in example (1) above) might pose questions other than that about Paul's love for Ann, e.g., questions about Ann's love for Paul, the availability to Ann of alternative suitors, and so on.

To summarize up to this point, it has been assumed that the epistemic episode is guided by an intention to assess the possibility of a knowledge increment relevant to the lay knower's underlying motive. Specifically, a question is formulated for which some potential answer(s) have relevance to the motive at issue.

Generation and Assessment of Answers, Consistency Loigc, and the Problem of Demarcation.   Once a question has been formulated by the lay in vestigator, the next step in the epistemic episode consists of the generation of one or more pertinent answers. Such answers are subsequently subject to the process of assessment. It is at this juncture that the validational logic elaborated earlier is invoked such that various implications deduced from a given answer are examined for consistency with pertinent evidence. For example, the hypothesis that "Mickey stole the cookies from the cookie jar" implies among other things that Mickey was present at the scene of the crime, had motive, opportunity, etc. If the evidence is consistent with these implications, the case against Mickey may be believed with some conviction. But if some of the implications are flagrantly violated by the evidence — say, if it is demonstrated that at time of the stealing Mickey was a long distance away from the jar, this should weaken considerably any suspicion against this individual.

In instances where several competing hypotheses are assessed simultaneously, those implications of a given hypothesis that *demarcate* it from the competing alternative would be of particular interest to the knower. For instance, suppose that both Mickey and Donald were suspects in the cookie stealing. Suppose further that only Donald had the capacity of reaching high places – not Mickey. In this instance, the investigator might be particularly interested in knowing the specific location of the cookie jar. On the other hand, if it were known that Mickey but not Donald had a particular passion for oatmeal cookies, our investigator might be especially curious about the kind of cookies stolen.

Termination of the epistemic episode. The epistemic episode completes its course when the epistemic intention is fulfilled (i.e., whenever the question posed by the knower receives an adequate answer). Such an answer must be consistent with the available evidence, and, more so than possible alternative hypotheses considered by the knower. Thus, our investigator may conclude that Donald (not Mickey) was guilty of the stealing (e.g., if the cookies were out of Mickey's reach, if Donald had been particularly fond of such cookies, if eye-witnesses saw Donald leaving the scene of the crime, etc.).

Knowable Versus Unknowable Aspects
of Content Evocation:
On Relevance and Availability

As noted earlier the course of the epistemic episode and the validational logic of consistency are presently assumed universally applicable accross all types of epistemic problems. By contrast, the contents of knowledge are assumed potentially variable from person to person and from situation to situation. In the present section, we explicate the boundaries within which situational evocation of contents seems accessible to systemic characterization and identify a domain assumed *in principle* inaccessible to such characterization.

In general, it is possible to characterize the cognitive contents evoked by the layman in a given situation as *relevant*[6] to the problem at hand. To this extent, at least, epistemic behavior is selective: It does not admit irrelevant contents. Based on previous learning and socialization the layman generates contents (questions and answers) that are broadly appropriate to this person's sequential

---

[6] As in Proposition 1, given earlier, by relevance is meant the existence of an implicational relation between two cognitive elements.

objectives in the epistemic situation. The several components of the epistemic episode are assumed mutually relevant in the following ways:

1. *Teleological relevance.* The question formulated and conjectural answers generated in response to the question will be relevant[7] to the layman's underlying motive if the confirmation of an answer (in the focal case) or the disconfirmation of an answer (in the complementary case) affects the motive in question. Thus, granting a specific, underlying motive salient to the attributor at a given moment, this person will not likely consider questions or answers irrelevant to the motive at issue. For example, in the circumstance where Ann sees Paul with a beautiful blonde (granted that Ann's immediate aim [motive] is to assess Paul's marital suitability), she is unlikely to pose irrelevant questions, e.g., about the price the blonde paid for her blouse. Furthermore, given that Ann did formulate a relevant question like, "Who is the blonde with Paul?" she is unlikely to consider teleologically irrelevant answers such as "a college graduate," "a member of the Democratic Party," etc.

2. *Deductive relevance.* This type of relevance pertains to the next stage in the epistemic sequence, notably the stage of assessment. Deductive relevance constrains the type of evidence sought by the lay knower in order to evaluate a potential answer under consideration. Specifically, such evidence *would need* to bear on the validity (vs. invalidity) of implications *following deductively* from the answer at stake, as opposed to other deductively unrelated information. As we have seen, the proposition that "Mickey stole the cookies" implies that he must have been present at the scene of the crime, so such evidence might well be of interest to someone attempting to determine Mickey's guilt or innocence.

3. *Demarcational relevance.* This type of relevance pertains to the case in which the layman's objective is to decide between competing answers to some question. Under such conditions, the specific deductions (from an answer) for which evidence would be sought would be those that are *noncommon* with the competing alternative, hence allowing rational choice (critical test) between the alternatives. As we have seen, an investigator with the objective of deciding whether Mickey or whether Donald snatched the cookies from the jar might be particularly interested in evidence pertinent to implications that demarcate the "Mickey" vs. the "Donald" hypothesis.

Availability. So far we have attempted to show that the use of cognitive (mental) contents in the course of the epistemic sequence is constrained by relevance requirements. In this sense it is selective. This may allow the outsider (e.g., the experimental psychologist) to make certain broad predictions about

---

[7]To show that teleological relevance is a special case of Definition 1, it is easy to rephrase the proposition just given into the implicational statement "*If* answer A is confirmed or disconfirmed *then* motive M is affected in way X."

contents that the layman would consider in specific circumstances, the necessary assumption being that the outsider is sufficiently familiar with the layman's conceptual repertory to judge content relevancies in an epistemic episode.

But relevance constraints afford only a rough boundary of predictability; within it, there exists a broad variety of permissible contents whose specifc elicitation seems best characterized as unpredictable or unjustified (cf. Campbell, 1974). Thus, for any epistemic goal there may exist numerous relevant questions and potential answers; for each potential answer there may exist numerous relevant deductions; and for each set of competing alternatives there may exist several demarcational implications. According to the present view, beyond the broad constraints of relevance, the formulation of the particular question, answer, or testable deduction depends on the momentary *availability* to the layman of the specific content. Which contents will be available seems to depend on such a particularistic constellation of factors as to render precise prediction outside the feasible powers of a theory of knowledge. The idea that evocation of concepts in the course of epistemic behavior is largely blind and haphazard has been stressed by major epistemological theorists (in particular by Donald T. Campbell, 1974; Bain, 1874; James, 1880; Jevons, 1974; Mach, 1895; Poincaré, 1913; Souriau, 1881).

In the remainder of this paper, the present concepts of lay epistemology will be applied to attribution theory. As a first step, the ANOVA model of attribution will be partitioned into the elements of content and logic. Second, experimental evidence will be reviewed for the idea that the content elements of the ANOVA cube are applied by the layman in *specific* circumstances suggested by the present analysis of epistemic course, rather than applied *generally.*

CONTENT AND LOGIC IN THE ANOVA CUBE

The central idea developed in the following section is that the ANOVA cube formulation of the attribution process may be conveniently parsed into elements of content and of logic. The identification of content elements within attribution theory should have important implications for their universal applicability: Because content elements are assumed to be invoked according to their *situational relevance,* they *may not* be *universally* applicable across attributional situations.

Content elements of the ANOVA cube.   The content elements of the ANOVA cube may be partitioned into: (1) the four *identity categories* that constitute the dimensions of the matrix, notably the categories "person," "modality," "time" (occasion), and (external) "entity"; and (2) the *relation* of causality that may connect some of the identity categories (e.g., person) to an effect (e.g., enjoyment of a movie) the explanation of which is sought by the attributor. According to the present view, then, the *specific* ANOVA representation of attribution may only afford tests of hypotheses about causality involving the *specific* content categories of "persons," "modalities," "times," and "external entities"; the specific ANOVA cube does not afford the evaluation of any other causal categories that may interest the attributor. Now one might wish to suggest that all possible causal categories may be mapped within the proposed ANOVA dimensions. However, this suggestion does not seem too appealing: though it may be innocuously true in some sense that any potential cause must reside somewhere between (or within) the person, the environment, the modality, or the occasion, this does not meaningfully convey the cause's *specific identity*. Yet it is this specific identity rather than the characterization of the cause within some broad system of coordinates that frequently is of interest to the attributor. Thus, an attributor might wonder whether some effect produced by another person was caused "intentionally" or "unintentionally" (Weiner, 1974) or whether another's action was an "end in itself" vs. being a "means" to a further goal (Kruglanski, 1975, 1977a, 1977b). But neither the specific concepts of "intentional" vs. "unintentional" causality nor "ends" vs. "means" are readily distinguishable within the ANOVA cube.

Secondly, it is of interest to note that the causality relation to which attribution theory has been deliberately restricted, though undoubtedly important to the lay knower, is merely one among a myriad of possible relations in which persons express interest. To mention just a few such possible relations, there are various logical, algebraic, or geometric relations, biological relations (or social relations), spatial and temporal relations, and so on.

In sum, the foregoing analysis suggests that the ANOVA formulation of attribution is a special case of the more general framework of lay epistemology. In particular, the ANOVA formulation speaks to the attainment of knowledge about a specific subset of identity categories and about a single, specific relation (causality), whereas the present epistemological theory speaks to the attainment of knowledge about all possible identity categories and all possible relations. In the following section we demonstrate in addition that the proposed *method* of gaining subjective knowledge within the ANOVA formulation is a special case of the present validational logic. To accomplish this objective, we now examine closely the attributional criteria (Kelley, 1967, p. 197) presumably employed by the layman to reach subjectively valid knowledge about the external world.

Consistency logic and the attributional criteria. Recall the Kelley's four criteria of external attribution were: (1) *distinctiveness* [i.e., covariation of effect with an external entity and *consistency* (of the effect)] across (2) times, (3) modalities, and (4) persons (consensus). Note further that within the ANOVA cube *person-attribution* (Kelley, 1967, p. 203; McArthur, 1972, p 172) requires *distinctiveness* (or covariation) between the specific person and the effect (low consensus in McArthur's terminology) and *consistency* across external entities, times, and modalities with respect to the effect. Extrapolating from this, it may be assumed that confident attribution to the specific "time" (or occasion) would require *distinctiveness* between *it* and the effect and *consistency* across the remaining categories. Similarly, confident attribution to the specific modality would require distinctiveness between *it* and the effect and consistency (of the effect) across the remaining categories.

It is possible to subsume all the foregoing examples under the general *distinctiveness—consistency* rule that *inference about the causal role of a given factor would be indicated to the extent that there was evidence regarding distinctiveness of the effect with respect to the factor in question, and to the extent that the effect did not manifest distinctiveness (i.e., that it manifested consistency across factors considered as plausible alternative causes thereof).*

We proceed to show that the attributional criteria (reduced now to the general *distinctiveness—consistency* rule) are actually a special case of the validational logic of consistency elaborated earlier. Recall that according to this logic, the layman would conclude that a given belief of his (hers) is valid if implications deduced from the belief in question turned out to be consistent with the evidence. In the case dealt with by attribution theory, the layman assesses *whether the causality relation between some factors and an effect may be considered valid.* Now consider just what the deductive implications of the causality relation are. Certainly one such implication is *covariation (distinctiveness)* between the putative cause and the effect. For instance, the conjecture that some identity category, say an external entity, is the cause of some effect is *consistent* with evidence that the effect covaried with this entity or manifested distinctiveness therewith, while maintaining constancy across variation in plausible alternative causes (say, persons, modalities, times, etc.).

It is particularly noteworthy that distinctiveness or covariation is *not the sole implication* of the causality relation. As we all know correlation need not mean causation. Thus, temporal antecedence of the presumptive cause relative to the effect is generally recognized as another fundamental implication of the causality concept; yet, it is not reflected in the extant version of the attributional criteria. Finally, note that mere association (co-presence) of the presumptive cause and the effect is also an implication of causality. Although the latter implication is subsumed under the idea of covariation (that includes alike co-absence *and* co-presence), in some instances association alone might be judged by a knower as sufficient for a causality verdict with respect to a given factor as opposed to possible alternative categories. Generally, which implication of the causality relation is selected for the test should depend on its *availability* to the lay knower and on its *demarcational relevance* in given circumstances.

To summarize, in this section we have demonstrated that Kelley's attributional criteria of subjective validity constitute a special case of the validation logic of consistency. According to the present analysis, the criteria express the idea that causality of a given factor would be concluded if the "distinctiveness" implication of the causal relation were confirmed with respect to the factor in question and were confirmed with respect to competing alternative factors. It was pointed out that the causality relation contains implications other than distinctiveness that could become of interest to the attributor. Finally, it is reiterated that although the validational logic of consistency applies to the assessment of causality, it applies also to the assessment of innumerable alternative relations and identities that might interest the lay knower in various circumstances.

EMPIRICAL EVIDENCE FOR THE PRESENT THEORY

In the following sections we examine novel experimental evidence pertinent to the present analysis. The evidence bears mainly on the idea that the ANOVA model of attribution is a special case of the theory of lay epistemology outlined earlier. Pursuant to this, the data examined subsequently will be interpreted as consistent with the hypothesis that the ANOVA categories and the related attributional criteria are not universally applicable but rather are applicable only in circumstances specified by the present model. In the following review, the empirical evidence is considered in accordance with the several types of content relevancies in the epistemic sequence. We first examine an experimental study on *teleological relevance* as it might affect the applicability of the ANOVA attributional categories.

Experiment 1:
Teleological Relevance of Answers
and the ANOVA Categories

In previous studies on which major support for the ANOVA attribution model has been based (e.g., McArthur, 1972; Orvis, Cunningham, & Kelley, 1975), the subject's interest in the ANOVA attributional categories was established by experimental fiat: subjects were explicitly instructed to infer whether one or more of the constitutive ANOVA categories was causally involved in the appearance of some event. But would subjects, *of their own accord,* generally find interest in such categories? That is, will subjects attempting to causally explain sombody's response predominantly consider as potential answers the possibilities that "something about the person,"[8] "something about the stimulus," or "something about the circumstance" produced the response?

That the traditional ANOVA partition between person and environmental stimulus (the internal—external distinction) need not be universally invoked by the layman was suggested in a recent article by the senior author (Kruglanski,

---

[8]This particular phrasing was employed by McArthur (1972) in her classic attribution study.

1975). In that paper it was argued that the internal—external distinction may not be appropriate when it comes to accounting for actions or voluntary behaviors. Instead, inferences previously linked with the internal—external distinction (e.g., the inferences of enjoyment, freedom, sincerity, benevolence) were shown to follow from an alternative explanatory partition between the endogenous—exogenous categories (for a recent debate about the relative merits of the internal—external and the endogenous—exogenous partitions see Calder, 1977; Zuckerman, 1977; and Kruglanski, 1977b).

According to the present theory, ANOVA categories (like "person," "stimulus," and "circumstance") or, for that matter, Kruglanski's (1975) "endogenous" categories are merely specific contents, or more precisely, specific identificational categories. Whether or not they should figure in the knower's causal hypotheses should depend on their *teleological relevance* to the knower's underlying motive and on their momentary availability to this person. In the experiment to be described presently, the question of availability was resolved *via direct presentation* to subjects of several potential answers to a causal question; this rendered the various answers (equally) available. Mainly, however, the present experiment manipulated the subject's underlying motive such that in the one case the categories of "person" and "stimulus" were more teleologically relevant thereto than were various alternative categories; and in the other case, they were less relevant than were various alternative categories. The subject's task in each case was to choose between a set that contained the ANOVA categories of "person" and "stimulus" and a set of alternative categories that variously comprised the categories of "means and ends;" "ability and effort," and "intentionality and unintentionality." The alternative categories just listed were chosen merely because they have been mentioned in the general attributional literature (e.g., in Weiner et. al., 1972; Weiner, 1974; Kruglanski, 1975). Consistent with the present viewpoint, they were *not* treated as an exhaustive set of alternatives to the ANOVA categories.

Each subject was given 12 problems coupled with 12 behavioral events modeled after those of McArthur (1972). In the terminology of the present model of the behavioral event constituted the epistemic background furnished to the attributor, and the specific problem defined the attributor's underlying motive. For example, one of the events was "John plans to see movie X on Saturday night." This event was coupled once with the problem "whether to invite John to a party given on Saturday" and another time with the problem "whether it is worthwhile to buy tickets to movie X for yourself and your friend." In each case, the subject's task was *to select the set of answers (hypotheses) that he or she would want validated* in order to solve the problem at hand. For each of the 12 problems there were two sets of potential answers from which to choose. One set always comprised the person—stimulus categories; the other set contained answers referring to either the means—ends, ability—effort, or intentionality—unintentionality categories.

For each of the 12 events presented to the subject, one of the two associated problems was made relevant to the person–stimulus categories and the remaining problem to the alternative set of categories. Thus, in the example above, the means–ends categories are relevant to the problem of deciding whether to invite John to a Saturday night party: To the extent that John's contemplated visit to the movies is a mere means of combating loneliness, he might accept instead an invitation to the party. But John's acceptance of the invitation seems much less likely if the intended visit to the movies is an end in itself, i.e., if watching the particular movie is what he specifically wants to do on the occasion in question.

By contrast, the person–stimulus categories are obviously relevant to the problem of deciding whether to purchase tickets for oneself to the movie selected by John. To the extent that John's plan to attend the movie reflects something about the movie rather than about John, other people – including the attributor – can be expected to share John's positive opinion of the movie.

In actual course of the present experiment, each subject was given four events in which the choice was between the person–stimulus and the means–ends sets, four events in which the choice was between the person–stimulus and the ability–effort sets, and four events in which the choice was between the person –stimulus and the intentional–unintentional sets. For one-half of the subjects, a given event appeared with one of the two problems constructed around this event; for the remaining half of the subjects, the same event appeared with the second associated problem.

The experimental hypothesis presently tested was that subject attributors attempting to causally explain an event would prefer the teleologically relevant over the irrelevant answers. More specifically, it was predicted that the alternative categories would be preferred over the ANOVA categories of "person" and "stimulus" in situations where the former would be more teleologically relevant to the attributor's problem (underlying motive). Similarly, where the ANOVA categories would be more teleologically relevant, they would be preferred by the attributor over the alternative categories.

Results pertinent to the above predictions are summarized in Table 11.1. As expected, an overall analysis of variance indicated that the frequency of preferences for the alternative (over the ANOVA) categories was higher when the former (vs. the latter), were more relevant to the attributor's motive ($p <$ .001). The *type* of alternative categories (i.e., means–ends vs. ability–effort, vs. intentionality–unintentionality) did not yield a main effect; nor did it interact with the problem presented to subjects.

More specifically, the mean frequencies of preference for the alternative (over the ANOVA) categories was significantly *above* unity[9] ($p <$ .005) when the alternative categories were teleologically relevant for the attributor, which in-

---

[9]Given the particular scoring system employed in this study, unity represents a lack of preference among the two sets of categories.

TABLE 11.1
Mean Frequency[a] of Preferences for Alternative
Over ANOVA (Person—Stimulus) Causal Categories

Teleologically Relevant Categories	Type of Alternative Categories		
	Means—Ends	Ability—Effort	Intentionality—Unintentionality
Alternative	1.53	1.46	1.46
ANOVA (Person—stimulus)	.73	.67	.80

[a]Scored on a 3-step scale (0, 1, 2) with unity indicating an equal number of preferences (one) for alternative vs. ANOVA categories and vice versa.

dicates that in this situation they were indeed preferred over the ANOVA categories by most subjects. Those mean frequencies were *below* unity ($p <$ .06) when the ANOVA categories were teleologically relevant for the attributor; that is, in this situation the ANOVA categories tended to be preferred over the alternative ones by most subjects.

To summarize, the results of this study support the idea that the attributor's interest in specific causal categories is a function of the degree to which the categories in question are teleologically relevant to this person. The specific constitutive categories of the ANOVA cube may be teleologically relevant to some underlying motives but not to others. Hence, they may not be considered universally or uniformly applicable in all attributional circumstances.

Experiment II:
Deductive Relevance of Information
and the Attributional Criteria

According to the present theory, the layman's validation of any hypothesis (including hypotheses about causality) involves the check for consistency between implications deduced from the hypothesis and information pertinent to these implications. Thus, information regarding distinctiveness across entities and consistency across time, modalities, and persons should suggest a confident attribution to an external entity, because the hypothesis "entity X was cause of effect E" *implies* among other things a distinctiveness outcome between X and E, as well as constancy (consistency) of the effect E across times, modalities, and persons. The latter categories are viewed as competing alternatives to the external-entity hypothesis. Hence the constancy outcome (i.e., the lack of distinctiveness) associated therewith renders their causal role less plausible and lends support to the external-entity hypothesis. Data consistent with the above interpretation were obtained in the classic attribution experiment by McArthur (1972).

Among other things, McArthur (1972) found that the frequency of person attributions was greater with *low* than with *high* consensus information (or in the present terms with high distinctiveness of the response with respect to a given person), with *low* than with *high* distinctiveness information across stimuli (or in the present terms with high consistency of the response across different stimuli), and with *high* than with *low* consistency information across different times. The frequency of stimulus attribution was greater with *high* vs. *low* distinctiveness information, with *high* vs. *low* consensus information (i.e., high consistency across persons), and with *high* vs. *low* consistency over times.

McArthur's (1972) results strongly suggest that distinctiveness, consistency, and consensus information could be useful in the generation of causal inferences. According to the present theory, however, *such information should be useful only for the assessment of (causal) hypotheses to which it may be deductively relevant and should not be useful otherwise.* More specifically, consensus, consistency, and distinctiveness information should only be of interest when validating the causal categories of person, stimulus, and circumstance and should be relatively devoid of interest when evaluating numerous possible alternative categories that can be involved by the layman.

Using a methodology similar to that of Experiment 1, subjects were presented with 12 events (e.g., John laughed at the comedian) and with causal questions concerning the events at stake. Each event appeared once with a causal question phrased in terms of the person—stimulus categories, or with a causal question phrased in terms of the now-familiar alternative categories; means—ends, ability —effort, or intentionality—unintentionality. The presentation of the stimulus materials was so arranged that for a given subject each event was coupled with only one of the two causal questions constructed for this event, i.e., either with the person—stimulus question or with the alternative question.

For each event the subject was asked to decide which among two sets of information was more useful in providing an answer to the particular causal question posed with respect to this event. One set contained information about the attributional criteria of consensus, distinctiveness, and consistency, whereas the alternative information set contained information *deductively relevant* to the alternative causal categories associated with the event (e.g., to the means—ends categories).

For example, consider one of the events employed in this research — "John laughed at the comedian." Some subjects were instructed to explain this event in terms of the person—stimulus question; i.e., they were instructed to decide whether "something about John probably caused him to laugh at the comedian," or "something about the comedian probably caused John to laugh at him." Other subjects were instructed to explain the same event in terms of an alternative pair of categories (e.g., the means–ends categories). These subjects were instructed to decide, e.g., whether "something about the comedian's jokes probably caused John to laugh at him," or "something about the comedian's status probably caused John to laugh at him." Regardless of the causal question posed for the subject in connection with a given event, he (or she) was required to choose among two sets of information. One set consisted of the criterial information of consensus, distinctiveness, and consistency, modeled after McArthur (1972). In the example at hand, this set included information that "almost everyone who hears the comedian laughs at him [high consensus];" "John does not laugh at almost any other comedian [high distinctiveness];" and "in the past John has almost always laughed at the same comedian [high consistency]." The second set contained information pertinent to the alternative causal categories, e.g., to the means–ends categories. In the present example the information was: "John does not usually enjoy humor;" "The comedian is John's supervisor at work;" and "John is afraid of losing his job."

As a check on the relevance manipulation, the subjects were asked to rate the extent to which each informational set was relevant to the causal question posed. Analysis of these data indicated that subjects indeed perceived the criterial set (of consistency, consensus, distinctiveness) as more relevant to the ANOVA (person–stimulus) than to the alternative categories and that they perceived the alternative set as more relevant to the alternative than to the ANOVA categories ($p < .05$ for both comparisons).

Mean frequencies of preferences for the alternative over the criterial information sets are displayed in Table 11.2. An overall analysis of these data disclosed, as expected, that the alternative information sets were preferred over the criterial sets more often when the alternative categories were being assessed than when the ANOVA categories were being assessed ($p < .01$). The type of alternative categories under assessment did not significantly interact with the relevance of informational sets, though it did yield a significant main effect ($p < .01$)

TABLE 11.2
Mean Frequency[a] of Preferences for Alternative
Over "Criterial" Informational Sets

Deductively Relevant Information	Type of Alternative Categories		
	Means–Ends	Ability–Effort	Intentionality– Unintentionality
Alternative	1.42	1.26	1.05
Criterial	1.21	.84	.31

[a]Scored on a 3-step scale (0, 1, 2) with unity indicating an equal number of preferences (one) for alternative vs. criterial sets and vice versa.

explicable primarily in terms of the contrast in magnitude of preferences for the alternative sets between the intentionality—unintentionality vs. the means—ends conditions.

Furthermore, the mean frequency of preferences for the alternative informational sets was 1.24 in the conditions where the alternative causal categories were being assessed; this is significantly ($p < .05$) above unity and suggests that under these circumstances subjects predominantly favored the alternative vs. the criterial informational sets. In the conditions where the ANOVA categories were being assessed, the mean preference frequency for the alternative sets was .78; this is marginally ($p < .10$) below unity and implies a preference for criterial information when the ANOVA causal categories were being assessed.

In sum, the results of Experiment II support the idea that the attributor's interest in given information is a function of the degree to which this information is deductively relevant to the hypothesis under assessment; the specific attributional criteria are deductively relevant to some causal categories (the ANOVA categories) and not to others. Hence they may not be considered universally applicable across varying attributional circumstances.

Experiment III:
Deductive Relevance and the Impact of
Consensus Information

If the layman's interest in criterial information (of the consensus, consistency, and distinctiveness type) is contingent on this person's interest in assessing the traditional ANOVA categories, *the provision* of criterial information should have no appreciable impact on attributions to alternative causal categories. The foregoing implication is relevant to the recent discussion as to whether consensus information should have considerable impact on attributions, given the mounting evidence that it does not universally have such effects (cf. Nisbett, Brogida, Crandall, & Reed, 1977). Nisbett et al. write on this point as follows:

> It comes as a surprise to discover that one of ... [the] fundamental axioms [of attribution theory] has found virtually no support in subsequent research. This is the notion that people respond to consensus information in allocating cause. Theory and common sensse notwithstanding, there is mounting evidence that people are largely uninfluenced in their causal attributions by knowledge of the behavior of others [p. 2].

According to Nisbett et al. (1977), the McArthur experiment (1972) mentioned earlier is among those studies that show a remarkable weakness of consensus information, because in this research the consensus variable accounted for less than 3% of the variance, whereas consistency and distinctiveness information accounted for considerably greater proportions of the variance.

The present theory suggest that provision of consensus information should affect attributions *only when* the attributor's interest is in causal categories to which consensus or dissensus is *deductively relevant.* Could it be, then, that where consensus information had failed to affect attributions this was so because the deductive relevance of consensus was not sufficiently apparent to the subject? The following experiment was designed to provide evidence regarding the above possibility, in particular reference to the relative weakness of consensus in the McArthur (1972) experiment.

For instance, in McArthur's study consensus information was expected to make a difference with respect to the question of whether a given effect could be attributed to "something about a person (e.g., "something about John"). Low consensus was expected to render such attribution more likely, and high consensus (i.e., information that most other persons have produced responses similar to the actor's), less likely. But the category "something about the person" is somewhat ambiguous and amenable to two divergent interpretations: "something about the person" seems to imply "something *unique* about the person" that distinguishes him (her) from most other people, or it could imply *any* definable characteristic of this person (e.g., his or her sex or nationality), not necessarily a unique characteristic. Now consensus information is deductively relevant only to the former interpretation and not to the latter. If the attributional hypothesis being assessed concerns a category unique to the person, it follows deductively that uniqueness of the response should be expected; so that evidence of high consensus (or non-uniqueness) is inconsistent with the hypothesis, whereas evidence of low consensus is consistent therewith. However, when the attributional hypothesis being assessed concerns *any* definable characteristic of the person, this has no implications regarding consensus. A person could emit a response *because* of something about him or her — say, because he is male, American, or overweight — without this having, any clear bearing on how many other persons should have emitted this response.

Subjects in the present experiment received a booklet with descriptions of eight events in which a person performs a response toward an object. With each event, there appeared an additional statement conveying low- or high-consensus information, i.e., that few other persons or that most other persons performed a similar response toward the object. For example, one of the events was, "Peter disliked the history professor," and the accompanying information was that hardly anyone disliked the history professor (low consensus) or that almost everyone disliked the history professor (high consensus). The subject received four of the eight events in each of the two consensus conditions; the experimental task was to make an inference about what probably caused the event. In one condition, henceforth referred to as the *standard condition,* subjects were asked whether "something about the actor (e.g., Peter) caused the response." This phrasing was identical to McArthur's (1972) and, presumably, was open to the two interpretations mentioned earlier. The latter interpretations were explicated in the remaining conditions of the present research. In one condition, subjects were asked whether "something *unique* about the actor" caused the response. In another condition, subjects were asked whether some (definable) "identity" of the actor caused the response, where the term "identity" was explained to mean any well-defined category (e.g., of Americans, females, stamp collectors) to which a person might belong.

Toward the end of the experiment, subjects in the "standard condition" were asked to rate on a 7-point scale the extent to which they interpreted "something about the actor" to mean "something *unique*" about the actor and the extent to which they interpreted it to mean an "identity" of the actor as defined above. As expected, subjects interpreted "something about" both as "something unique" and as an "identity" to a significantly greater degree than the low point of the 7-step scale labeled at the ends "not at all" and "very much" ($p < .01$). However, they perceived themselves making the "something unique about" interpretation significantly more than the "identity" interpretation ($p < .01$).

The effects of consensus information on attributions to the three causal categories presently investigated are displayed in Table 11.3. As expected, the overall ANOVA yielded a significant interaction ($p < .001$) between the consensus variable and the causal category assessed. More detailed tests indicate that consensus information had no significant effect on attributions in the "identity" condition ($F < 1\%$); it had a marginally significant effect (low consensus giving stronger attributions) on attributions in the "something about" condition ($p < .06$); and it did have a significant effect in the "something unique about" condition ($p < .01$).

It is also noteworthy that the mean difference between the two consensus conditions was significantly greater in the "something unique about" condition than in the "identity" condition ($p < .025$, one-tailed), but it was *not* significantly greater in the "something about" than in the "identity" condition. The difference between low and high consensus was marginally greater ($p < .11$, one-tailed) in the "something unique about" than in the "something about" condition.

TABLE 11.3
Attributions[a] to Personal Categories
Under High- and Low-Consensus Conditions

		Causal Category Assessed		
		Personal Identity	*Something About Person*	*Something Unique About Person*
Consensus	High	3.91	4.05	2.29
	Low	4.63	5.64	5.31

[a]Higher numbers indicate more confident attributions.

In sum, the data reviewed so far are consistent with the argument that the layman's interest in the attributional criteria, including consensus, is restricted to situations to which criterial information is deductively relevant. In Experiment III, just reviewed, consensus information had greatest impact when the causal category being assessed had an unambiguous implication involving consensus (as in the "something unique" condition), had somewhat lesser effect when the category being assessed had an ambiguous implication concerning consensus (as in the "something about" condition), and had *no* appreciable effect when the category being assessed ("identity") had no implication concerning consensus.

Experiment IV:
Demarcational Relevance and
Attributional Criteria

The research reported above suggests that traditional attribution criteria (consensus, consistency across times, and modalities, and distinctiveness across external entities) may not be universally applicable across attributional situations but may be restricted to the assessment of causal categories to which such criteria are deductively relevant. But according to the present analysis, even when deductively relevant, attributional criteria will not be of *uniform* interest to the lay knower. First they may not be uniformly available to the attributor. But availability aside, in situations where the attributor chooses among several competing hypotheses, this person's interest in specific criterion of information would be affected by its *demarcational relevance.* More specifically, a deductively relevant attributional criterion would be of greater interest to the lay knower when it demarcates among plausible competing hypotheses than when it does not. The experiment reported subsequently tests the demarcational-relevance hypothesis just stated.

In previous experiments included in this chapter, naive hypotheses based on traditional ANOVA categories were the only ones the subject was instructed to assess. By contrast, in the present experiment, hypotheses based on the ANOVA categories were rival alternatives to a different hypothesis of some special interest to the attributor.

Specifically, the subjects were presented with information about an event and about the causal attribution made by an expert. The expert's attribution was of interest to someone whose role it was the subject's task to assume. In the proportion of the experiment pertinent to the present discussion, there were three conditions. In the M condition (for modality), additional information given to subjects suggested that the modality of the attributor's (expert's) interaction with the event might have rendered false this person's causal attribution. In the P condition (for person), the additional information was that the expert had a strong vested interest in the attribution, thus casting doubt on his objectivity. Finally, in the T condition (for time), the additional information was that the attribution was rendered in unusually disruptive circumstances, implying that this (time-contingent) factor may have strongly biased the attribution. To summarize then, in the M, P, and T conditions respectively, the traditional ANOVA categories of modality, person, and time provided a basis for competing alternatives to the hypothesis of interest.

In each condition, the subject responded to three specific questions that inquired whether, before accepting the expert's recommendation (implied by his/her attribution), he (she) desired information regarding: (1) consensus, in the form of an attribution made by another, similar expert; (2) consistency across modalities, in the form of an attribution reached *via* a different modality; (3) consistency across times, in the form of an attribution reached at a different time. As should be clear, these questions represent three of the attributional criteria specified by Kelley (1967).

In present terms, each of the questions is *deductively relevant* to the focal hypothesis: *if* the expert's attribution is indeed valid, it should not be the case that a *different* attribution would be reached via a different modality, by another similar expert, or at a different point in time. Moreover, each of these questions is *demarcationally relevant* in a different experimental condition. Evidence regarding consistency across modalities demarcates the attribution rendered by the expert in condition M against an alternative hypothesis based on faulty modality. Evidence regarding consensus demarcates this attribution in condition P against an alternative hypothesis of personal bias. Evidence regarding consistency across times demarcates the attribution in condition T against an alternative hypothesis based on time-bound circumstances. The above, somewhat abstract, notions are illustrated below by a sample event employed in the present research.

The physician.   In these event frame the expert attributor is a physician, Dr. Jones, who shares an office with a colleague, Dr. Scott, and who one morning is approached by a patient with a skin rash. Dr. Jones examines a sample of the skin via a microscope and concludes that *milk* is the causal agent involved in the rash and that the patient must therefore refrain from milk for some time (the recommendation).

The M condition.   Unbeknownst to Drs. Jones and Scott, the new electronic microscope used to examine the skin sample contained a defective electronic circuit.

The P condition.   Dr. Jones had recently published an article in which he identified milk as a major cause of many allergies. The article was attacked by many critics, and Dr. Jones was hurt very much.

The T condition.   On the particular morning during which Dr. Jones conducted the pertinent skin examination, things in the office were unusually hectic.

A dentist occupying an adjoining suite had been robbed and beaten, and the police were investigating. Understandably, Drs. Jones and Scott were extremely upset.

Specific questions inquired into the extent to which the subject, before accepting Dr. Jone's recommendation, would insist on: (1) having Dr. Jones repeat the test via a *different microscope* (consistency across modalities); (2) having *Dr. Scott* examine the skin sample (consensus); (3) having Dr. Jones repeat the test on *another day* of the week (consistency across times).

Five event frames were created, each with the three conditions (M, P, T) above and with two additional conditions D (for distinctiveness) and N (for normal); presentation of these latter two conditions is beyond the scope of the present discussion. This resulted in 25 combinations of events and conditions. The 25 combinations were divided into 5 groups of 5 combinations (randomly ordered) such that all subjects viewed a given event frame and a given condition *only once*.

It was hypothesized that in circumstances where a given causal category provides a basis for a complementary hypothesis, subjects would exhibit greater interest in the *demarcationally relevant* criterion than in the remaining criteria. Data pertinent to the foregoing hypothesis are summarized in Table 11.4

As expected, in the M condition, interest in consistency over modalities exceeded interest in the remaining criteria ($p < .0005$). In the P condition, interest in consensus information exceeded interest in the remaining criteria ($p < .0005$). In the T condition interest in consistency across times exceeded interest in the remaining criteria ($p < .025$).

In sum, the results of Experiment IV lend support to the notion of demarcational relevance. It appears that even when several kinds of information are deductively relevant to a given causal category, they need not be of uniform

TABLE 11.4
Mean Amounts[a] of Criterial Information Requested

Criterion of Information	Basis for Competing Complementary Hypothesis		
	Modality	Person	Time
Consistency over modalities	5.18	2.71	3.56
Consensus	2.67	5.69	4.09
Consistency over times	2.69	2.96	4.64

[a]Higher numbers indicate greater amounts of information requested.

interest to the attributor. Rather, interest in deductively relevant information should be a positive function of the extent to which it demarcates the central hypothesis from an alternative one entertained by the attributor in the course of a particular epistemic episode.

Demarcational Relevance, Availability, and Spontaneous Utilization of Attributional Criteria

According to the present theory, an attributor who attempts to assess the causal role of a given category will actively seek out validational information in accordance with: (1) the *deductive relevance* of such information to the cateogry assessed; and (2) the *mementary availability* to the attributor of the specific deduction. Furthermore, an attributor whose objective is to choose among several competing hypotheses would seek out information about those deductive implications that are *demarcationally relevant* as well as momentarily *available.*

Suppose now that a subject is asked to assess the causal role of a given ANOVA category (say, of "person" or "external stimulus") to which the "attributional criteria" are deductively relevant and is free to specify the information he (or she) would need to accomplish the assessment. Whether a given attributional criterion is specifically sought in these circumstances should depend on: (1) what (if any) competing hypotheses are invoked by the attributor and need to be ruled out, and (2) the momentary availability to the attributor of deductions involving the criteria.

In light of the above analysis, consider the recent experiment by Garland, Hardy, and Stephenson (1975). These authors presented subjects with statements modeled after McArthur (1972). The statements by Garland et al. (1975) described "the responses of various people to different stimuli (e.g., Mary got an "A" on the chemistry exam) [p. 613]." Each statement was followed by an open-ended question asking what further information the subject would require to make a person attribution (e.g., "What further information would you require in order to say that Mary is intelligent? . . . ") or a stimulus attribution. Requests for information were classified by independent coders into the three informational categories of consensus, consistency, and distinctiveness.

Garland et al. (1975) noted with surprise the

> "relatively low percentage of total information requested which was assignable to Kelley's three informational categories. Only 23 percent of the data could be coded into consistency, consensus and distinctiveness leaving 77 percent in a residual category. Expressed as a percentage of the total information requested, consistency accounted for 3 percent, consensus for 7 percent and distinctiveness for 13 percent [p. 615]."

The authors concluded that their "inability to code so much data suggests that Kelley's attribution model is less than complete in its ability to describe the kinds of information which will be sought by attributors in a natural situation where they are free to search for any information they choose [p. 615]."

But within the present theoretical framework, the low percentage of information falling into the criteria of consensus, consistency, and distinctiveness is not surprising. The mere specification of a category to be assessed (e.g., the "person" category) should not uniquely determine the kinds of information sought by the attributor. To reiterate, given some category to be assessed, the information actually sought should depend on: (1) whatever competing hypotheses the attributor might invoke and hence seek to rule out; and (2) the momentary availability of deductions involving the criteria.

For instance, Garland et al. (1975) describe a subject who desires to determine whether Mary's success on the chemistry exam warrants a person attribution (i.e., is explicable in terms of Mary's intelligence). Now, it should be clear that McArthur's (1972) criteria for person attribution (low consensus, high consistency, and low distinctiveness) are implications of the identificational concept "stable personal property" (hence the requirement of temporal consistency), *unique* to the person (hence the requirement of low consensus), and *general* (hence the requirement of low distinctiveness across specific stimuli like the specific chemistry exam). But although the common notion of high intelligence does imply *stability, uniqueness,* and *generality,* this hardly exhausts the implications of the intelligence concept. Besides being unique, stable, and general, intelligence is also inborn rather than learned, concerns ability rather than motivation, does not require cheating to yield success, etc. Under the appropriate conditions of demarcational relevance and availability, information about each of the above implications rather than about the "attributional criteria" might well be sought by the lay knower. Thus the results of Garland et al. (1975) are quite compatible with the present model.

The Experimental Evidence: Summary

The experimental evidence reviewed in the preceding sections warrants the following conclusions:

1. Specific causal categories may be of interest to the lay knower in accordance with their teleological relevance (Experiment I above) and availability. The traditional ANOVA categories need not be universally *relevant* to the attributor's epistemic goal; nor need they be universally *available* to this person. Hence, they need not be invoked by the lay knower across all types of attributional circumstances.

2. Specific evidence may be of interest to the lay knower in accordance with its deductive relevance to the categories validated. Deductive relevance of evidence will determine: (a) whether or not its provision will *have impact* on attributions (Experiment III); and (b) whether or not the attributor will *seek* such evidence

when attempting to validate specific causal hypotheses (Experiment II). The attributional criteria (of consistency across times and modalities, consensus, and distinctiveness across environmental stimuli) are not deductively relevant to all possible causal categories. Hence, provision of criterial information need not have universal impact on attributions; nor need criterial evidence be universally sought by the attributor.

3. Even when deductively relevant (and available), specific evidence may not be sought by the attributor when it lacks *demarcational relevance* to the attributor's objective to decide among various competing hypotheses. A given criterion of information may sometimes be deductively, but not demarcationally relevant. Under those circumstances criterial information would not be particularly sought by the attributor (Experiment IV). This suggests the following corollary.

4. The layman's desire to validate a given ANOVA category (e.g., something about the person) need not evoke spontaneous search for the attributional criteria. The latter may not appear demarcationally relevant to the specific competing alternatives considered by the attributor, and/or it may not be momentarily available (as may have been the case in research by Garland, et al., 1975).

More broadly, the experimental evidence considered earlier supports the present partition between the content, logic, and course aspects of lay epistemology. Specifically this research suggests that the lay knower's interest in specific cognitive contents (e.g., specific causal categories or specific validating information) may vary considerably from one situation to the next and from one person to the next. Furthermore, the situational elicitation of contents seems to occur in accordance with the relevancies that interrelate the different constitutive elements of the epistemic course.

Finally, the extant evidence seems thoroughly supportive of the proposal that the validational logic of lay epistemology rests on the principle of consistency. For example, in experiments by McArthur (1972), Orvis, Cunningham, and Kelley (1975), or Experiment III in this paper, specific causal hypotheses were assumed more valid when the implications derived from such hypotheses (e.g., that an effect was caused by something unique about the person) were consistent with the evidence provided (few other persons produced a similar effect) than when the evidence was inconsistent with such implications. In passing, it is mentioned that the extensive evidence regarding Kelley's discounting rule (reviewed in Kelley, 1971, 1973; Kruglanski, 1975; Kruglanski, Schwartz, Maides, & Hamel, in press), whereby the assumed validity of a given hypothesis is lowered by the presence of a competing alternative hypothesis, is also interpretable in terms of the consistency logic. The presence of two mutually exclusive explanations of some effect suggests inconsistency (or logical contradiction), since they may not both be correct at the same time.

CONCLUSION

This paper featured a novel attributional perspective of which the extant ANOVA formulation was assumed a special case. In essence, the present attributional framework is a conceptualization of naive epistemology founded on the crucial partition between the contents of knowledge, the logic used to validate such knowledge, and the course of the epistemic episode.

It has been noted that the contents of lay knowledge are highly diversified, are continually evolving, and include in a given instance of knowledge-seeking behavior a restricted, situationally specific portion of the knower's conceptual repertory. By contrast to the situational specificity of contents, the course of an epistemic episode and the validational logic were assumed invariant across situations. The course of an epistemic episode has been characterized in terms of the knower's *underlying motive,* one that operates against a certain *background of knowledge* and gives rise to an *epistemic intention.* This leads to the *formulation of questions* and the generation of *conjectural answers* validated against relevant evidence. In line with recent developments in the conception of science, the validational logic of knowledge was assumed to rest on the *principle of consistency;* the individual's experience of valid knowledge was assumed to be a function of the degree to which his or her network of interrelated beliefs did not contain contradictions.

It was argued that the ANOVA representation of attribution is a special case of the present theory in the following senses:

1. The ANOVA model deals specifically with the problem of causality, whereas within the present framework causality is merely one among many possible relations of interest to the layman;

2. The attributional categories implicit in the ANOVA model (notably the categories of "person," "entity" [or stimulus], "modality," and "time") are merely four among numerous possible categories potentially considered as causes of various events.

3. The attributional criteria (of consensus, consistency across times, modality, and distinctiveness among entities) represent, at root, the validational logic of consistency whereby a given hypothesis is believed whenever its implications are consistent with the evidence and is disbelieved whenever its implications are inconsistent with the evidence.

More specifically, the criteria confirm the covariance (distinctiveness) implication of the causality relation with respect to a given ANOVA category (e.g., external entity) and disconfirm the same implication (by showing constancy or non-distinctiveness) with respect to the competing ANOVA categories (time, modality, person). According to this analysis, significance of the specific criteria is restricted to the case in which an attributor has: (1) interest in one of the ANOVA categories (say, an external entity); and (2) entertains as competing alternatives all of the remaining ANOVA categories (i.e., time, modality, person). To the extent that an attributor lacks interest in a given ANOVA category, the corresponding "criterion" in which distinctiveness or consistency is conjoined to the category in question (say, modality) is rendered irrelevant.

In the following final paragraphs, we discuss in turn two potentially fruitful directions for future research within the present framework. One such line of inquiry involves extending the present analysis to several areas of attributional theoreizing that have not been dealt with presently. A second line of inquiry involves exploring the place of error and bias in the present epistemological scheme.

1. Because of limited space, the present essay has dealt exclusively with the ANOVA model of attribution. This particular model is certainly a major organizing perspective for viewing attribution, but it is not the sole influential conceptualization of the phenomena involved. Important attributional theorizing is represented by Jones and Davis's (1965) model of correspondent inferences (synthesized by Jones & McGillis, 1976, with the ANOVA cube), in Kelley's ideas (1971, 1972, 1973) about attributional principles and schemata, and in Ajzen's and Fishbein's (1975) formulation of attributional inference in terms of the Bayesian rule. We believe that the present theoretical framework allows a novel interpretation of these various ideas and their integration under a common set of principles; but in order to cogently demonstrate this, further exegetic and empirical work is required.

2. The question of bias and error and its place in a rational attribution model has not received a fully adequate answer within existing attributional formulations. To be sure, major attribution theorists (e.g., Jones & Davis, 1965; Kelley, 1967, 1971) have acknowledged the problem and have made some intriguing stabs at a resolution (as in Jones and Davis's (1965) hypothesis of "hedonic relevance" or in Kelley's (1971) postulate of "control" motivation). Presently, however, discussions of bias and error seem loosely attached to the rational attribution models, without clear specification of the divergent circumstances in which biased vs. unbiased inferences might obtain.

Several interesting questions arise in this regard. For example, insofar as attribution theory (or the present theory of lay epistemology) lays claim only to the explanation of *subjective* validity, is it at all appropriate to view errors (i.e., departures from *objective* truth) as exceptional? Secondly, it is not clear whether possible motivational influence on epistemological processes (e.g., ego-defensive influences) should be *contrasted* with rational attributional models (cf. Ajzen & Fishbein, 1975; Miller & Ross, 1975) as opposed to being *interpret-*

able within such models. For instance, in the present framework, the "rational" aspects seem restricted only by the validational logic of *consistency* and the several *relevancy* relations among elements of the epistemic sequence. But motivation might influence only the momentary *availability* to the attributor of various cognitive contents (e.g., by selective attention or repression), and hence such influence need not be considered irrational. Clearly, further analysis and research are needed to elucidate these issues, but the potential value of such clarifications seems to be quite considerable and to have broad ramifications.

REFERENCES

Ajzen, I., & Fishbein, M. A Bayesian analysis of attribution processes. *Psychological Bulletin,* 1975, *82,* 261–277.

Bacon, F. *Essays.* London: Dent, 1962.

Bain, A. *The senses and the intellect.* New York: Appleton, 1874.

Bartley, W. W. III. *The retreat of knowledge.* New York: A. A. Knopf, 1962.

Calder, B. D. Endogenous–exogenous versus internal–external attribution explanations. *Personality and Social Psychology Bulletin,* 1977, *3,* 400–406.

Campbell, D. T. Evolutionary epistemology. In P. A. Schilpp (Ed.), *The philosophy of Karl Popper* (Vol. 14, I and II), *The libarary of living philosophers.* La Salle, Ill.: Open Court Publishing Co., 1974.

Campbell, D. T., & Stanley, J. C. Experimental and quasi-experimental designs for research on teaching. In N. L. Gage (Ed.), *Handbook of research on teaching.* New York: Rand McNally & Co., 1963.

Duhem, P. [*The aim and structure of physical theory.*] *La théorie physique: Son objet, sa structure* (2nd ed.). 1906. Translation of second edition by Princeton University Press, 1954.

Festinger, L. *A theory of cognitive dissonance.* New York: Harper & Row, 1957.

Feyerabend, P. Consolations for the specialist. In I. Lakatos & A. Musgrave (Eds.), *Criticism and the growth of knowledge.* Cambridge England: Cambridge University Press, 1970.

Garland, H., Hardy, A., & Stephenson, L. Information search as affected by attribution type and response category. *Personality and Social Psychology Bulletin,* 1975, *1,* 612–615.

Harvey, J. H., Ickes, W. J., & Kidd, R. F. (Eds.) *New directions in attribution research* (Vol. 1). Hillsdale, N.J.: Lawrence Erlbaum Associates, 1976.

Heider, F. *The psychology of interpersonal relations.* New York: Wiley, 1958.

Hempel, C. G. The empiricist criterion of meaning. In A. J. Ayer (Ed.), *Logical positivism.* New York: The Free Press, 1959.

James, W. Great men, great thoughts and the environment. *The Atlantic Monthly,* 1880, *46,* 441–459.

Jevons, S. *The principles of science.* London: Macmillan, 1974.

Jones, E. E., & Davis, K. E. From acts to dispositions: The attribution process in person perception. In L. Berkowitz (Ed.), *Advances in experimental social psychology* (Vol. 2). New York: Academic Press, 1965.

Jones, E. E., & McGillis, D. Correspondent inferences and the attribution cube: a comparative reappraisal. In J. H. Harvey, W. J. Ickes, & R. F. Kidd (Eds.), *New directions in attribution research* (Vol. 1). Hillsdale, N.J.: Lawrence Erlbaum Associates, 1976.

Kelley, H. H. Attribution theory in social psychology. In D. Levine (Ed.), *Nebraska Symposium on Motivation,* 1967. Lincoln, Neb.: University of Nebraska Press, 1967.

Kelley, H. H. *Attribution in social interaction.* Morristown, N.J.: General Learning Press, 1971.

Kelley, H. H. *Causal schemata and the attribution process.* Morristown, N.J.: General Learning Press, 1972.

Kelley, H. H. The processes of causal attribution. *American Psychologist,* 1973, *28,* 107–128.

Kruglanski, A. W. The endogenous–exogenous partition in attribution theory. *Psychological Review,* 1975, *82,* 387–406.

Kruglanski, A. W. Endogenous attribution and intrinsic motivation. In D. Greene & M. R. Lepper (Eds.), *The hidden costs of reward.* Hillsdale, N.J.: Lawrence Erlbaum Associates, 1977, in press. (a)

Kruglanski, A. W. The place of naive contents in a theory of attribution: Reflections on Calden's and Zuckerman's critiques of the endogenous–exogenous partition. *Personality and Social Psychology Bulletin,* 1977, *3,* 592–605.

Kruglanski, A. W., Schwartz, J. M., Maides, S. A., & Hamel, I. Covariation, discounting and augmentation: Toward a clarification of attributional principles. *Journal of Personality,* in press.

Kuhn, T. S. *The structure of scientific revolutions.* Chicago: University of Chicago Press, 1962.

Kuhn, T. S. Logic of discovery or psychology of research. In I. Lakatos & A. Musgrave (Ed.), *Criticism and the growth of knowledge.* Cambridge, England: Cambridge University Press, 1970.

Lakatos, I. Criticism and the methodology of scientific research programmes. *Proceedings of the Aristotelian Society.* 1968, *69,* 149–186.

Lakatos, I. Falsification and the methodology of scientific research programmes. In I. Lakatos & A. Musgrave (Eds.), *Criticism and the growth of knowledge.* Cambridge, England: Cambridge University Press, 1970.

Mach, E. On the part played by accident in invention and discovery. *Monist,* 1895, *6,* 161–175.

McArthur, L. The how and what of why: Some determinants and consequences of causal attribution. *Journal of Personality and Social Psychology,* 1972, *22,* 171–193.

Mill, J. S. *Collected works.* Toronto: University of Toronto Press, 1963.

Miller, D. T., & Ross, M. Self-serving biases in the attribution of causality: Fact or fiction? *Psychological Bulletin,* 1975, *82,* 213–225.

Nisbett, R. E., Borgida, E., Crandall, R., & Reed, H. Popular induction: Information is not necessarily informative. In J. Caroll & J. Payne (Eds.), *Cognition and social behavior.* Hillsdale, N.J.: Lawrence Erlbaum Associates, 1977.

Orvis, B. R., Cunningham, J. D., & Kelley, H. H. A closer examination of causal inference: The roles of consensus, distinctiveness and consistency information. *Journal of Personality and Social Psychology,* 1975, *32,* 605–616.

Poincaré, H. *The foundations of science,* New York: Science Press, 1913.

Popper, K. R. *The logic of scientific discovery.* New York: Harper, 1959.

Popper, K. R. *Conjectutes and refutations.* New York: Harper, 1963.

Popper, K. R. *Objective knowledge.* Oxford: Oxford University Press, 1972.

Reichenbach, H. *Experience and prediction.* Chicago: The University of Chicago Press, 1961.

Schachter, S., & Singer, J. E. Cognitive, social and physiological determinants of emotional state. *Psychological Review,* 1962, *69,* 379–399.

Souriau, P. *Théorie de l'Invention.* Paris: Hachette, 1881.

Watkins, J. W. N. Against normal science. In I. Lakatos & A. Musgrave (Eds.), *Criticism and the growth of knowledge.* Cambridge, England: Cambridge University Press, 1970.

Weiner, B. *Achievement motivation and attribution theory.* Morristown, N.J.: General Learning Press, 1974.

Weiner, B., Frieze, I., Kukla, A., Reed, L., Rest, S., & Rosenbaum, R. *Perceiving the causes of success and failure.* Morristown, N.J.: General Learning Press, 1972.

Zuckerman, M. On the endogenous–exogenous partition in attribution theory. *Personality and Social Psychology Bulletin,* 1977, *3,* 387–399.

12

Attribution and Misattribution of Excitatory Reactions

Dolf Zillmann
Indiana University

In this chapter, principal theories of emotion, especially the two-factor theory of emotional state, are examined in terms of assumptions and propositions regarding the self-perception of excitatory reactions and its projected implications for emotional behavior in man. Research findings that have been viewed as implicating the operation of specific attributional processes are critically evaluated, and their decisiveness and generalizability are called into question. A three-factor theory of emotion is then outlined, and it is shown how this model accommodates the propositions of the two-factor theory as a special case. It is also shown how the three-factor model reconciles conflicts between different theoretical approaches to emotion. Finally, research findings on the transfer of excitation in emotional behavior are drawn upon to support the three-factor model in general and its assumptions regarding attributional processes in particular.

TWO-FACTOR THEORY OF EMOTION AND ATTRIBUTIONAL PROCESS

In the two-factor theory of emotional state (Schachter, 1964), the interaction between cognitive factors and physiological arousal is specified in the following three propositions:

1. "Given a state of physiological arousal for which an individual has no immediate explanation, he will 'label' this state and describe his feelings in terms of the cognitions available to him."
2. "Given a state of physiological arousal for which an individual has a completely appropriate explanation, no evaluative needs will arise, and the individual is unlikely to label his feelings in terms of the alternative cognitions available."

3. "Given the same cognitive circumstances, the individual will react emotionally or describe his feelings as emotions only to the extent that he experiences a state of physiological arousal [p. 53]."

The crucial assumptions upon which these propositions are founded are:

1. Emotional states are not associated with unique arousal states. In other words, physiological reactions are nonspecific or diffuse in emotion (Schachter, 1964, 1970).
2. The individual is acutely aware of nontrivial elevations in the level of diffuse physiological arousal he experiences, mainly through interoceptive cues.
3. Awareness of a heightened arousal state actuates "evaluative needs" in the individual; that is, it produces motivational pressures that "act on an individual in such a state to understand and label his bodily feelings [Schachter, 1964, p. 52]."

The latter assumption is of particular importance for an attributional analysis. In attribution theory (e.g., Heider, 1958; Jones & Davis, 1965; Kelley, 1967, 1971a, 1971b), it is generally assumed that man seeks to comprehend the causal relationships that govern his interaction with the environment. The very proposal that the individual *makes causal attributions* — that is, that he isolates conditions in his physical surroundings, in the activities of others, and in his own actions, and holds them responsible for having brought about certain "consequences" — presupposes the operation of an epistemic motivation. It has been suggested that such epistemic motivation is most pronounced under conditions of uncertainty (Berlyne, 1960). It has further been proposed that in human behavior response selection is associated with a degree of ambiguity that makes it advantageous, in terms of adaptation, for the individual to rely on inferential processes in determining an appropriate response mode (e.g., Festinger, 1954; Schachter, 1959; Bem, 1972). The individual is viewed as trying to understand the causal dependencies in his environment, including his own responses to environmental situations. The knowledge gained through this analysis of causal relations is seen as being employed in setting forth a suitable course of action. Schachter (1964) extended this line of reasoning so to include covert activities along with overt behavioral manifestations: The individual's interoceptive feedback, primarily from a nonspecific state of physiological arousal, is thought of as initiating the quest for a causal accounting that will ultimately provide the individual with an explanation of his emotional reaction and guide his response selection.

In terms of an attributional analysis, then, the two-factor theory of emotion can be rephrased as follows:

1. In the absence of an awareness of heightened autonomic activity, a need for the causal attribution of such activity cannot arise. As a consequence, emo-

tional behavior cannot eventuate. Needless to say, this condition should generally be met when actual autonomic activity is low or normal. (This is the equivalent to Schachter's third proposition.)

2. Awareness of heightened autonomic activity produces an epistemic need that motivates the individual to generate attributions of his autonomic reaction to an inducing condition (or to a set of inducing conditions). The epistemic need is satisfied with the individual's adoption of a plausible causal account — within his subjective frame of reference. (This is the equivalent of Schachter's first proposition.)

3. The adoption of a plausible causal account arrests the generation of attributions. The individual is therefore unlikely to change the assessment of his emotional state. (This is the equivalent of Schachter's second proposition.)

These transformed propositions of the two-factor theory of emotional state should make it clear that in this model an emotional state is in principle preceded by the excitatory reaction associated with it. According to the theory, an emotion cannot be experienced until an excitatory state materializes, is recognized, and is labeled. It might appear that this projected chain of events is necessarily time-consuming and that emotional experiences would be possible only with a considerable latency following the introduction of an emotion-inducing stimulus. In recognition of the apparent immediacy of an emotional experience, Schachter acknowledged that causal attributions are usually suggested and even forced upon the respondent by the emotion-inducing stimulus itself. "Cognitive or situational factors trigger physiological processes, and the triggering stimulus usually imposes the label we attach to our feelings [Nisbett & Schachter, 1966, p. 228]." A plausible account of an excitatory reaction thus is available as soon as the individual becomes aware of this reaction. Technically speaking, because of the latency of sympathetic arousal, the label for an emotional experience can be viewed as being provided before the excitatory reaction takes and reaches awareness — the latter being a condition that, according to Schachter, is necessary for an emotional state to occur.

The above qualification notwithstanding, two-factor theory generally stipulates that the experience of heightened diffuse arousal *requires an explanation* through some form of cognitive appraisal. The stipulation of such a requirement apparently derives from Schachter's assumption of nonspecificity of arousal. This principal assumption of two-factor theory asserts that the individual who becomes aware of his excitatory reaction learns only about the fact of his physiological arousal and potentially the intensity of his arousal reaction (cf. Latané & Schachter, 1962; Singer, 1963). Since the arousal is said to be nonspecific, the individual cannot tell from his excitatory reaction alone what exactly caused him to become aroused. This ambiguity regarding a reaction of his would exist even if it were assumed that the individual is equipped with a sensitive, interoceptive system that provides him with adequate feedback of his reaction: Since the reaction is said to be diffuse, even a sensitive feedback system could convey only a diffuse arousal reaction. According to this reasoning, then, the individual is virtually forced into employing attributional processes to "understand" his reaction and to be able to select an adaptive response.

Schachter's two-factor theory certainly stresses the individual's reliance on environmental cues in the determination of emotional experiences and emotional behavior. But it should be recognized that it does so in a highly specific manner: The model requires that the excitatory reaction sensed by the individual be *explained* as resulting from particular conditions, characteristically from exposure to stimuli in the immediate environment. It will be recalled that the concept of explanation is directly employed in the principal propositions regarding labeling. This point is made in illustrations as well: It is suggested that a person generally labels "his bodily feelings in terms of the situation in which he finds himself. Should he at the time be watching a horror film, he would probably *decide* that he was badly frightened. Should he be with a beautiful woman, he might *decide* that he was wildly in love or sexually excited. Should he be in an argument, he might explode in fury and hatred. Or, should the situation be completely inappropriate, he could *decide* that he was excited or upset by something that had recently happened [Schachter, 1967, p. 120, italics added]." Clearly, the labeling of an emotional experience is construed as the result of a *decisional process* that involves the deliberation of possible causal attributions. Emotional behavior, especially its motor component, is not conceived of as an immediate response to a stimulus but as one that is mediated by this attributional deliberation. The proposal of such a procedure may appear quite compelling when ambiguous response conditions are being considered, but it seems to be somewhat forced for conditions in which ambiguity does not exist (cf. Leventhal, 1974). Is it meaningful, for example, to assume that upon unexpectedly encountering a snake, a person with a phobia of snakes must contemplate the significance of his arousal reaction before he can display an adaptive reaction? Although it can be argued in defense of an attributional analysis that the critical emotion-labeling attributions have been established in the past and are virtually pre-existent in such a case, the two-factor model remains challenged by the type of behavior depicted in this illustration because the latency of adaptive motor reactions is characteristically shorter than the latency of excitatory reactions. It appears that emotional behavior can be displayed before an emotional experience can manifest itself, a sequence of events that conflicts with two-factor theory. We will discuss this issue in more detail later. At this point, let it suffice to say that the two-factor model posits that *all* emotional experiences are based on a labeling process in which a recognized state of diffuse arousal is causally attributed to an inducing condition. This process is assumed to operate even in those cases where response ambiguity does not exist (i.e., when emotional reactions such as "fight or flight" occur quasi-instantaneously). However, the proposal that heightened diffuse arousal of which the individual becomes aware need be causally attributed in order to produce an emotional experience has usually been applied to conditions in which some ambiguity exists. Under such conditions, the *search* for plausible causal accounts has been viewed as consuming a considerable period of time (Nisbett & Valins, 1971). In experimental work on the attribution process (Barefoot & Straub, 1971), a period of 10 seconds, for example, has been considered inadequate for necessary attributional deliberations. A period of 25 seconds, in contrast, has been judged as more adequate for the presumed search for the most plausible explanation. Such operationalizations leave no doubt that some investigators assume some form of explicit, covert reasoning process to serve the labeling function. In two-factor theory proper, however, a particular mode for the representation of the proposed causal attributions has not been specified. Yet the emphasis of the concept of explanation would seem to suggest very strongly that the attributions said to be made by an individual to account for his emotional experiences are conceived of as covert linguistic processes.

ALTERNATIVE APPROACHES TO EMOTIONAL BEHAVIOR

Clearly, two-factor theory projects man as greatly dependent upon the use of his cognitive skills. It suggests that the individual's overt and covert heightened (re)activity in any given situation does not sufficiently determine an emotional experience, and that the individual is forced to resort to inferential techniques to resolve a potentially ambiguous stimulus–response relationship and thereby "make sense" of his state of being. Proposing such an extreme plasticity of emotional behavior has been a point of departure from earlier theories of emotion. In earlier theories, emotional experience and emotional behavior were viewed as entirely determined by the somatic reactions produced by an arousal-inducing stimulus. It thus was unnecessary to postulate a dependency of experiential states on inference and attribution. The theories posited different loci for the emotion-determining information or the emotion-controlling mechanism, however. Depending on the placement of the proposed emotion-determining events and agents, these approaches are commonly classified as peripheral vs. central theories of emotion.

Peripheral Theory

The peripheral theory of emotion, also called the body-reaction theory, is usually credited to James (1884) and Lange (1887), although it was originally formulated by Henle (1876). James, in particular, was accused of plagiarism (Titchener, 1914), but the fact that he has been the most vehement advocate of the theory might explain why it is associated mainly with his name.

Peripheral theory as expounded by James asserts that "bodily changes directly follow the *perception* of the exciting fact, and that our feeling of the same changes as they occur *is* the emotion [Lange & James, 1922, p. 13]." James thought he had turned common sense upside-down in suggesting that emotions are not directly triggered by an arousal- and action-producing stimulus event, but that instead they derive from the afferent impulses occasioned by the overall bodily reaction to the stimulus event. Arguing that "we feel sorry because we cry, angry because we strike, afraid because we tremble, and not that we cry, strike, or tremble, because we are sorry, angry or fearful [p. 13]," he stressed the dependency of an emotional experience on feedback from the directly induced peripheral manifestations of a reaction. James insisted that emotions could not materialize without such afferent feedback. "If we fancy some strong emotions, and then try to abstract from our consciousness of it all our feelings of its bodily symptoms, we find we have nothing left behind [p. 102]." "Emotion dissociated from all bodily feelings is inconceivable [p. 103]." In equating emotions with the awareness of peripheral sensations associated with excited reaction, James went far beyond the requirement of sensory feedback as such. He assumed: (a) that excited reactions are uniquely patterned; (b) that afferent sensory pathways are highly differentiated and provide specific feedback of uniquely patterned reactions; and (c) that the afferent impulses deriving from this highly differentiated structure produce awareness of a potentially large number of distinct, peripheral activities.

James conceived of excited reactions as patterned responses with unique internal and external manifestations. He envisioned the various conceivable emotions as differing in their visceral response structure as well as in their motoric manifestations. Specificity in visceral reactivity was considered especially important. Bodily changes were seen as induced "by a preorganized mechanism [p. 101]" and as forming an unlimited number of complex and unique patterns. "The various permutations and combinations of which these organic activities are susceptible make it abstractly possible that no shade of emotion, however slight, should be without a bodily reverberation as unique, when taken in its totality, as is the mental mood itself [p. 101]." Most importantly, however, James went on to propose that this virtually unlimited variability in response patterns is fully represented at the awareness level. *"Every one of the bodily changes, whatever it be, is felt, acutely or obscurely, the moment it occurs* [p. 101]."

It should be clear at this point that the peripheral theory of emotion does not project the experiential plasticity of two-factor theory. Although both theories involve afferent feedback of excitatory changes as a crucial mediator of emotional state, the emotional mechanics are entirely different. They diverge, because of the assumption of specificity of somatic reactivity in emotion in peripheral theory in contrast to the assumption of diffuse excitatory reactions in two-factor theory. The two assumptions have enormously different theoretical implications.

With regard to the plasticity of emotional experience, the James–Lange version of peripheral theory entails no response ambiguity whatsoever. The emotional experience is obviously entirely fixed with the immediate reaction to an arousing event. There is no leeway for cognitive modification – no room for attributions or misattributions of excitation. According to peripheral theory per se, the excited reaction in all its somatic manifestations is predetermined by the inducing stimulus; and awareness of these manifestations (i.e., visceral responses in the gut, cardiovascular changes in the chest and in the extremities, and muscular activity throughout the body and in the face) is in turn completely determined by a sensitive and reliable system of afferent pathways.

Peripheral theory thus posits a one–to–one relationship between emotional experience and the reaction that produces the experience. It will be recalled that the reaction, in and of itself, is not to be considered an emotion. Is it an emotional reaction? Is it emotional behavior? Yes and no! As James so tenaciously argued, emotions are our feelings of an excited reaction. If so, the reaction necessarily precedes the experience of emotion, and consequently it is not (or was not) the emotion. But James also argued that the afferent impulses that produce the emotion are provided instantaneously. The emotional experience should thus follow the excited reaction without any appreciable latency. For all practical purposes, it would be simultaneous, and the reaction in question might be considered an element of the emotional experience. Running away from a bear, to use one of James's examples, might therefore be construed as an emotional reaction. But even such a somewhat forced, possible interpretation does not alter the behavior-determines-feelings sequence of peripheral theory. Emotional experience is always seen as the result of behavior. The experience of emotion, on the other hand, is not viewed as having any significant subsequent effect on behavior.

This is very much in contrast to two-factor theory. According to the two-factor model, the emotional experience arrived at through causal attribution is a most crucial mediator between experience and reaction. The emotional experience is viewed as guiding the selection of appropriate responses. Two-factor theory, then, is a theory of both emotional experience and emotional behavior. The James–Lange version of peripheral theory is not. It presupposes meaningful, adaptive "emotional" reactions and then outlines a way in which these reactions might produce an emotional experience, an experience that does not act back on the associated reactions. Peripheral theory thus neither explains the initial behavior nor predicts alterations of it. It merely asserts that feelings mechanically follow behavior – like a shadow cast by an object. Peripheral theory is therefore a theory of emotional experience only. It does not concern the action component of emotional behavior. Strictly speaking, the theory posits that emotional experiences derive from emotional behavior, regardless of how such behavior is induced or what purpose it serves. Put another way, it asserts that in order for people to experience emotions they must behave emotionally. It thus would appear to be unwarranted and potentially misleading to treat peripheral theory as a general theory of emotion or of emotional behavior.

The reliance of peripheral theory on the interoception of visceral changes has prompted strong criticism. Cannon (1927, 1929) examined the physiological viability of the proposed role of the viscera in emotional states and concluded:

1. The total separation of the viscera from the central nervous system does not alter emotional behavior.
2. The viscera are relatively insensitive structures.
3. Visceral changes are too slow to be the source of emotional feeling.
4. The same visceral changes occur in very different emotional and nonemotional states.
5. The artifical induction of visceral changes typical of strong emotions does not produce them.

This criticism is often presented as devastating. Actually, it is at best troublesome to the James–Lange version of peripheral theory. This is not because the points raised have lost their validity. Only the first objection needed modification: Removal of visceral activity is of little or no consequence for emotional *behavior* only after the behavior in question is well established (cf. Wynne & Solomon, 1955; Grossman, 1967). It is because Cannon's criticism of peripheral theory implies that James declared visceral interoception to be the absolutely crucial, if not the only, determinant of emotion. Clearly, James did not make such a declaration. As will be recalled, James conceived of complex patterns of interoception. If visceral reactions should prove to be slow, muscular reactions (e.g., in the face) might be fast. If the viscera should prove insensitive (something James certainly did *not* anticipate), the remaining emotion-determining structures might not be. If there should be overlap in visceral reactivity, different emotions could still result from differences in other structures. And if the artificial induction of strong visceral changes fails to produce specific emotions, it would be because the entire, unique interoceptive structure associated with particular emotions has not been replicated. James (Lange & James, 1922) discussed the mockery of emotions at length; and he pointed out that as long as the entire, complex interoceptive pattern associated with a particular emotional state is not accomplished, the resulting feelings will be different. James could thus have readily averted Cannon's criticism by assigning somewhat less significance to the viscera and somewhat more to structures such as the face. In fact, with such shift of emphasis, elements of peripheral theory have survived. Izard (1971) and Leventhal (1974), for example, have assigned a crucial role in the determination of emotions to the interoception of changes in the facial musculature.

The above discussion should make it clear that the James–Lange version of peripheral theory, because it encompasses *all conceivable* elements of interoception in the determination of feeling states, is so broad as to elude decisive falsification. But research evidence that is highly inconsistent with the principal assumptions underlying peripheral theory has steadily accumulated. As Cannon had proposed, the viscera proved to be comparatively insensitive structures that fail to display emotion-specific activity patterns with any reliability (cf. Oken, 1967; Grossman, 1967; Kety, 1970). Interoception of specific somatic changes also proved to lack the precision James and Lange had assumed (cf. Ádám, 1967; Mandler, 1975). The research of Schachter and his co-workers (cf. Schachter, 1964), which demonstrated the dependency of an emotional experience on the processing of externally provided information (and thus the insufficient determination of such experiences by the interoception of complexly patterned bodily states), produces further damaging evidence. Peripheral theory, then — especially as espoused by James — is severely challenged by more recent research findings.

Central Theories

The common characteristic of central theories of emotion is that they assign little or no significance in the determination of emotional behavior or emotional experience to feedback from peripheral manifestations. All manifestations of the emotions are considered centrally controlled, with various structures of the brain held responsible for this control. Central theories differ mainly with regard to the specific structures designated as controlling emotional behavior. In the so-called Cannon—Bard theory of emotion (Cannon, 1929; Bard, 1934), for example, the thalamus is implicated as the primary seat of the emotions. In Papez's (1937, 1939) theory, the hypothalamus is considered the principal structure that controls the emotions in its interaction with the anterior thalamic nuclei, the gyrus cinguli, and the hippocampus. MacLean's (1949, 1970) theory essentially agrees with that of Papez, but places less emphasis on the involvement of the cingulate gyrus. The conscious experience of the emotions is explained as an upward discharge of excitation from these lower centers into the cerebral cortex. Characteristically, the consciousness-generating process is not considered to entail peripheral information, but some efforts have been made to involve interoception of peripheral manifestations of emotional behavior in the determination of feeling states (e.g., Arnold, 1960; Gellhorn, 1964).

The contribution of such neurological analyses to the understanding of emotional behavior and emotional experience has often been branded as negative. Leventhal (1974), for example, suggested that central theory is potentially misleading because it proposes "that body responses are not responsible for emotional feelings, but fails to tell us what *cues* or reactions *are* responsible for feelings [pp. 13–14]." Although the first part of this argument (because some central theories — as just indicated — acknowledge the influence of information from the afferent network) is not entirely correct, the second part points out the problem one encounters when — through the use of neurological theories – one seeks to learn about the evocation of emotional reactions and feeling states. What external events and internal conditions initiate the processes detailed in central theories? In general, neurological theories of emotion do not systematically address the induction issue. The purpose of such theories is not so much to explore particular S–R relationships in emotion as it is to map out R with the greatest precision possible. The delineation of specific neural (and associated endocrine) processes in emotional *reactions* is important in its own right, and it has the potential of aiding the analysis of emotional behavior in an S–R context. But, in and of itself, it fails to explain *why* an organism responds emotionally when it does. It merely specifies *how* it does when it does. Neurological theory, then, offers little assistance in explaining the occurrence of specific emotions under particular environmental and constitutional conditions.

TWO-FACTOR FORMULATIONS IN BEHAVIOR THEORY

In behavior theory, the concept of emotion has been employed alongside the concept of motivation. It is therefore not surprising to find that crucial underlying assumptions of the two concepts have undergone similar changes. In the treatment of emotion, it became clear that the consideration of internal determinants alone was insufficient to explain emotional behavior and emotional experience. Similarly, in behavior theory, early conceptions of the fixed internal determination of behavior proved inadequate. The view that behavior in higher species resulted from a set of rigid instincts yielded to the view that behavior was controlled by a set of highly specific drives. The latter view, in turn, was more and more abandoned in favor of the projection of behavior as a complex interplay of internal *and external* forces (cf. Marler & Hamilton, 1966). Most importantly, with the possible exception of deprivation-based drives such as hunger and thirst (cf. Morgan, 1959), drive states came to be conceived of as potentially *nonspecific* (cf. Cofer & Appley, 1964; Nevin, 1973). The organism was no longer thought of as being internally driven to perform specific activities, seemingly in defiance of prevailing environmental conditions. Instead, nonspecific drive was considered an energizing force that needed guidance, and the necessary guidance was viewed as being provided by external stimuli. This point of view has been succinctly expressed by Hebb (1955): "Drive is an energizer, but not a guide; an engine but not a steering gear [p. 249]." Hebb stressed the *cue function* in the guidance of behavior; and he declared arousal, presumed to be nonspecific, to be equivalent with a "general drive state." The notion of a general drive state was further suggested by activation theory (e.g., Duffy, 1962; Lindsley, 1951, 1957), an approach in which diffuse arousal was more or less equated with measured activity in the reticular formation. The most influential use of the concept of nonspecific drive, however, proved to be that in motivational theory with a learning orientation (e.g., Hull, 1943, 1952; Spence, 1956; Brown,

1961). The behavior theory of Hull, in particular, promoted the view that arousal is a state of acute excitation that energizes the performance of conditional or unconditional responses made to stimuli in the environment. Hull proposed a *generalized drive;* that is, a nonspecific, undifferentiated drive state that integrates components of drive from various sources. In his view, the strengths of simultaneously active drives combine into an *effective* drive state. The entire accumulated force of this state then *energizes the behavior that is prepotent in the habit structure.* There thus is no one—to—one linkage between a particular drive and associated behaviors. Depending upon the prevailing stimulus conditions that control habit, elements of one drive come to facilitate behavior associated with another drive. In principle, any behavior can be facilitated by *irrelevant drive;* that is, by energy that in the past has not been connected with the behavior it comes to energize.

Hull's reasoning regarding "generalized drive," a concept he treated as a hypothetical construct, has been taken to mean that "generalized arousal," a concept operationalized in various indices of physiological arousal (e.g., activity in the reticular formation, skin conductance, heart rate, blood pressure, or vasoconstriction), serves to potentiate and intensify the behavior that holds a prime position in the response hierarchy (e.g., Geen & O'Neal, 1969). With the prime position being mainly a function of the stimuli provided in the immediate environment, arousal is viewed, then, as the principal determinant of the intensity of behavior triggered by these stimuli. It has generally been implied that the intensity of behavior is proportional to the level of arousal prevailing at the time of the behavior enactment. Accordingly, behavior performed at high levels of physiological arousal should be highly intense; and since "emotional *behavior"* is commonly conceived of as behavior associated with high levels of arousal, highly intense behavior may be treated as emotional.

Clearly, the reasoning presented above can be construed as a two-factor model of emotional behavior. This model, analogous to the two-factor theory of emotion (cf. Schachter, 1964), projects the emotional reaction as a joint function of cognitive and excitatory processes. As in the two-factor theory of emotion, arousal is assumed to be diffuse and incapable of providing response guidance. In both models, the specifics of the response are considered to be determined by environmental cues. The crucial difference is with regard to the attributional process. In the behavior-theoretical paradigm of an emotional reaction, the response is either unconditional or conditioned. In the two-factor theory of emotion, it is mediated by attributional processes. Furthermore, it should be clear that the behavior-theoretical paradigm does not address the experiential component of emotional behavior and its potential implications for the emotional behavior itself (cf. Lazarus, 1966). The paradigm simply does not offer any rationale that deals with conscious feeling states. It thus cannot be considered a theory of emotional *experience,* but it appears to be a viable theory of emotional *behavior* or *reactivity.*

As a theory of emotional reactivity, the behavior-theoretical, two-factor approach has one fundamental advantage over the two-factor theory of emotion: It can explain the immediacy of an adaptive emotional reaction. In the case where a person is surprised by a snake, for example, the likely motor reaction is abrupt withdrawal. This reaction is displayed before the individual can become aware of his arousal response, mainly because pertinent peripheral manifestations of such a response have a characteristic latency of 3 to 15 seconds (cf. Newman, Perkins, & Wheeler, 1930; Grossman, 1967). According to the behavior-theoretical approach, the motor response to an emotion-inducing stimulus is unconditional or conditioned and thus quasi-instantaneous. It can be assumed that the covert concomitants of the reaction are analogously initiated and that after the discussed latency, arousal manifests itself and energizes the overt response. In the example, the immediate withdrawal from the snake should, within seconds, turn into potentially vigorous emotional behavior.

The two-factor theory of emotion is at a loss, it appears, when immediate emotional reactions — especially those that serve an adaptive function — are being considered. In the case of the unexpected encounter with a snake, it must be argued that the individual first awaits feedback from peripheral manifestations of his excitatory reaction to the situation, then causally attributes his reaction to the snake, and finally, having thus comprehended his predicament, decides on a presumably adaptive course of action. This projected process is simply at odds with the established immediacy of emotional behavior (cf. Cofer & Appley, 1964).

In contrast to the two-factor theory of emotion, then, the behavior-theoretical approach provides an adequate account of immediate emotional behavior. Obviously, since the response is considered unconditional or conditioned, such behavior occurs under conditions that are not associated with response ambiguity. As we will see shortly, behavior-theoretical explanations fail under conditions in which response ambiguity exists.

EPISTEMIC SEARCH AND ATTRIBUTION IN EMOTION

Research on the two-factor theory of emotion is largely associated with an ingenious experimental manipulation (cf. Schachter, 1964): The theoretically pertinent condition in which the individual notices a state of diffuse arousal that he then attributes to situational factors is accomplished in principle; (a) by elevating sympathetic excitation through the injection of epinephrine; and (b) by misinforming subjects about the response to the injection. In the initial investigation by Schachter and Singer (1962), for example, subjects received subcutaneously a dose of epinephrine that produced the following experiential symptoms: palpitation, tremor, accelerated breathing, and the sensation of flushing. These symptoms derive, of course, from the sympathetic reaction induced by the adrenalin: elevated blood pressure, accelerated heart rate, increased muscular blood flow, decreased cutaneous blood flow, accelerated respiration rate, heightened blood sugar concentration, etc. Regarding the symptoms the subjects would come to experience, subjects were: (1) appropriately informed; (2) not informed; or (3) misinformed. In the "informed" condition, they were told that the drug had some transitory side effects: It would probably make their hands shake, their hearts pound, and their faces get warm and flushed. These subjects were thus provided with an entirely appropriate explanation of their forthcoming sensations. They knew exactly what they would feel and why they would feel it. In the "uninformed" condition, subjects were led to believe that the drug had no side effects. As in the "informed" condition, subjects in the "misinformed" condition were told of transitory side effects of the drug. But in sharp contrast to the former condition, the misinformed subjects were led to expect sensations that were *not* forthcoming: feeling of numbness in the feet, itching sensations over parts of the body, and a slight headache. Both in the "uninformed" and in the "misinformed" condition, then, subjects were without an adequate explanation for the arousal reaction they would come to experience. These are, of course, precisely the circumstances that two-factor theory projects as motivating the search for an explanation that will ultimately lead to a causal attribution that determines the phenomenological status of the emotional state. The experiment employed placebo conditions as additional controls. To equate the procedure, a saline solution was injected, and subjects were not informed of side effects of the "drug." The placebo conditions thus parallel the "uniformed" arousal conditions. It was expected that — because subjects would fail to experience an arousal reaction — they would not be confronted with an attribution problem and not respond emotionally.

In general, the critical causal attributions are strongly suggested by cues that are obtrusively displayed in the subjects' immediate environment (cf. Schachter, 1964). In the investigation described, whether or not an arousal reaction was induced, these cues were provided by the behavior of a stooge. Subjects were either exposed to conditions under which they could come to label their feelings as annoyance or as mirth. In the "annoyance" condition, along with the experimental confederate they were to respond to a questionnaire that probed issues of great intimacy and embarrassment. The subjects witnessed the stooge burst into anger over the questions. In the "mirth" condition, they waited in the company of the stooge, who improvised a variety of "fun games" and sought to involve the subject in his mirthful activities.

The interaction with the stooge occurred, of course, at the time when the drug took effect. According to two-factor theory, subjects in the "informed" conditions — because they had an acceptable explanation for the arousal symptoms they experienced during the interaction period — should not attribute their arousal to alternative causes; and they should respond in a comparatively nonemotional manner both in the "annoyance" and in the "mirth" conditions.

Subjects in both the "uninformed" conditions and in the "misinformed" condition (the latter condition was employed for the "mirth" treatment only), in contrast, should come to accept the explanation of their arousal state that is provided with its causal attribution to the situational cues. They should regard their experience as an emotional one, and they should behave emotionally in accordance with their experience. In the "mirth" conditions, "uninformed" and "misinformed" subjects should develop feelings of mirth of greater intensity than those of subjects in the "informed" condition (or in the placebo control). Allowing for the missing comparisons due to the absence of the "misinformed" condition, the same differences regarding emotional intensity should be found in the feelings of annoyance in the "annoyance" conditions. Emotional reactivity was measured through self-report and unobtrusive observation of expressive nonverbal behavior. On these measures, the findings generally confirmed the theoretical expectations.

The demonstration that — depending upon the immediate presence of suggestive environmental cues — an unexplained, puzzling arousal reaction apparently led to feelings of annoyance *or* feelings of merriment is obviously damaging to the belief that specific "internal" responses determine emotional experiences. The sympathetic reaction induced was, after all, identical in both labeling conditions. The Schachter–Singer experiment, then, evidences that the individual confronted with ambiguous experiential circumstances does indeed employ externally provided information to disambiguate the situation and arrive at a specific emotional state. The findings seem to be in accord with the proposal that the individual searches for a viable explanation for his arousal reaction and then, in making a causal attribution, adopts the most plausible explanation he can find. Predictions based on such attributional assumptions were confirmed, and this confirmation can be taken to demonstrate the theoretical usefulness of these assumptions as hypothetical constructs. It cannot be taken as evidence for the actual occurrence of an epistemic search and the explicit, conscious execution of a causal attribution, however. The discussed experiment and similar investigations (e.g., Schachter & Wheeler, 1962) have left the empirical status of arousal attributions unclear.

Leventhal (1974) has recently challenged the attributional interpretation of investigations in which subjects were provided either with appropriate or with false information about an experienced arousal state. In essence, he has argued: (a) that the uninformed or misinformed subject who suddenly finds himself aroused is placed into a state of uncertainty; (b) that the experience of uncertainty is in itself arousing; (c) that uncertainty prompts an information search; and (d) that the state of uncertainty renders the subject vulnerable to suggestion. The appropriately informed subject, in contrast, would not experience uncertainty; hence, he would not be further aroused, he would not need to conduct an information search, and he would not be vulnerable to suggestion. Leventhal suggested that in the Schachter–Singer experiment the aroused, bewildered, and suggestible subject at first simply followed his inclination to *imitate* the stooge's behavior, then *expressed* similar feelings, and finally came to *experience* the feelings he expressed. This projected chain of events obviously does not necessitate assumptions about explicit cognitive processes that would serve the subject's inference of his own feeling state from external, environmental cues. In trying to provide an alternative explanation for the findings reported by Schachter and Singer, Leventhal further suggested that the low degree of emotionality observed in those subjects who were appropriately informed about their arousal reaction may have resulted from the lack of unexpectedness. Emotionality, according to Leventhal (1974; Johnson, 1973), cannot manifest itself when the individual is well prepared for forthcoming events – apparently independent of the potential impact these events might have. If no uncertainty is created, the individual should be comparatively calm. Applied to the Schachter–Singer findings, this reasoning again accounts for the crucial observed differentiation without requiring the assumption that subjects behaved nonemotionally, because – having been furnished with the causal connection between drug and reaction – they could not conceive of their arousal as a part of an emotional reaction induced by the surrounding circumstances.

Leventhal's contention that in the research seeking to support the two-factor model the information treatment (accurate vs. inaccurate or inadequate description of arousal symptoms) may have inadvertently effected a differentiation in actual arousal is not only unrefuted by data (measures of sympathetic excitation were not obtained in the initial investigations) but seems to be supported by the results in the placebo conditions of the Schachter–Singer investigation. According to two-factor theory, either annoyance or merriment in the placebo condition should have been as low as in the "informed" condition. Subjects should not have experienced arousal and, hence, should have behaved nonemotionally. Actually, both annoyance and merriment were found to be somewhat more intense in the respective palcebo conditions than in the respective "informed" conditions. Emotionality in the placebo conditions was clearly intermediate: It assumed a level between those associated with the "informed" and the "uninformed" conditions. This pattern of effects, it would seem, is readily explained on the basis of differential levels of arousal that produce differently intense reactions, with proportionality assumed between level of arousal and reaction intensity. Following Leventhal's argument, only the accurately informed subjects were at ease about the injection they had received (everything was as expected); the uninformed subjects, in contrast, were in the dark about the drug's effect and thus had ample reason to be apprehensive and to "get excited." The latter applies to the placebo treatment as well. Lacking the injection of an arousing agent, however, arousal resulting from uncertainty could only reach an intermediate level.

Leventhal's criticism (Calvert–Boyanowsky & Leventhal, 1975) has proved similarly troublesome to a modified version of two-factor reasoning. The initial investigations into the attribution of excitatory reactions (cf. Schachter, 1964) sought to demonstrate that arousal that, because of deception, could not be properly attributed to its cause (the injection of a stimulant) would be causally related to an apparent, plausible inducer — usually a set of stimuli in the respondent's immediate environment — and accordingly would produce a particular emotional state. Epistemologically speaking, the apparent and behaviorally binding relationship is of course erroneously inferred. Arousal deriving from a particular condition is being *misattributed* to another condition. This process of misattribution, independent of the artificial manipulation of arousal, attracted attention because it seemed to permit, within limits, the arbitrary control of emotional states (cf. Ross, Rodin, & Zimbardo, 1969). It was proposed that the intensity of emotional states can be reduced, essentially, through the induction of a misattribution of arousal. Through misattribution, the emotion would be deprived, so to speak, of its excitatory component. To be more specific, Ross et al. proposed that an emotional reaction caused by condition *A* will be experienced less intensely when the arousal response can be plausibly attributed to condition *B*, which covaried with *A*. Ti.e requirement of the covariation of two similarly plausible, inducing stimuli is certainly restrictive, but it was seen as readily met in a therapeutic setting — the context in which induction of misattributions was seen to hold great promise.

Ross et al. (1969) sought to establish the validity of this reasoning on misattribution through an investigation in which they induced fear: Subjects expected to receive electric shock. While in this state of apprehension, they were exposed to noise. Initially, the subjects had been told that the study explored the effects of noise on performance, and they had been presented with lists of "likely physiological side effects" of exposure to the noise. The outlined side effects were either symptoms commonly associated with fear (e.g., palpitations) or symptoms unrelated to fear (e.g., ringing of the ears). It was expected that those subjects who were made to believe that the noise induced the "fear symptoms" would *misattribute* their fear-associated reactions to the noise and hence experience less fear than those subjects who were not misled about the fear symptoms and thus would properly attribute them to their apprehensions regarding the reception of shock. Ross et al. provided experimental support for these predictions of the intensity of fear.

It is not the emotion-reducing effect reported in this or similar investigations (e.g., Barefoot & Girodo, 1972), but its (mis)attributional interpretation that proved to be controversial. Calvert–Boyanowsky and Leventhal (1975) explored the presumed misattribution process in great detail and found no direct evidence to support its existence. Furthermore, these investigators were able to show that a reduction of emotionality can be achieved by the provision of arousal symptoms alone, regardless of any particular attributional suggestion. This accords with Leventhal's (1974) suggestion that the subject who is not prepared for an arousing experience is placed into a state of apprehension that in itself is arousing. In the Ross et al. investigation, then, the subjects who received a list of symptoms commonly associated with fear may have been less fearful, simply because they were better prepared and thus less distressed — and not because they misattributed anything to anything. This alternative explanation potentially applies to all reports of misattribution effects based on investigations in which subjects were provided with different lists of "symptoms."

Two-factor theory has prompted yet another approach to attributional process: false feedback (cf. Valins, 1970; Nisbett & Valins, 1971). Generally speaking, subjects are confronted with stimuli that credibly could induce an emotional response, and they are simultaneously or immediately thereafter provided with purportedly "objective" information on their arousal reaction. Independent of the subjects' actual responses, they are fed signals of excitatory changes that are commonly associated with emotional reactivity or with the lack thereof.

Heartbeat has been most frequently employed as feedback of excitatory reactions. Subjects were led to believe that the heartbeat they would hear was their own, but they were actually exposed to recordings of fast, normal, or slow; accelerated, stable, or decelerated heart-beating. Subjects exposed to accelerated, fast pounding should — according to the two-factor model — take note of their apparent state of arousal and attribute its induction to obtrusive stimuli in the environment. The subjects should thus come to respond emotionally to these stimuli, regardless of the specific emotional state occasioned by them. The subjects' provision with stable or decelerated heart-beating, on the other hand, should prevent emotional reactions. In case an emotional reaction is evoked by the environmental stimuli alone, such "calmness-exhibiting" behavior of the heart should reduce, if not eliminate, emotional responsiveness.

A substantial amount of research evidence supports the above predictions (cf. Valins, 1970). The evidence is by no means unequivocal, however (cf. Goldstein, Fink, & Mettee, 1972). But granted that false feedback can affect emotional experience and emotional behavior, the attributional mechanics said to mediate this effect are very much in doubt.

The false-feedback paradigm has been taken to represent processes that "naturally" occur in emotional behavior. Nisbett and Valins leave little doubt about this when they say that "we infer our feelings about stimuli from information about the degree and source of autonomic arousal, *even when* that information is verbally supplied by an experimenter and *even when* it is false [1971, p. 74; italics added]." These investigators go on to suggest that "such inferences are not necessarily passive or immediately accepted," and that people "actively attempt to validate their inferences before encoding them as truth [p. 74]." It is evidently implied that at least in waking hours, the individual: (1) carefully monitors his autonomic excitation; (2) probes for an explanation of increased activity; and (3) adopts the causal connection that best resists attempts at falsification. The epistemic search is apparently conceived of as a set of conscious operations. It seems that because of this, it is of little moment to the investigators whether a possible attribution is self-generated or provided by an outside agent:

Both types merely constitute trial attributions that the individual considers and evaluates.

At the operational level, it obviously has been assumed that the individual monitors the activity of his heart and that he does so with considerable interoceptive sensitivity. This sensitivity assumption is not only at variance with pertinent data (e.g., Mandler & Kremen, 1958; Mandler, Mandler, Kremen, & Sholiton, 1961) but also leads to contradiction within the discussed research. On one hand, the individual is assumed able to discriminate autonomic changes and to habitually employ the information gained on the basis of his discriminatory capacity in determining the intensity of his affective reactions. On the other hand, the very fact that arbitrarily manipulated, interoceptive feedback was externally presented through artificial means (i.e., playback of heart-beating that did not correspond with the subject's actual heartbeat) shows that it also must have been assumed that the individual's discriminatory abilities are rather limited and readily overpowered by the suggestion of "objective" feedback. Giving credence to the suggestive force of false feedback, the question arises whether similar effects could be obtained with feedback from something other than a specific autonomic reaction. What if the subject were simply told that he got excited (whether or not he did)? What if he were shown the deflection of a needle on an apparatus and were told that this indicated his "gut level" response? Procedures such as these have in fact been employed, and they proved to be effective manipulators of emotional experience (e.g., Berkowitz, Lepinski, & Angulo, 1969). It would appear, then, that it is mainly the "power of suggestion" that produces the false-feedback effects. More specifically, in the somewhat ambiguous research situation, the subject may have been ambivalent about his feelings and, thus vulnerable, may readily have accepted any purportedly reliable information about his reaction. The so-called false-feedback studies may illuminate this state of information dependency, but they fail to shed light on the individual's interoceptive sensitivity regarding autonomic reactions and on the use of eventually obtained feedback in presumed attributional deliberations (cf. Mandler, cited in Oken, 1967).

To summarize, it can be said that although the discussed type of research on the two-factor model has convincingly demonstrated the great plasticity of emotional behavior (i.e., its relative independence from specific somatic reaction patterns together with the individual's reliance on environmental cues), it has not generated evidence that would support the assumption of a conscious epistemic search that ultimately produces highly specific causal attributions or misattributions. The assumptions in question may be regarded as useful constructs, but the empirical status of their projections is still in doubt. This is the case, because the effects that can be accurately predicted on the basis of these assumptions can also be explained without them — which is obviously more parsimonious. The evidence regarding attributional processes, then, may be regarded as suggestive; but it certainly cannot be considered compelling.

The indecisiveness of inferences about attributional process is mainly the result of the deceptive manipulations used in the discussed research: Subjects in whom an arousal reaction was induced were (indirectly) told to believe that it did not result from an apparent inducer (e.g., Schachter & Singer, 1962; Ross et al., 1969); and independent of their actual arousal or nonarousal, subjects were (indirectly) told that they responded emotionally to certain stimuli (e.g., Valins, 1966). This procedure, because it created ambiguities and offered solutions, invited the alternative accounts that were discussed; these accounts, as will be remembered, were based on the consideration of uncertainty and suggestion. But because the subject was virtually told what to make and what not to make of a situation, the deceptions involved may also be seen as having *artifically stimulated* an epistemic search and *forced* attributional deliberations. The misled subject may have been puzzled because he could recognize that the suggested emotional reactions did not conform to his "normal" manner of reacting. In his bewilderment, he may well have conducted the presumed search for an explanation, and only to resolve his bewilderment may he have come to accept the causal attribution obtrusively delineated in the situation. What if the subject had not been thus puzzled? What if he had not been told how and how not to attribute his arousal reaction? Should we assume that he still would have conducted an epistemic search?

It can be argued that in the research discussed thus far, the subject — through deceptions about his excitatory reactions and the response ambiguity thereby created — *was coerced into thinking about his behavior.* Had no ambiguity been created, an epistemic search would not have been necessary, and a conscious causal attribution should not have been made. In two-factor theory it is generally assumed, however, that any and every arousal reaction triggers the quest for an explanation. Response ambiguity is apparently thought of as ever present. But is it? If a person runs into a venomous snake, for example, is it meaningful to assume that his excitatory reaction is so ambiguous to him that he must go through a formal reasoning process in which he causally connects snake and response? As discussed earlier, Schachter has acknowledged that the arousal-inducing stimulus usually provides and even imposes an explanation. This quasi-simultaneous awareness of arousal response and explanation does not, however, change the proposed principal chain of events. According to two-factor theory, the puzzling awareness of an arousal reaction always precedes its attribution to an inducing condition. The time discrepancy between "question" and "answer" may be considered trivial, but in theory, it exists. Also, it is apparently inconceivable that the answer precedes the question.

In spite of the lack of decisive support for the proposal of an emotion-determining epistemic search, this element of two-factor theory holds great intuitive appeal for the explanation of emotional experience and emotional behavior under ambiguous response conditions (and presumably this is why it is so strongly adhered to and so vehemently defended). The proposal must appear awkward and in violation of parsimony of explanation, however, when applied to response conditions devoid of ambiguity. Why should one, for example, assume that a person who has just been blatantly insulted must consciously go through the motion of causally attributing his arousal to the assaultant's action and that only after that does he know what he feels? Why could it not be assumed that, based on the recall of his reactions in similar, past situations, he knows more or less immediately that he is annoyed and that he will be angry? The latter assumption implies that causal connections can be made before the individual becomes aware of peripheral manifestations of arousal. In a sense, he knows of his anger (and acts accordingly) before he is able to feel it. Most importantly, however, as peripheral arousal manifests itself and the individual becomes aware of excitatory changes, there is no ambiguity and, hence, no cause for an epistemic search. The need for the individual to reflect on his state of excitation simply does not exist.

The two-factor theoretical projection that the individual infers his emotional state in unambiguous response situations just as in ambiguous ones appears to be not only forced conceptually but also in terms of the generalization of research findings. As argued earlier, in the research paradigms used, attributional process has been explored only under ambiguous response conditions. Strictly speaking, nothing is known about unforced, spontaneous attributions of excitatory reactions in unambiguous situations. The projection, then, can only be treated as most tentative.

It should be clear from this discussion that the attributional analysis of excitatory reactions has been greatly compromised by deception manipulations in two-factor research. Not all research employing a two-factor approach has employed deceptions about arousal reactions, however. In research on excitation-transfer phenomena (e.g., Zillmann, 1971, 1972), subjects were not told what to make and what not to make of their excitatory reactions. In fact, these reactions never constituted a focus of attention for the experimental subject. If, under these conditions, subjects behave in accord with predictions from an attributional model, spontaneously generated causal connections — whatever their specific form may be — can be inferred. We will soon turn to the discussion of these possible inferences of attributions and misattributions. First, however, we will briefly outline: (1) a theory of emotion that avoids some of the deficiencies of two-factor theory; and (2) a theory of excitation transfer.

A THREE-FACTOR THEORY OF EMOTION

Aside from the already discussed problems regarding the empirical securement of the proposed epistemic search and the explanation of a puzzling arousal reaction it is to produce, the two-factor model suffers from a basic theoretical deficiency: It fails to explain why the individual responds in an aroused fashion to certain stimulus conditions in the first place. Two-factor theory simply *presupposes* that "cognitive or situational factors trigger physiological processes [Nisbett & Schachter, 1966, p. 228]," and it then deals with what happens after the individual becomes aware of these processes. As a theory of emotion, the two-factor model is thus incomplete, because it does not address the very origin of an emotional reaction.

If one takes the position that in two-factor theory it has been implied that excitatory reactions are unconditional or conditioned, one would expect the same mechanisms to govern motor reactions. As stated earlier, such an assumption has not been made, however; and consequently, the determination of immediate motor reactions remains as unclear as that of excitatory reactions.

The failure of two-factor theory to explain either the origin of excitatory reactions or the immediacy of motor reactions proves most restrictive. The requirement that the individual be able to reflect on a situation and execute causal attributions consciously or quasi-consciously further restricts the applicability of the model. Two-factor theory, it seems, applies only to the emotional experience of human beings who have developed the rational skills necessary to conduct a full-fledged epistemic search. To the extent that the development of these skills coincides with the development of linguistic skills, the prelinguistic child's emotional behavior goes unexplained. So does the emotional behavior of primates and generally speaking of all higher vertebrates capable of highly intense reactions based on pronounced, temporary activity increases in the sympathetic nervous system. And even in linguistically mature humans, as has been explicated earlier (under "Two-factor Formulations in Behavior Theory"), immediate emotional reactions — especially motor reactions — are left unexplained.

The three-factor model proposed in the following is designed to eliminate the discussed shortcomings of the two-factor approach and to expand its applicability. Essentially, the three-factor model seeks to accomplish this objective by integrating the valid features of the two-factor theory of emotional state and those of the behavior-theoretical, two-factor approach, and by avoiding the weaknesses of these models.

The three-factor theory of emotion projects emotional experience and emotional behavior as the result of the interaction of three principal components of emotional state: the *dispositional*, the *excitatory*, and the *experiential* component.

1. The dispositional component is conceived of as a response-guiding mechanism. As in behavior theory, the motor aspects of behavior are considered either unconditional (such as in startle reactions) or acquired through learning (such as in atypical phobias). It is obviously assumed that the ES—ER connection pre-exists or is established without the involvement of "cognitive operations at higher levels." The motor behavior associated with emotions is thus seen as an unmediated, direct response made without appreciable latency to the presentation of the emotion-inducing stimulus.

2. The excitatory component is conceived of as the response-energizing mechanism. It is assumed that excitatory reactions, analogous to motor reactions,

are either unconditional or acquired through learning. If the ES–ER connection is established through learning, it is again assumed that this happens without the intervention of higher-level cognitive processes. In accordance with Cannon's (1929) proposal of the "emergency" nature of emotional behavior, the excitatory reaction associated with emotional states is conceived of as heightened activity of the sympathetic nervous system, primarily, that prepares the organism for the temporary engagement in vigorous motor activities such as needed for "fight or flight."

3. The experiential component of emotional behavior is conceptualized as the conscious experience of either the motor or the excitatory reaction or of both these aspects of the response to a stimulus condition. Both exteroceptive and interoceptive stimuli are thus assumed to reach the awareness level. The individual's awareness of his ER (i.e., of its motor and/or its excitatory components), the so-called feeling state, is viewed as potentially initiating further cognitive operations at the awareness level. Under conditions yet to be specified, these operations may serve the appraisal of the emotional reaction; and this appraisal, in turn, may effect changes in the motor and/or excitatory manifestations of the initial emotional reaction. In principle, then, the experiential component of emotions is viewed as a modifier or a corrective that, within limits, controls the more archaic, basic emotional responsiveness governed by unlearned and learned S–R connections.

The principal interactions and interdependencies of these three components of emotional behavior are described in the following propositions:

1. A stimulus condition that evokes a specific motor response without evoking an excitatory reaction will produce nonvigorous behavior, which the individual is unlikely to experience as emotional.

2. A stimulus condition that evokes a specific motor response and an excitatory reaction will produce vigorous, "emotional" behavior. As the individual becomes aware of his state of elevated excitation, he will appraise his reaction. If he deems his behavior appropriate, he will continue to respond emotionally. If he deems it inappropriate, he will change his mode of reacting, both motorically and excitationally. He will experience his reaction as emotional to the extent that it is thus characterized by the social community whose value criteria he has adopted.

3. A stimulus condition that evokes an excitatory reaction without evoking a specific motor response will produce a state of acute response ambiguity marked by motor restlessness. As the individual becomes aware of his aimless, excited behavior, he is likely to perform an epistemic search directed at the comprehension of the inducement of his state of elevated excitation. The adopted explanation will guide further activities both at the motor and the excitatory levels. Again,

the individual will experience his behavior as emotional to the extent that it is thus characterized by the social community whose value criteria he has adopted.

In these propositions, it is generally assumed that: (1) most complex motor reactions are under voluntary control; (2) excitatory reactions are usually not thus controlled by the individual; (3) the linguistically mature individual habitually appraises the appropriateness of his behavior; and (4) unless motor reactions are voluntarily controlled as the result of an appraisal process, their vigor is a simple function of the level of sympathetic excitation prevailing at that time. It is further assumed that the intensity of feelings, that is, the intensity of the subjective experience of emotional actions and reactions, is approximately proportional to the perceived magnitude of increases in excitation.

It should be clear that the above propositions were formulated with the human respondent in mind. The potentially behavior-modifying appraisal, a crucial function in the experiential component of emotion, obviously presupposes attributional and judgmental capabilities (cf. Lazarus, 1966) that only man has developed. This function cannot be assumed operative in subhuman species. Emotional behavior in these species consequently must be explained as an inter-action between the dispositional and the excitatory components only. The three-factor model is reduced in this case to the behavior-theoretical elements contained in it. The explanation of emotional behavior in humans, on the other hand, requires all three components. The interaction between the dispositional and the excitatory components may be regarded as a "primitive heritage of man." The basic response tendencies that are delineated by the interaction in question have had obvious survival value (cf. Cannon, 1929). Only in recent history, under the conditions of living in modern society, have they largely lost their adaptive value. However, along with the evolution of rational capabilities in man has evolved the capacity to modify emotional behavior through the appraisal of its effectiveness and appropriateness. The experiential component of emotion, then, is viewed as more than the mere awareness of a behavioral state associated with elevated excitation ("I sense this; I feel that"). It is considered to serve as a *corrective* of emotional reactions that are appraised as inappropriate. It is further seen to *provide guidance* in situations in which a response disposition does not exist. These two functions of appraisal can be viewed as an evolutionary adjustment that assures that in man the more archaic response mechanisms (i.e., the interaction between the dispositional and the excitatory components) do not become overly maladaptive.

The mechanics of three-factor theory can be briefly illustrated with research findings on the effect of mitigating information on anger and hostile behavior (Zillmann & Cantor, 1976). Subjects were provoked through the extremely rude behavior of an experimenter, and they were later provided with an op-portunity to retaliate. In a no-mitigation condition, no excuse was offered for the experimenter's behavior. In two mitigation conditions, an outside agent offered an "explanation" for the experimenter's apparent irritability: The experimenter was described as being very uptight about exams. The subject received this mitigating information either (1) prior to or (2) after he was mistreated by the experimenter. Under these experimental conditions, the time course of excitatory reactions and the ultimate, retaliatory actions were assessed.

In terms of three-factor theory, the mistreatment is expected to induce a severe excitatory reaction in the subject together with an impulsive inclination for fight or flight. The excitatory reaction can be seen as preparing the organism to act on the situation so as to avert a potential threat. After a trivial latency, the subject should receive exteroceptive and interoceptive feedback of his state of excitation. No epistemic search is necessary, however, to find an appropriate explanation for his reaction. The type of his emotional experience is virtually forced upon the subject by the apparent contingency of situational and response cues. Labeling, then, need not be assumed to take the form of an explicit inference — as many advocates of the two-factor model have suggested or implied. It possibly manifests itself in covert verbal reactions such as "This guy is crazy," "You S.O.B.," or "Boy, I'll get you for this!" Presumably, the flow of covert verbal *responses* to a situation, highly idiosyncratic as it may be, constitutes the cognitive side of emotional experience and determines the subjective typology of these experiences ("I hate him," "I feel sorry for him").

If no mitigating circumstances are known, the subject can only appraise the mistreatment he has received as deliberately inflicted on him by the experimenter. The cognitive processes involved in this appraisal can certainly be expressed in attributional terms. In these terms, the subject would come to attribute malicious intent to his annoyer, and this attribution should lead him to view the interaction with his annoyer as a continued threat. This appraisal requires him to maintain his preparedness to act on the situation so as to avert further harm or to terminate the state of acute annoyance. A high level of excitation thus needs to be perpetuated, and intense feelings of annoyance and strong retalitatory measures are likely.

If, on the other hand, mitigating circumstances are known prior to provocation, the subject can attribute the experimenter's misbehavior to his state of irritability. As a consequence, he cannot really perceive the experimenter as malicious, and his actions cannot be construed as a personal attack or a threat. Such a *preattribution* of forthcoming events should prevent intense excitatory reactions from developing. Since the situation is not appraised as a threat, the organism need not prepare for taking action. Also, because excitation should remain normal, intense feelings of annoyance — which would motivate hostile actions — cannot materialize. Under these circumstances, then, retaliation should be minimal.

Finally, if mitigating circumstances become known only after the subject has been mistreated, he will have intensely suffered from the treatment. Up to the point where he receives the mitigating information, he perceives the experimenter's behavior as a threat, and he thus should have maintained a high level of excitation that produced feelings of annoyance of great intensity. The appraisal is then suddenly changed: The interaction becomes comparatively nonthreatening. This should immediately initiate the decay of excitation. Retaliation, however, should be relatively severe because of the intensity of feelings of anger that the subject had suffered and for which he seeks retribution.

The findings were in accord with these predictions. Both dispositional and excitatory reactions were thus shown to be mediated by the appraisal function of the experiential component of emotion. The control that this function can exert over emotional reactions should not be considered unlimited, however. There is some evidence (Zillmann, Bryant, Cantor, & Day, 1975) that suggests that the cognitive processes involved in appraisal are greatly impaired at high levels of sympathetic excitation. It appears that the intensely aroused individual in his "impulsive" emotional behavior is forced to rely on the more archaic response mechanisms.

A THEORY OF EXCITATION TRANSFER

The excitation-transfer paradigm has been detailed elsewhere (e.g., Zillmann, 1971, 1972; Zillmann & Bryant, 1974; Zillmann, Johnson, & Day, 1974; Cantor, Zillmann, & Bryant, 1975) and therefore is presented here in summary form only. Essentially, the paradigm applies the three-factor approach to the experience of emotional states in sequence, and it specifies the conditions under which residues of sympathetic excitation from a preceding emotional state will (or will not) intensify a subsequent emotional state.

Excitation-transfer theory is based on the following assumptions:

1. With respect to emotions, the interoception of excitatory reactions is generally nonspecific. This interoceptive nonspecificity is largely due to a high degree of nonspecificity of the excitatory reactions themselves.

2. The individual can determine the intensity of his excitatory reaction through interoception. However, only comparatively gross changes in the level of excitation will draw the individual's attention and produce an awareness of his state of excitation.

3. The individual relates an excitatory reaction of which he becomes aware to the apparent inducing condition and may recall this connection at later times.

4. The individual generally does not partition excitation compounded from reactions to different inducing conditions. More specifically, the individual does not identify all factors that contribute to an experienced state of excitation; nor does he apportion his excitation to the various contributing factors. Instead, he tends to ascribe his entire excitatory reaction to one particular, inducing condition.

5. Intense excitatory reactions do not terminate abruptly. Because of slow, humoral processes involved in the control of sympathetic excitation, excitation decays comparatively slowly. Residues of this slowly decaying excitation, then, may enter into subsequent, potentially independent experiential states.

The implications of residual excitation for emotional behavior are specified in the following propositions:

1. Given a situation in which (a) an individual responds to emotion-inducing stimuli and appraises his reactions, (b) he experiences a level of sympathetic arousal that is still elevated from prior stimulation, and (c) he is not provided with apparent extero- and/or interoceptive cues that would indicate that his arousal results from this prior stimulation, excitatory residues from prior arousal will combine inseparably with the excitatory response to present stimuli and thereby intensify emotional behavior.

2. Emotional behavior and emotional experience will be enhanced in proportion to the magnitude of the prevailing residual excitation.

3. Both the period of time in which transfer can manifest itself and the magnitude of transferable residues are a function of: (a) the magnitude of the preceding excitatory response, and/or (b) the rate of recovery from the excitatory state.

4. The individual's potential for transfer is: (a) proportional to his excitatory responsiveness, and (b) inversely proportional to his proficiency to recover from excitatory states.

The validity of the transfer paradigm has been demonstrated under a wide range of emotional circumstances. It has been shown, for example, that residues of excitation from physical exertion can intensify feelings of anger and aggressive behavior (Zillmann, Katcher, & Milavsky, 1972; Zillmann & Bryant, 1974) or the experience of sexual excitement (Cantor et al., 1975). Furthermore, it has been shown that residues of sexual arousal can potentiate aggression (Zillmann, 1971; Meyer, 1972; Zillmann, Hoyt, & Day, 1974) and that residues from either sexual arousal or from disgust reactions can facilitate such diverse experiences as the enjoyment of music (Cantor & Zillmann, 1973), appreciation of humor (Cantor, Bryant, & Zillmann, 1974), and dysphoric empathy (Zillmann, Mody, & Cantor, 1974). There is also evidence in support of the predictions regarding individual differences in the propensity for transfer (e.g., Zillmann, Johnson, & Day, 1974). We will not review these investigations here but rather concentrate on the findings that are particularly pertinent to the attribution or misattribution of excitation in transfer situations.

ATTRIBUTION IN EXCITATION TRANSFER

Excitation-transfer phenomena are widespread in animal behavior. In ethological analyses they have been extensively discussed under the concept of "displacement activity" (cf. Hinde, 1970). Evidently, as a stimulus condition for specific responses changes and the responses in question become inappropriate and lose their adaptive value, residual excitation intensifies reactions to other stimuli or – in the absence of unambiguous stimuli – fosters strong motor habits (e.g., preening in birds). Laboratory investigations on animals have further substantiated the transfer phenomenon, even in domains of behavior that have been generally held to be extremely S–R-specific. It has been shown, for example, that residual arousal from conditioned fear can greatly potentiate the sexual behavior of rats (e.g., Crowley, Popolow, & Ward, 1973). In such instances, attributional considerations appear to be misplaced. They might also be considered misplaced when a similar behavioral facilitation occurs in man. In accordance with Cannon's (1929) reasoning on the emergency function of emotion, one might readily grant that as an emergency condition ceases to exist, residual excitation from the state of emergency favors intense motor reactions in the postemergency period. No matter what particular responses are called for by the environmental stimuli to which the individual is exposed during this period, residual arousal should intensify behavior – at least at the motor level. This, essentially, is the behavior-theoretical position discussed earlier. The intensification of behavior by residual arousal from preceding activities is explained as resulting from the combination of "irrelevant" and "relevant" drive, with the irrelevant drive component being a residue from prior arousal states. Assumptions about the conscious or quasi-conscious attribution or misattribution of arousal are obviously *not required*. The mere demonstration of excitation transfer in the emotional behavior of man, then, cannot be considered to prove anything about attributional process. The facilitation of mirth through prior sexual arousal or through disgust (Cantor et al., 1974), for example, could be explained in drive terms without resorting to attributional conceptualizations. In fact, the involvement of attributional concepts in the explanation of such transfer effects can be viewed as violating epistemological parsimony.

Regarding the attribution and misattribution of excitatory reactions, it is not so much the analysis of excitation transfer itself but the analysis of the conditions under which transfer fails that proves revealing and instructive. These conditions are critical in that they permit a separation of the predictions from behavior theory and the transfer paradigm – that is, from a model that does not entail assumptions about attribution and a model that involves attributional considerations. According to the behavior-theoretical view, the facilitation of behavior through residual excitation (i.e., elements of irrelevant drive) is *always* proportional to the magnitude of these residues at the time. Excitation-transfer theory, in contrast, stipulates that the behavior-enhancing transfer of excitation depends upon certain attributional conditions: Transfer will occur only if the individual does *not* causally connect a residual state of excitation to a prior inducer. If the individual links a state of excitation, rightly or wrongly, to a prior inducer, behavior-facilitating transfer is *not* expected to occur. It is at this point that the predictive accuracy of the two approaches can be evaluated and potentially that a case can be made for the involvement of attributional concepts.

A substantial amount of research implicates attributional processes in excitation transfer (e.g., Cantor & Zillmann, 1973; Zillmann & Bryant, 1974; Zillmann, Katcher, & Milavsky, 1972). The investigations that pertain most directly to the issue, however, are those by Zillmann, Johnson, and Day (1974) and Cantor, Zillmann, and Bryant (1975). We will very briefly discuss these latter investigations.

The first of the two investigations examined the effects of residual excitation from strenuous, physical exercise on aggressive behavior. Subjects were provoked, engaged in exercise, and then provided with an opportunity to retaliate against their annoyer. The provision of this opportunity was timed differently: Subjects either retaliated immediately after exercise or, similar to earlier research (e.g., Zillmann et al., 1972), after a brief recovery period. It was argued that the recovery period not only lowers the level of excitation but — more important for an attributional analysis — *removes obtrusive exteroceptive and/or interoceptive cues* (e.g., trembling of hands, heaving breathing, sweating, heart-pounding) that would attract the subject's attention and virtually force him into linking his reaction to exercise. The subject, then, should become vulnerable to "overreact" aggressively only after the recovery period: As he is reconfronted with his annoyer, anger is cognitively reinstated, and residual excitation from exercise can potentiate anger and aggression because it is not linked to its appropriate source. Immediately after exercise, on the other hand, the subject is cognizant of the fact that he is still aroused from exercising; consequently, in the reconfrontation with his annoyer, he is unable to misattribute residual excitation to his reaction to that person. In other words, residual arousal — because it is properly attributed to its actual source — cannot transfer into anger and aggression. The findings corroborated this reasoning. Motivated aggressive behavior was more intense after partial recovery than immediately after exercise. It should be clear that this result is not only nonsupportive of, but *opposite* to, behavior-theoretical predictions. Residual arousal from strenuous exercise was, of course, of a much greater magnitude during aggression immediately after exercise than during aggression after a recovery period. Aggression, then, should have been far more "energized" in the former than in the latter condition.

The second of the two investigations explored transfer from exercise-induced arousal into the experience of sexual excitement. In contrast to earlier studies in which attributional processes were only presumed, in this experiment the attribution of residual excitation from exercise was empirically assessed by a pretest. Subjects were instructed to report the degree to which they considered themselves still aroused from the exercise. These reports were recorded at regular intervals. The subject's recovery was also monitored in indices of sympathetic excitation (heart rate, blood pressure). On the basis of the measures of *perceived* and *actual* excitation, three phases were constructed:

1. In *Phase 1,* measured residual excitation was pronouncedly elevated, and subjects were aware that they were still aroused from prior stimulation.
2. In *Phase 2,* measured excitation was still significantly elevated, but subjects reported feeling that they had recovered from their prior excitatory response.
3. In *Phase 3,* measured excitation had returned to base level, and subjects perceived that they had recovered. Because residual excitation that could affect behavior is absent during Phase 3, this phase constitutes a control condition in comparison to which transfer effects are assessed.

Taking a no-attribution, behavior-theoretical perspective, behavior during Phase 1 — because residues are at a maximum — should be greatly enhanced. During Phase 2, behavior enhancement should occur but to a lesser degree. Predictions from excitation-transfer theory are the opposite. During Phase 1, transfer should not occur, because the individual is unable to misattribute residues of excitation from prior stimulation. If anything, the individual should underestimate his excitatory reactions to the stimuli he is exposed to during Phase 1, because he will tend to perceive them as a part of his state of excitation induced by prior stimulation. Phase 2 constitutes the transfer period, and only during this phase should emotional behavior and emotional experience in response to environmental stimuli be facilitated by residual excitation.

In the main experiment, these predictions were tested by exposing subjects in Phase 1, in Phase 2, or in Phase 3 to erotica and by measuring the degree to which they perceived themselves as sexually aroused. As predicted from the theory of excitation transfer, and in sharp contrast to behavior-theoretical expectations, reported sexual excitement was found to be markedly elevated only during Phase 2. In spite of greater residual excitation, sexual excitement during Phase 1 failed to exceed that in the control condition. In fact, it was slightly lower.

These and related findings show quite compellingly that it is advantageous to involve attributional considerations in the explanation of emotional behavior and emotional experience. It is the involvement of such considerations in the excitation-transfer paradigm that — compared to the behavior-theoretical reasoning on response-facilitation through arousal — gives it superior predictive accuracy. In theories of emotion, then, attributional process *must* be recognized and incorporated in the construction of theory — at least as hypothetical constructs — in order to achieve a satisfactory level of explanatory adequacy.

The reported findings not only stress the need to involve attributional concepts in theories of emotion, but permit some rather specific inferences about the attribution and misattribution of excitatory reactions. First, the comparative analysis of the time course of an excitatory reaction in terms of actual and perceived arousal revealed a lack of correspondence: Subjects reported recovery long before they had recovered. Put another way, they correctly reported relatively high levels of sympathetic activity as arousal but failed to sense and recognize lower levels of arousal. This result corroborates the view that the interoception of excitatory reactions is largely unreliable and that the individual distinguishes only rather gross changes in level of excitation (cf. Mandler, 1975). The individual apparently does not skillfully monitor his excitatory reactions. He does not reliably trace back excitatory residues. More specifically, *he does not engage in a continuous causal accounting of his sympathetic activity induced by earlier stimulation.* As a consequence, he is prone to "overreact" emotionally, as predicted by the excitation-transfer paradigm. In this overreaction, he de facto misattributes residual excitation; that is, he *behaves as if he believed* that these residues had been induced by a prevailing stimulus condition.

Second, and most importantly, in the face of obtrusive exteroceptive and interoceptive cues of an excitatory reaction whose *prior* induction was immediately apparent and unambiguous, subjects were unable to misattribute residual excitation. This must mean that *the individual who recognizes a causal connection between an inducing condition and his arousal reaction is aware of this connection, at least for the period of time in which the symptoms of his reaction are in obtrusive evidence.* During this period of awareness, the individual apparently is unwilling to consider alternative sources for his arousal. Such a state of affairs may be viewed as supporting the second proposition of the two-factor theory of emotion (Schachter, 1964). But though *some form of recognition* of a causal connection can be inferred from behavioral consequences, the specific form of such a recognition remains entirely unclear. Does the individual, as suggested by two-factor theory, really ponder the circumstances and select a suitable explanation for his arousal reaction? Or is the process more primitive? Could it be that the individual does not *search* for viable attributions at all, but instead readily *accepts* an obtrusive stimulus condition that immediately precedes an arousal reaction as a "causal" connection? After all, such a possible confusion between S—R contiguity and causal relation would statistically speaking be quite adequate and sufficient in terms of adaptation. Could it be, then, that man is rather uncritical in adopting "explanations" for his excitatory reactivity and that in this context the notion of "epistemic search" is misplaced? Unfortunately, the data at hand have no bearing on this issue. They evidence that causal connections between excitatory reactions and inducing conditions are made and have immediate and delayed behavioral implications (cf. Zillmann & Bryant, 1974). The specific cognitive manifestations of these causal connections remain to be determined, however.

ACKNOWLEDGMENTS

The author's research discussed in this chapter has been supported by Grant MH −19456 from the National Institute of Mental Health and by Grants GS−35165 and SOC−7513431 from the National Science Foundation.

REFERENCES

Ádám, G. *Interoception and behavior: An experimental study.* Budapest: Publishing House of the Hungarian Academy of Sciences, 1967.

Arnold, M. B. *Emotion and personality.* New York: Columbia University Press, 1960.

Bard, P. The neuro-humoral basis of emotional reactions. In C. A. Murchison (Ed.), *A handbook of general experimental psychology.* Worcester, Mass.: Clark University Press, 1934.

Barefoot, J. C., & Girodo, M. The misattribution of smoking cessation symptoms. *Canadian Journal of Behavior Science,* 1972, *4,* 358–363.

Barefoot, J. C., & Straub, R. B. Opportunity for information search and the effect of false heart-rate feedback. *Journal of Personality and Social Psychology,* 1971, *17,* 154–157.

Bem, D. J. Self-perception theory. In L. Berkowitz (Ed.), *Advances in experimental social psychology* (Vol. 6). New York: Academic Press, 1972.

Berkowitz, L., Lepinski, J. P., & Angulo, E. J. Awareness of own anger level and subsequent aggression. *Journal of Personality and Social Psychology,* 1969, *11,* 293–300.

Berlyne, D. E. *Conflict, arousal, and curiosity.* New York: McGraw–Hill, 1960.

Brown, J. S. *The motivation of behavior.* New York: McGraw–Hill, 1961.

Calvert–Boyanowsky, J., & Leventhal, H. The role of information in attenuating behavioral responses to stress: A reinterpretation of the misattribution phenomenon. *Journal of Personality and Social Psychology,* 1975, *32,* 214–221.

Cannon, W. B. The James–Lange theory of emotions: A critical examination and an alternative theory. *American Journal of Psychology,* 1927, *39,* 106–124.

Cannon, W. B. *Bodily changes in pain, hunger, fear and rage: An account of researches into the function of emotional excitement* (2nd ed.). New York: Appleton–Century, 1929.

Cantor, J. R., Bryant, J., & Zillmann, D. Enhancement of humor appreciation by transferred excitation. *Journal of Personality and Social Psychology,* 1974, *30,* 812–821.

Cantor, J. R., & Zillmann, D. The effect of affective state and emotional arousal on music appreciation. *Journal of General Psychology,* 1973, *89,* 97–108.

Cantor, J. R., Zillmann, D., & Bryant, J. Enhancement of experienced sexual arousal in response to erotic stimuli through misattribution of unrelated residual excitation. *Journal of Personality and Social Psychology,* 1975, *32,* 69–75.

Cofer, C. N., & Appley, M. H. *Motivation: Theory and research.* New York: Wiley, 1964.

Crowley, W. R., Popolow, H. B., & Ward, O. B., Jr. From dud to stud: Copulatory behavior elicited through conditioned arousal in sexually inactive male rats. *Physiology and Behavior*, 1973, *10*, 391–394.

Duffy, E. *Activation and behavior*. New York: Wiley, 1962.

Festinger, L. A theory of social comparison processes. *Human Relations*, 1954, *7*, 117–140.

Geen, R. G., & O'Neal, E. C. Activation of cue-elicited aggression by general arousal. *Journal of Personality and Social Psychology*, 1969, *11*, 289–292.

Gellhorn, E. Motion and emotion: The role of proprioception in the physiology and pathology of the emotions. *Psychological Review*, 1964, *71*, 457–472.

Goldstein, D., Fink, D., & Mettee, D. R. Cognition of arousal and actual arousal as determinants of emotion. *Journal of Personality and Social Psychology*, 1972, *21*, 41–51.

Grossman, S. P. *A textbook of physiological psychology*. New York: Wiley, 1967.

Hebb, D. O. Drives and C.N.S. (conceptual nervous system). *Psychological Review*, 1955, *62*, 243–254.

Heider, F. *The psychology of interpersonal relations*. New York: Wiley, 1958.

Henle, J. *Antropologische Vorträge*. Braunschweig: Vieweg, 1876.

Hinde, R. A. *Animal behavior: A synthesis of ethology and comparative psychology* (2nd ed.). New York: McGraw–Hill, 1970.

Hull, C. L. *Principles of behavior*. New York: Appleton, 1943.

Hull, C. L. *A behavior system: An introduction to behavior theory concerning the individual organism*. New Haven, Conn.: Yale University Press, 1952.

Izard, C. E. *The face of emotion*. New York: Appleton–Century–Crofts, 1971.

James, W. What is emotion? *Mind*, 1884, *9*, 188–204.

Johnson, J. E. Effects of accurate expectations about sensations on the sensory and distress components of pain. *Journal of Personality and Social Psychology*, 1973, *27*, 261–275.

Jones, E. E., & Davis, K. E. From acts to dispositions: The attribution process in person perception. In L. Berkowitz (Ed.), *Advances in experimental social psychology* (Vol. 2). New York: Academic Press, 1965.

Kelley, H. H. Attribution theory in social psychology. In D. Levine (Ed.), *Nebraska Symposium on Motivation, 1967*. Lincoln, Neb.: University of Nebraska Press, 1967.

Kelley, H. H. Attribution in social interaction. In E. E. Jones, D. E. Kanouse, H. H. Kelley, R. E. Nisbett, S. Valins, & B. Weiner (Eds.), *Attribution: Perceiving the causes of behavior*. Morristown, N.J.: General Learning Press, 1971. (a)

Kelley, H. H. Causal schemata and the attribution process. In E. E. Jones, D. E. Kanouse, H. H. Kelley, R. E. Nisbett, S. Valins, & B. Weiner (Eds.), *Attribution Perceiving the causes of behavior*. Morristown, N.J.: General Learning Press, 1971. (b)

Kety, S. S. Neurochemical aspects of emotional behavior. In P. Black (Ed.), *Physiological correlates of emotion*. New York: Academic Press, 1970.

Lange, C. *Über Gemütsbewegungen: Eine psycho-physiologische Studie*. Leipzig: Thomas, 1887.

Lange, C. G., & James, W. *The emotions* (Vol. 1). Baltimore: Williams & Wilkins, 1922.

Latané, B., & Schachter, S. Adrenalin and avoidance learning. *Journal of Comparative and Physiological Psychology*, 1962, *65*, 369–372.

Lazarus, R. S. *Psychological stress and the coping process*. New York: McGraw–Hill, 1966.

Leventhal, H. Emotions: A basic problem for social psychology. In C. Nemeth (Ed.), *Social psychology: Classic and contemporary integrations*. Chicago: Rand McNally, 1974.

Lindsley, D. B. Emotion. In S. S. Stevens (Ed.), *Handbook of experimental psychology*. New York: Wiley, 1951.

Lindsley, D. B. Psychophysiology and motivation. In M. R. Jones (Ed.), *Nebraska Symposium on Motivation, 1957*. Lincoln, Neb.: University of Nebraska Press, 1957.

MacLean, P. D. Psychosomatic disease and the "visceral brain": Recent developments bearing on the Papez theory of emotion. *Psychosomatic Medicine*, 1949, *11*, 338–353.

MacLean, P. D. The limbic brain in relation to the psychoses. In P. Black (Ed.), *Physiological correlates of emotion.* New York: Academic Press, 1970.

Malmo, R. B. Activation: A neurophysiological dimension. *Psychological Review,* 1959, *66,* 367–386.

Mandler, G. *Mind and emotion.* New York: Wiley, 1975.

Mandler, G., & Kremen, I. Autonomic feedback: A correlational study. *Journal of Personality,* 1958, *26,* 388–199.

Mandler, G., Mandler, J. M., Kremen, I., & Sholiton, R. D. The response to threat: Relations among verbal and physiological indices. *Psychological Monographs,* 1961, *75*(9, Whole No. 513).

Marler, P., & Hamilton, W. J. *Mechanisms of animal behavior.* New York: Wiley, 1966.

Meyer, T. P. The effects of sexually arousing and violent films on aggressive behavior. *Journal of Sex Research,* 1972, *8,* 324–331.

Morgan, C. T. Physiological theory of drive. In S. Koch (Ed.), *Psychology: A study of a science* (Vol. 1). *Sensory, perceptual and physiological formulations.* New York: McGraw –Hill, 1959.

Nevin, J. A. (Ed.). *The study of behavior: Learning, motivation, emotion, and instinct.* Glenview, Ill.: Scott, Foresman, 1973.

Newman, E. B., Perkins, F. T., & Wheeler, R. H. Cannon's theory of emotion: A critique. *Psychological Review,* 1930, *37,* 305–326.

Nisbett, R. E., & Schachter, S. Cognitive manipulation of pain. *Journal of Experimental Social Psychology,* 1966, *2,* 227–236.

Nisbett, R. E., & Valins, S. Perceiving the causes of one's own behavior. In E. E. Jones, D. E. Kanouse, H. H. Kelley, R. E. Nisbett, S. Valins, & B. Weiner (Eds.), *Attribution: Perceiving the casues of behavior.* Morristown, N.J.: General Learning Press, 1971.

Oken, D. The psychophysiology and psychoendocrinology of stress and emotion. In M. H. Appley & R. Trumbull (Eds.), *Psychological stress: Issues in research.* New York: Appleton–Century–Crofts, 1967.

Papez, J. W. A proposed mechanism of emotion. *Archives of Neurology and Psychiatry,* 1937, *38,* 725–743.

Papez, J. W. Cerebral mechanisms. *Journal of Nervous and Mental Disease,* 1939, *89,* 145–159.

Ross, L., Rodin, J., & Zimbardo, P. G. Toward an attribution therapy: The reduction of fear through induced cognitive–emotional misattribution. *Journal of Personality and Social Psychology,* 1969, *12,* 279–288.

Schachter, S. *The psychology of affiliation: Experimental studies of the sources of gregariousness.* Stanford: Stanford University Press, 1959.

Schachter, S. The interaction of cognitive and physiological determinants of emotional state. In L. Berkowitz (Ed.), *Advances in experimental social psychology* (Vol. 1). New York: Academic Press, 1964.

Schachter, S. Cognitive effects on bodily functioning: Studies of obesity and eating. In D. C. Glass (Ed.), *Neurophysiology and emotion.* New York: Rockefeller University Press, 1967.

Schachter, S. The assumption of identity and peripheralist–centralist controversies in motivation and emotion. In M. B. Arnold (Ed.), *Feelings and emotions: The Loyola Symposium.* New York: Academic Press, 1970.

Schachter, S., & Singer, J. Cognitive, social and physiological determinants of emotional state. *Psychological Review,* 1962, *69,* 379–399.

Schachter, S., & Wheeler, L. Epinephrine, chlorpromazine, and amusement. *Journal of Abnormal and Social Psychology,* 1962, *65,* 121–128.

Singer, J. E. Sympathetic activation, drugs and fright. *Journal of Comparative and Physiological Psychology,* 1963, *56,* 612–615.

Spence, K. W. *Behavior theory and conditioning.* New Haven, Conn.: Yale University Press, 1956.

Titchener, E. B. An historical note on the James—Lange theory of emotion. *American Journal of Psychology,* 1914, *25,* 227–247.

Valins, S. Cognitive effects of false heart-rate feed-back. *Journal of Personality and Social Psychology,* 1966, *4,* 400–408.

Valins, S. The perception and labeling of bodily changes as determinants of emotional behavior. In P. Black (Ed.), *Physiological correlates of emotion.* New York: Academic Press, 1970.

Wynne, C. C., & Solomon, R. L. Traumatic avoidance learning: Acquisition and extinction in dogs deprived of normal peripheral autonomic function. *Genetic Psychology Monographs,* 1955, *52,* 241–284.

Zillmann, D. Excitation transfer in communication-mediated aggressive behavior. *Journal of Experimental Social Psychology,* 1971, *7,* 419–434.

Zillmann, D. The role of excitation in aggressive behavior. In *Proceedings of the Seventeenth International Congress of Applied Psychology, 1971.* Brussels: Editest, 1972.

Zillmann, D., & Bryant, J. The effect of residual excitation on the emotional response to provocation and delayed aggressive behavior. *Journal of Personality and Social Psychology,* 1974, *30,* 782–791.

Zillmann, D., Bryant, J., Cantor, J. R., & Day, K. D. Irrelevance of mitigating circumstances in retaliatory behavior at high levels of excitation. *Journal of Research in Personality,* 1975, *9,* 282–293.

Zillmann, D., & Cantor, J. R. Effect of timing of information about mitigating circumstances on emotional responses to provocation and retaliatory behavior. *Journal of Experimental Social Psychology,* 1976, *12,* 38–55.

Zillmann, D., Hoyt, J. L., & Day, K. D. Strength and duration of the effect of aggressive, violent, and erotic communications on subsequent aggressive behavior. *Communication Research,* 1974, *1,* 286–306.

Zillmann, D., Johnson, R. C., & Day, K. D. Attribution of apparent arousal and proficiency of recovery from sympathetic activation affecting excitation transfer to aggressive behavior. *Journal of Experimental Social Psychology,* 1974, *10,* 503–515.

Zillmann, D., Katcher, A. H., & Milavsky, B. Excitation transfer from physical exercise to subsequent aggressive behavior. *Journal of Experimental Social Psychology,* 1972, *8,* 247–259.

Zillmann, D., Mody, B., & Cantor, J. R. Empathetic perception of emotional displays in films as a function of hedonic and excitatory state prior to exposure. *Journal of Research in Personality,* 1974, *8,* 335–349.

IV CURRENT PROBLEMS AND FUTURE PERSPECTIVES

In Chapter 13, Edward Jones and Harold Kelley discuss historical influences in their attribution work, the status of and some problems with extant attribution approaches, the future of attribution research, and some possible applications of attributional ideas. Jones and Kelley express considerable optimism about the burgeoning amount of work on attribution and the potential development of this area of research in social psychology. However, they also express some concern about the area. In particular, they view with concern the lack of theoretical advances in understanding basic attributional processes — understanding that they feel is needed to balance the extensive amount of research currently being conducted on attributional phenomena.

After Fritz Heider pioneered attributional ideas in social psychology, Jones and Kelley contributed important early statements that provided a foundation for further systematic theorizing and research. They are currently both active workers in the attribution area. Nevertheless, it is clear that they already stand as scholars of historical significance in the emergence of the area. Jones and Davis' 1965 "From Acts to Dispositions: The Attribution Process in Person Perception" and Kelley's 1967 "Attribution Theory in Social Psychology" represent classic contributions to the attribution area and to the field of social psychology.

We believe that because of Jones and Kelley's status as major attribution theorists and because of the unique historical perspectives they can provide on the development of attributional approaches, this conversation represents an appropriate final piece for this volume and a fitting complement to the first manuscript in Volume 1 involving an interview with Fritz Heider. For both the Heider and Jones—Kelley interviews, Larry Erlbaum provided enthusiasm and financial backing. We are most grateful for his help and perceptiveness in recognizing the possible import of these discussions for the field. The Jones—Kelley conversation occurred at Duke University in October 1976 and was graciously hosted by Ned and Ginny Jones. What follows is a collection of excerpts edited from two sessions of conversation recorded at that time.

13

A Conversation With Edward
E. Jones and Harold H. Kelley

Editors:

John H. Harvey
Vanderbilt University

William Ickes
University of Wisconsin

Robert F. Kidd
Boston University

HISTORICAL INFLUENCES

Editors: Maybe a good way to begin would be for you to tell us how your work with attribution started.

Kelley: I was first stimulated with these kinds of ideas by John Thibaut and from his experiments with Henry Riecken and then Lloyd Strickland. John and I did those studies on the minimal social situations in which two persons control one another's outcomes but have no knowledge of exactly how they affect each other (Kelley, Thibaut, Radloff, & Mundy, 1962). There was a little attributional problem in the middle of these studies, having to do with the kind of logical processes that tend to catch my attention. The problem was one of a person analyzing his effect on the other by doing something like a covariation analysis. They couldn't both be doing this at once, or their mutual analytic efforts would interfere with either one finding out anything. Subsequently, a number of my students did follow-ups of the Thibaut work: Ken Ring did an extension of the Thibaut and Riecken studying (Ring, 1964), and Arie Kruglanski did a follow-up of Strickland (Kruglanski, 1970). They were just getting into doing these things in the mid-60s, a little before I wrote that paper for the *Nebraska Symposium on Motivation* (Kelley, 1967). I had had an interest in social perception all along.

Jones: In fact, you did your dissertation on first impressions in social perception.

Kelley: Yes. In part I became interested in social perception problems because Kurt Lewin wasn't dealing with social perception, yet he was always talking about the results of perception. It seemed to me that there needed to be an analysis of the processes involved. I remember that as being the rationale for my thesis.

Editors: Was Heider much of an influence?

Kelley: Very much so. I should not have forgotten that. When I was at Minnesota at the Laboratory for Research in Social Relations, the whole group would occasionally read a book and really study and discuss it. We did this for Heider, so I read Heider more closely than I might have otherwise. I also was killing two birds with one stone, because I reviewed it for *Contemporary Psychology* in 1960. In my review of Heider, I emphasized the balance and the attribution ideas and distinguished between them. I am almost certain that Fritz, and I think he is right, regards that as overlooking vast chunks of the book. It's a very rich work and undoubtedly has some other kinds of themes that can be drawn out.

Editors: Do you think you may have been responsible for separating those two notions more then Heider originally intended? Somewhere back in the history of social psychology, people started talking about balance and not bringing along with it the ideas involved in attribution.

Kelley: In some way, the two streams that are divided seem to be due to the separate treatment Heider gave them in his various papers.

Jones: Especially the one on phenomenal causality and the other on cognitive organization.

Kelley: Just one other small point on that. Really the main exception to balance is an attributional effect. It's when you expect to agree about something with someone you don't like. That's a very powerful effect, an exception to balance almost any way you think about it. It is an illustration of an assumed consensus effect where you and this other person, even though you don't like him, are assumed to be looking at the same external stimulus and responding to it. The assumed agreement implies you are both responding to the same entity property. This exception is another kind of thing that got me interested in the middle 60s.

Editors: Ned, would you like to tell us how you got interested in attribution?

Jones: I think Jerry Bruner started my interest in person perception by referring me to Asch's (1946) warm—cold study. It struck me as very intriguing because it got away from the accuracy problem that was inhibiting developments in perception — questions concerning who makes the best judge or what kind of person makes the most perceptible target. Then John Thibaut also put me onto Heider through the causality paper. John was thinking about the Thibaut and Riecken (1955) study when I was with him at Harvard in the early 50s. I was terribly intrigued by the causality notion, by the fact that people have to truncate the causal sequence somehow. At some fairly early point we have to stop and say, "He is responsible" or "It is responsible" or something like that. Let me give you an example.

When I was a 4th-year student in clinical psychology, we had a group therapy seminar run by Elvin Semrad at Boston State Hospital. Although it was supposed to be a seminar, there were only two ground rules. One was that somebody had to take notes every time, and the other was that we had to write a term paper. Everything else was up for grabs. It turned into an early encounter group. Semrad had been doing this for some years and was very skillful at bringing out hostilities between group members, turning them on himself, and then having us analyze the authority problem. The problem that kept arising in this group was, is there any such thing as legitimate hostility? Under what circumstances can you get mad at somebody and really justify it, especially if you are a clinical psychologist and concerned with the infinite regress of causation and the determinants of behavior? That sort of stuck in my craw. Many of the early things I did were an attempt to work out this problem. Under what circumstances is hostility ameliorated or mitigated by some knowledge of the hostile person or the situation under which that hostile remark was produced?

In an early, complex, and not entirely successful study (Jones, Hester, Farina, & Davis, 1959), we had a subject verbally attacked by someone who was either presented as well adjusted or maladjusted. We were attempting to see if a person being attacked actually felt better about the attack and therefore dealt with it better, knowing it came from a maladjusted person.

The Jones and Davis (1965) paper really came out of an attempt in the introductory social psychology classroom to talk about person perception in terms of choice. I had a little model I used to put on the board. I don't know if the students ever took much away from it, but when Berkowitz invited me to do a chapter for *Advances in Experimental Social Psychology,* it was just a logical unfolding of all the problems involved in the simple-minded notion that you can tell what a person is like if he has choice when he behaves. I think the word "attribution" got in the subtitle of that article.

When Heider was at Duke in 1962 and 1963 he talked a lot about attribution and was concerned with extensions of the Heider and Simmel study of animated geometric figures. I don't remember any special emphasis on the term in the book, but I later went back and saw that it was in the 1944 article; and I'm sure it was in the book, too. It seemed like a happy term to deal with a fundamental process in person perception.

I think I came to attribution from a narrower perspective than Hal, because I had been concerned with two paradigms. One was person perception; the other was self-presentation. I think at the outset I didn't really see clearly the connections between these things; however, now I see them as very closely related if only because self-presentation provides the figural stimulus for person perception in the real world. Initially, my interest in ingratiation was quite separate from my interest in person perception, but that didn't last very long. I remember being stunned and a little envious and delighted when Hal's 1967 paper came out. It was a comprehensive statement which tied together a lot of things that I had not thought about as being tied together. The social comparison process, for example, I hadn't really related to attribution in any way. The pictorial representation of an analysis of variance cube was very heuristic.

Kelley: Actually, I added on those pictures of the cube and the ANOVA analogue after I had written the paper. In some ways I am not happy that I did it. It was a thought-aiding device, not something I wanted to feature. I should have known that the analogue, being very concrete, would provide the label.

Jones: I think it's very important to have a label. In this case it integrates a set of variables that otherwise seem unrelated.

ON ATTRIBUTION AS A CAUSE
VERSUS ATTRIBUTION AS A PROPERTY

Editors: Is attribution, then, a process, or is it really a conceptual domain, or is it just a way of talking about a particular kind of perception? — that is, perception which is related to causality and motive?

Kelley: I sense that the way we could all agree to use the term is in relation to causality. Clearly, when I drew in the social comparison ideas, I started talking about entity attribution. I was stretching the notion of causation and starting to include the more vague processes of attributing properties.

Editors: There seems to be a strong implicational link there.

Kelley: I think the field suffers from some vagueness or some uncertainty about where it should be focused. I'd like your comments on whether it should be focused on causal explanations or whether it is wise to keep a lot of property imputations. Maybe it's unwise to judge these things prematurely. Maybe it's best to keep the boundaries a little vague.

Editors: Ned focused on that issue a minute ago when he was talking about the fascinating question of what is a sufficient cause for a behavior. The Jones and Davis paper indicated that if most people can point to some trait or underlying disposition in another's behavior, that is as far back as they need to take it. That certainly has the characteristics of attribution in the sense of describing reality in terms of its attributes and characterological features.

Kelley: There the focus is on causation.

Editors: Yes, but the two are combined when you're dealing with attributions of motives or traits.

Jones: I think maybe it is better not to restrict a domain prematurely. What would have happened if in the early days of dissonance research, somebody really got serious about defining a "cognition" or "dissonance"? The theory rested on a whole set of very badly or vaguely defined terms, yet nobody really had a terrible problem using those terms. I suspect some of the same thing is true here.

Kelley: I think in one way the label [attribution] has been applied too generously. It has been applied to two very different sets of phenomena which should really be sharply distinguished. One of them is the theory about the causal attribution process. There isn't a great deal there. There are other things to be drawn on, things floating in the background, like Duncker's (1945) ideas. I think we ought to discuss this more. I think you were dead right, Ned, in that paper with McGillis (Jones & McGillis, 1976), that the development of the theory has been slow — the theory about data, rules, inferences, the cognitive inferential part of the process. And in some ways that's really quite a small part of the attribution research. Most of the material I think should be called *attribution-based* theory. That's what most of the research is about — attribution-based theories of emotion, achievement motivation, affiliation, helping, revenge, equity. Weiner's analysis would epitomize it. Given an attribution, what are its implications for motivation, persistence, or affect? He is not very much interested in all the prior attribution process issues. He did a little dabbling in that, but it was really not his interest. So he's got an attribution-based theory of achievement motivation. Perhaps, as John Arrowood has suggested, the distinction here should be between *attribution* theory and *attributional* theories.

Editors: It seems that attribution as a word has been applied to anything having to do with the cognitive process in social psychology, very carelessly. People are adopting the language of attribution without also adopting some of the rigorous thinking that has gone along with the development of the theory. The theorist, on the other hand, has to work either at the level of the antecedents of the attribution or with the motivational consequences of arriving at an attribution.

Kelley: I suppose it's all right to put them under the attribution domain; but it is stretching it, and it leaves things muddy. Someone might have the impression that there's more coherence over a broader domain than there really is. Having two rather different things hooked together makes some people uncomfortable with the umbrella term "attributional."

Editors: The tendency simply is to assume what you and Ned and other people have said about the process.

Kelley: One problem with the attributional process part (i.e., that data, rules, etc.) is that it is available to all of us, because we are all intuitive, man-on-the-street psychologists. As a consequence, we can set up studies without being very explicit about the attribution process or revealing anything about the process itself. This research just takes some version of the process for granted. That's OK, but it is difficult thinking of how to put these two bodies of literature [the literature primarily concerned with process and the literature not primarily concerned with process] together, because in some ways they don't belong together. They're just contiguous to each other.

Jones: One sense in which they can belong together and rub up against each other is the sense in which we provide rational models by painting a prescriptive set of what the logical person would do if he had plenty of time and lots of resources. I think a lot of the things that I've done and the things that I've seen done have to do with looking at bias, identifying bias with the use of these logical models. In other words, you get data that don't fit the model, and they become interesting; so people pay some attention to them. For instance, you shouldn't attribute an opinion statement to an attitude cause if it was made under no-choice conditions. But people do. In that sense, the model has been used as a foil or a kind of logical backdrop against which to look at biases when they become interesting in their own right and lead to new problems and new attempts to explain.

ON THE DEVELOPMENT OF ATTRIBUTION THEORY
AND RESEARCH

Kelley: You hear the question raised, is attribution research a fad or a passing fancy like much of the cognitive dissonance work?

Jones: I'm not struck by the prevalence of such questions. This is a little bit of a digression, but it has interested me how few enemies attribution people have made.

Editors: Are you talking in relative terms, say compared to dissonance theory?

Jones: Compared to dissonance in particular, attribution hasn't really generated a lot of controversy.

Kelley: But a good deal of disdain.

Jones: I wonder how much of the disdain that exists isn't really a kind of minimal disdain and perhaps some envy, the disdain everyone feels toward something that has caught on and is so prominent in the literature?

Kelley: Anything will be a fad.

Editors: We wonder if you invite people to seriously challenge your models and thinking.

Kelley: One of the reactions to the field is that attribution is colorless. It doesn't have the old conflict, the issues.

Editors: Do we need some of that in attribution?

Kelley: It's not something to be taken as needed, per se, but it's rather symptomatic of imprecision in defining opposing views.

Editors: But do you welcome it?

Kelley: I welcome it if it's symptomatic that things are becoming clear enough and specified enough that it can be intelligently done. That's another difference between this area and cognitive dissonance. As dissonance theory developed it became rather tightly specified, and the experimental procedures were rather clearly defined. The possible variations in experimental paradigms were indicated specifically by the theory. In that setting, you could get a rather sharp juxtaposition of issues generating conflicting data — confirmation and disconfirmation.

I think the absence of that in our field is perhaps symptomatic of the looseness of thinking. Attribution is still exploratory — defining boundaries and classes of phenomena and so on.

Jones: It's almost in a sense an antitheory, too. You're almost saying that no set of specifications can possibly be accurate as long as they don't take into account the permutations and combinations of attributional possibilities. One of the implications of what you are saying, Hal, is that you can't get premature closure; somehow you have to keep it open, keep the complexity there.

Part of what is indicated here is that it's very desirable to have a division of labor in the field so that some people are thinking more broadly, and of necessity more vaguely, keeping at the public forefront various kinds of phenomena. Others will not be happy with that and will want great specificity, great clarity of methodology. They will burrow into these things. The logical possibilities of what they can study are endless. The function of the framework setters is to keep the bigger set of problems in front of them so that they would have some realization that what they were working on may be very much a part process. This is exactly how Bem describes your impact on him, Hal. He talks about a specific occasion when you took him by the scruff of the neck and said, "Lift yourself from this controversy and realize that you're dealing with something much broader with many more implications than you are making clear at this point."

ON ATTRIBUTIONS AND BEHAVIOR

Editors: What are your views about the relationship between attributions and behavior?

Jones: I think some people may be buying the assumption that people always behave in line with their attributions. If you can find out their attributions, you can, in a sense, explain why they behave the way they do. But we know that's not really true. There are many occasions where there is a slippage between the way in which people explain reality and how they respond to that reality. It's not always inherent in the cognition of what that reality is.

Wouln't it be interesting if people who overattribute by some criterion, traits, dispositions to others, actually don't act in line with that attribution at all? In some sense they know better; their behavior toward other people takes into account very clearly that this was situationally caused. I'm not sure that would happen if one could find a decent way of checking it out, but it would be terribly interesting if that were true. In some sense it has to be true. People can't live with severe distortions all the time. You ought to be able to find at least some occasions where the behavior is more rational than the attribution, or the behavior is more sensitive to what's actually going on in the situation than you would ever infer if you asked the person to expose his logical processes. That's just a hunch I have.

Editors: Does the behavior then tune the cognitions to some extent?

Jones: Part of it gets into the distortion that's introduced by the attributional questions you ask. We're not adept enough as yet to ask the right kinds of questions, and to be sensitive to the *sequence* of questions. This is particularly a problem in the perceived freedom and attribution of responsibility areas. You can get almost anything you want, depending on how you phrase the questions. In simple terms, when a guy tells you he has a particular attribution, he may not really have that attribution. Somehow you gave him too broad a category; but he doesn't have anything else to use, so he uses it.

Editors: How then do people project these elaborate decision trees, select among them, and then choose a response immediately?

Kelley: There are three possibilities: One, it's done very fast. Two, it doesn't happen. Three, it's done by rote. The corollary of the third one is that it has happened before.

That reminds me of something I've been puzzling about. How can figural notions be brought into the theory; for example, the assumption of similarity between the effect and the cause? If you hear a big bang you look for something big that banged, or if you see a big "heavy" footprint you can look for a big, fat elephant. It becomes clear to me that if we are going to deal with that kind of thing as part of the attribution process, then we have to distinguish between some process of simply perceiving properties like size or intensity and another process, perhaps partly related to and supported by the first, of inferring causes.

Jones: Another aspect of this is the extent to which too much thinking about the process interferes with the process. We had a graduate student who was actually an ex-Russian spy. One day in class I was talking about ingratiation. He said that it reminded him of an old Russian fable of a man who had a long beard. A child asked him one day whether he slept with his beard outside the covers or inside the covers, and the man said he never slept again! Obviously, a natural athlete is natural, because he doesn't have to think about what he's doing with his arms and legs. The same must be true in some sense about interpersonal behavior. It can be very disruptive to be caught in this reflexive process.

ATTRIBUTION AND SELF-PERCEPTION

Editors: Ned, you mentioned Bem earlier. Do you think Jones and Davis (1965) has implications for self-perception?

Jones: We didn't have that in mind when we wrote it. I have thought about it since. Do people use the theory to present themselves if they want to convey something about themselves, get across a particular point, or emphasize their uniqueness? I think there's a particular reservation, a proviso, and that has to do with the actor–observer differences emphasized by Jones and Nisbett. You're much less likely to formulate dispositional characterizations for yourself than you are for others. It's something that you don't necessarily need to do. I think you do and can ask yourself whether you did something that anybody else would have done under the circumstances. There we are talking about normative behavior that doesn't lead you to any new information about yourself or to any new decision about what kind of person you are. With Bem, you would get either an attribution to the demands of the situation, in a mands sense, or an attribution to the reality of the situation, in a tacts sense. Neither of these applies to dispositions. You can't get to dispositions from tacts and mands. Perhaps an exception might be when you tact your own feelings. But if you think about tacting an environmental object or describing what's out there, I don't see that as a disposition.

Editors: What's missing in that is the notion that people have a self-concept. In the cognitive-interactionist approach, people have very clear perceptions of self.

Kelley: The situation Bem's analysis is designed around is an experimental setting in which you're being asked about something you're experiencing for the first time; so you have no self-concept.

ACTOR—OBSERVER DIFFERENCES

Editors: Ned, do you have anything to say now about the actor—observer difference phenomenon, since it has gotten so much publicity?

Jones: The interesting question is why the actor—observer difference was so appealing. Even people who had reservations about it didn't question its fundamental reasonableness.

Editors: The idea of control is important. People want to render others more predictable, and one way to do that is to view others as objects.

Jones: One of the things that wasn't clarified by Jones and Nisbett (1971) has to do with the motive or need to see oneself in control, even if that in some way is not objectively true. You can phrase the attributional question in such a way that the observer will deny or resist the implication that he is being pushed around by the situation. We missed the 2 x 2 bifurcation into causes and reasons on the one hand, crosscut by the locus of attribution on the other. You can be commanded by or determined by a personal disposition, and you can be commanded and determined by the environment. In the original statement, we implied that you were much more likely to feel commanded by a disposition (and we want to resist that implicatioin of being controlled) than you would feel you were commanded by the situation. I think that's not necessarily true. You have to make a further differentiation. If you ask people, did the situation make you do this, they'll say no. But if you ask them if they wanted to do this because of the nature of the situation, they'll say yes, because they have control then. I think that's an important distinction. Many of the discrepancies in the data that followed from working with the Jones—Nisbett hypothesis can be accounted for by making allowances for this need for control.

Editors: We wonder about the actor—observer hypothesis in the future. Will there be continuing intense work on the hypothesis in the future?

Jones: I think there's a misleading implication in that paper that actors are always more accurate than observers. I know that's implied in Jones and Nisbett from time to time, but there are undoubtedly circumstances under which observers may be more accurate. I think the most intriguing and compelling kinds of evidence in the actor—observer area have to do with divergent perceptual perspectives. Those data are quite solid, and I think that's going to stay alive for awhile. The practical implications just hit you in the face. In a group situation, if you are sitting across from somebody, or if you are in some way looking at a guy, you are going to have a different conception of his contribution to the group. The implications and limiting conditions of this need to be very carefully explored.

APPLICATIONS: ATTRIBUTION AND SELF-PRESENTATION

Editors: An important point to raise now that people are beginning to think about applying attribution to approaches in other areas and to practical problems is that there may be only so far that you can apply them. Do you think attribution can be applied, or is it too simplistic?

Kelley: There must be some differences between application and dealing with nonlaboratory, complicated, real-world problems?

Jones: We're involved in a study that has potential applications. It could be called a study of what leads people to drink or take certain kinds of drugs. In a broader context, it comes out of my strong feeling that we should move in the direction of exploring the self-presentational aspects of attribution. In the interests of controlling attributions, how do people construct their own situations, and how do people set up their choices? Wachtel (1973) says that social psychologists only deal with a very limited part of human behavior, because we construct a situation for the subject and he follows that situation. One of the most important things that takes place in real life is that the individual maneuvers himself into situations or helps to define the situation he's in by making various prior choices.

We finished a simple study which needs several control groups to rule out some alternative explanations. It has to do with the conditions under which a person will choose to take a drug which allegedly inhibits his performance versus a drug which will facilitate his performance. We can think of the drug in this case as a performance "situation" that is being chosen. Putting together several notions from Seligman's helplessness and noncontingent performance work with some of the intrinsic—extrinsic motivation ideas, we came up with a hypothesis. Imagine somebody who has had a success experience but can't identify his own contribution to that success; that is, one who gets success feedback but doesn't have a feeling of control, who works a set of insoluble problems and is told that he did very well. This is a little bit like Seligman's success-depression notion. Such a person — when in a situation where he takes the test, expects to take a drug, and then expects to take the test again after experiencing the effects of the drug — should take a performance-inhibiting rather than a performance-facilitating drug. On the other hand, a person coming out of the contingent success experience should take no drug or the performance-facilitating drug. It turns out this is clearly true, at least with males. This seems to be a case, then, where the person acts on his environment to control the attributions that can be drawn from his behavior. We have a condition where the subject's performance is not known to the guy who is administering the drug. It apparently doesn't make any difference whether it's a public self-presentation or a private one, even though it's very difficult ensuring that something is really private. So maybe we are saying something about even the solitary drinker. People put themselves in a situation where whatever feedback they get, whether success of failure, they can't really lose. If it's failure, there's a clear reason for it. If it's success, they succeeded in spite of this presumably inhibitory drug and should feel very good about it.

I think in general the attributionists, especially Kelley, Nisbett, and others in our group that met in 1969 at UCLA, have tried to extend the domain of purely cognitive explanations as far as they could and to stay away from defensive attributions as much as possible. The game is to do whatever you can to explain what goes on without resorting to that kind of throwback to the "new look in perception." I think we are simply going to have to face it. As I read your couples' chapter [Orvis, Kelley, & Butler, 1976], Hal, it seems to me that you're getting back to defensive attribution, the rationalization of behavior. What I'm doing now is taking that whole defensive attribution thing very seriously and seeing what kinds of implications it has for different kinds of self-presentational behaviors.

Editors: You are suggesting that if we could extrapolate from the kind of result you are talking about, it might be a motivation for drinking.

Jones: Yes. Obviously, I have to qualify it seriously by saying that there are probably lots of reasons why people drink, but that this may be one reason which is particularly likely to be involved in problem drinking.

Editors: In your study, you seem to be getting close to the notion of fear of success.

Jones: I don't think it's a fear of success. In this case, it's a fear of failure which has unequivocal implications for one's own sense of competence. Berglas and I wrote a papter (as yet unpublished) in which we try to relate this "fear" to underachievement. Underachievement is another kind of strategy where you avoid putting all your effort eggs in one basket. If you hold back some motivation, then you can always, say, well, I could have done it if I really tried. If you opt for a situation which has constraints, impediments, difficulties, then there's an external explanation for not doing well, if in fact you don't do well. But again, if you do well and didn't try, or if there are external constraints, that's very reassuring to a person overly concerned with his competence.

Kelley: So they're put in the same category — people who don't try and others who try an unrealistically difficult task.

Jones: I see underachievement and overachievement as paradoxically very closely related. Under- and overachievers are taking different paths toward solving the problem of competence feedback. The overachiever is essentially saying, "I'm never going to take the risk of failing." It's kind of a vicious circle, because he really has to keep it up.

Editors: And the other guy is saying he'll never take the risk of succeeding. He may take the risk, but he's going to provide himself with an out.

Jones: I think the overachiever with a certain amount of luck and a certain amount of talent can survive.

Editors: The underachiever takes less of a risk because if the overachiever should fail, he hasn't got an out while the underachiever always has one.

Jones: Yes, I would agree.

APPLICATIONS: ATTRIBUTIONS IN INTERPERSONAL RELATIONS

Editors: What specific type of work with applied implications are you doing, Hal?

Kelley: My interest is in the study of couples and close relationships, not the study of attributional things per se. Attribution is coming in as it is needed. I find it useful to make sense of some of the kinds of things that are going on. I accept Ned's description of the things we uncovered in the attributional disagreement as reflecting defensive behavior, but then I would start thinking a little more about their function in preserving the relationship. It seems to me that maybe a lot of the attributional explanations given by couples revolve around concern for the relationship. The offender who engages in a bad behavior provides an explanation that is intended to gloss over the behavior. That puts the explanation process more in the context of being able to depart from the norms of the relationship, that is, get away with it, and not destroy the relationship. This process loosens up the structure and breaks down the constraints in the relationship that may be too severe for the people involved.

Editors: How does attribution fit in?

Kelley: I'm not entirely clear in my mind how it goes. In that one study that's in the book [Harvey, Ickes, & Kidd, 1976], I took the occasion to think about the possiblity that a great deal of the attribution process is perhaps learned and sustained in close relationships. In such relationships, it is important to learn to analyze why you have done something, why your partner has done something, figuring out the partner's commitment to the relationship, portraying your own commitment, etc. There may be some other motivation involved in the attribution process than being able to understand or control — such as an intense interest in keeping the relationship smoothed over and going. Sociologists tell us this in their analysis of "accounts" as justifying and excusing behavior.

As I've tried to become somewhat more subtle and comprehensive in the analysis of interdependence in a relationship, it seems to me that we are forced to the realization that people are interdependent on a variety of levels. Partners are not only interdependent on a specific behavior level — that tends to be the level that Thibaut and I have analyzed — but are also interdependent in the norm and role conceptions that they bring to the relationship. They are also interdependent in the attitudes and the personal characteristics that they bring to the relationship. Once I start thinking about the different levels of interdependence, to talk about these levels, I have to analyze what it means, for instance, that a person gets rewards at the normative level or at the level of personal characteristics and attitudes. You get rewards from being acknowledged to have certain characteristics, or you get rewards from the other person expressing some kind of attitude to you. You incur costs as your behavior departs from some norms that are valued and taken as appropriate to the relationship.

Once you start distinguishing these levels of interdependence, it becomes clear that some of the central processes in the relationship involve a movement back and forth among these levels behaviorally and cognitively. As participants in close relationships, we move from the specific to the general. We move from looking at the details, the specifics of interaction, to general inferences about what we are like, what rules we are following, what we're feeling about each other. On the other hand, there's the reverse process of moving from the general to the specific. For example, a person may have a desire to "come off" as a certain kind of person vis-à-vis another kind of person. Then you get into the processes where your specific behaviors are guided, in some loose way, by the general concepts corresponding to these ideals, rules, and so on. These include the self-presentational problems; but they are also cognitive, like deduction problems — specification, instantiation, exemplification. I wish to be seen as a loyal and capable group member; what do I do to show that? From this perspective the attribution process becomes simply one case of interest. I don't know what all the other cases are, maybe abstracting, generalizing about the relationship, as well as the specification processes.

At any rate, there aren't many social psychologists who keep their attention on interpersonal problems explicitly. It's partly our research methods and partly our theory.

Editors: Are the sociologists where we should be? What do we have to offer them?

Kelley: Cognitive theory. We start spelling things our on a more concrete level that they have talked about more vaguely.

Jones: I think we can also offer them the fruits of the experimental paradigm.

THE FUTURE OF ATTRIBUTION RESEARCH

Editors: There is one last topic that we would like for you to address briefly. Is attribution work going to peak like some other theoretical approaches, and five years from now we'll never hear of attribution?

Jones: I have a stock answer to the question about the future. If I knew what the future would be, I'd be there already.

Editors: But is social psychology such a haphazard enterprise that we can't make projections? Obviously, people who fund grants make decisions like this to some degree and are forced to project where the future of the field is by funding research that looks promising. Maybe it's an intuitive feeling that the attribution approach represents a fairly substantial framework that can be used and developed. The potential seems to be there. Is that belief naive?

Kelley: I share the feeling. I think the very way it developed — it wasn't some bright idea that somebody had, that somebody forced on somebody's data or tried to extend by brute force. It came out of a lot of phenomena that social psychologists have looked at and tried to interpret. I just can't imagine that the phenomena that are hooked into that kind of cognition will change or be modified. They'll never go away. We'll always have to have that kind of explanation.

Jones: Partly talking to the applications area and partly going back to the fact that we haven't made any enemies, I think there is a potential collision in sight with the psychotherapist, the dynamic psychologist. Maybe the collision has already occurred. It seems to be that the boundary of attributional notions is going to get established somewhere and that there is going to be a wave of returning to try to understand the irrational and the emotional, the phenomena that I don't think attribution theory is well designed to handle. In general social psychology doesn't have much to say about the intense affects, the strong emotions. Everything attributionists talk about works fairly well with a very intact organism. For example, I can't really buy the notion that schizophrenia is a proper attributional development from an improper initial inference. I think there is a lot more there that we can't come to terms with at all. From talking to clinicians who have seen a lot of anxiety patients and depressions, I would say here's a whole domain about which we really have very little to say.

Editors: This suggests a criticism along the lines that you were talking about earlier. A person may take the extreme point of view that only attributional theorists behave in an attributional fashion; everyday people don't have attributions. They're governed by factors that lie largely outside the realm of our theorizing.

Jones: It may be that what I'm suggesting reflects the kind of thing that Gordon Allport used to say. He had a real conviction that neurotics and normals did not lie on a continuum, that there was a real qualitative difference. It may have something to do with the failure of emotional responses to be contingent on what's happening in the environment. I think that's a fairly accurate description of many neuroses. This group of people, or everybody some of the time, acts in a neurotic fashion in the sense that they're not following any attributional principles; or if they do, the principles don't have any feedback control over the emotions they're experiencing.

Editors: Do you think that in the future there will be a convincing integration of cognitive—experimental approaches, such as the Bayesian approach and attributional approaches?

Jones: There does seem to be something happening that is going to happen more in the future: the integration of attribution with information processing, a more mathematical or Bayesian approach. This has already happened, but I think there is going to be a wing of attribution theory, quite technical, quite mathematical representing a convergence of two subdisciplines that initially were very different.

Editors: We wonder if all the basic process statements done by you two are going to be rewritten by people with mathematical expertise?

Kelley: Those basic process statements are gross, partial views, but there are going to be specifications by mathematicians. What does that imply? My work and Ned's framework, these basic things, will provide the impetus. They will point to some of the terms for dealing, for example, with the circular causal relationship that exists between information and conceptualization, the latter being partially derived from the former but also starting to determine it.

Editors: Baruch Fischhoff (1976) is coming to attribution from a very different kind of training. Do you think he might provide that insight for us?

Jones: Thus far there has been a division of labor. It's a caricature of that division to suggest that some people with a more mathematical, information-processing background have been scavengers, taking the research of other people and creating formulas to accommodate the results after the fact. What I think is going to happen, and has already happened a little bit, is that the information-processing people are turning more and more to designing their own problems in their own terms. The problem with much of the information-processing research is the reactiveness or obtrusiveness of the information the experimenter puts in and the kinds of recurrent decisions required of the subjects. Subjects become so self-consciously cognitive that the investigators end up studying a different process. That may be a danger in some of these approaches where the validation of the inputs is taken so seriously.

Kelley: Some really interesting future problems are the forms that the process takes, the ways it can be truncated, simplified, or dealt with in patterned terms. Related to this is the necessity of looking at the range of circumstances under which interpretation of intentions, for instance, might be usefully made. When does a person have time to think? When does he stop and think? What kinds of people are able to stop and think? I was thinking in terms of a cocktail party after which you go home and say, gee, why did I say that? You reminisce about the instigations and the possible consequences. Those kinds of obsessional times are very disturbing. They bring to mind the diplomat's looking with great deliberation at the intentions of the nations' leaders, thinking through things either in this kind of retrospective manner, but also in a prospective one. What have we done in relation to what they have done? What can I seem to infer about their intentions, their understanding of our intentions? Jervis (1976) has found good examples of that in diaries of diplomats.

It's just part of the latent question that gets addressed to us from outside. Do people really do it, how often, and to what degree do they do it explicitly, as your model is implying? What might be very useful is to look at patterns that people might recognize. Taking the chess master analogy, they apparently can recognize patterns very quickly, remember them well, and know their implications right away.

Editors: Does that suggest a paradigm for pattern recognition?

Kelley: We have to think in the context of the informational pattern, somehow finding a way to study the recognition of those patterns that would produce interpretations and response quickly. One that would, to some degree, map onto some full-blown, logical, step—by—step sequence.

Jones: There may even be some kind of individual difference dimension where some people are very sensitive to the ongoing situation; but they need the feedback, they need to get into it, they need to be controlled by what's going on right at the moment, whereas other people are better able to anticipate things.

One dimension of relevance for attribution research is what Mark Snyder (1974) calls self-monitoring. The little scale that he developed apparently separates very different kinds of people. The high self-monitor is really somebody who acts like he has read Kelley and Jones and Davis. He behaves like an attributionist says that people should behave, i.e., with sensitivity to the contingency between behavior and situation. The low self-monitor does not. There's no theory, no real conceptualization of what this variable is; but maybe just by luck or some kind of intuition, Snyder hit upon it. It's sort of a scale of the kind of person who monitors his own expressive behavior and is very concerned about whether it's appropriate in the situation. The high self-monitor is very attuned to what other people are doing and their expressive behavior. The self-monitoring variable is a package of things that seem to have a powerful influence on social behavior and social cognitions. I think another avenue that needs further exploration is the kinds of personality theories persons have about themselves.

Kelley: One of the things this reminds me of and that I think is needed is some kind of analysis or taxonomy of situational configuration. This is where I keep feeling that we need something to support the conceptual structure, something at the level of a taxonomy or detailed characterization of the stimuli.

Another side of the same thing would be characterizations of the action situations. The choice and its consequences would be one such characterization. You could have a taxonomy of choice situations with various alternatives or, what Thibaut and I have been using, the payoff matrix as a characterization of an action context. It's a way of defining choices and their consequences. I don't think we yet have the right terms. All of those analyses tend to be lacking in flexibility.

Editors: Maybe the Abelson scripts are types of little interaction stories.

Kelley: They have that in them, but they are stereotyped versions of interactions. If you flip around attributional problems from the problem of interpreting behavior to the self-presentational problem, you start looking more at the types of situations and how they are to be differentiated and classified. What do they support in the way of presentation of self?

REFERENCES

Asch, S. E. Forming impressions of personality. *Journal of Abnormal and Social Psychology,* 1946, *41,* 258–290.

Duncker, K. On problem-solving. *Psychological Monographs,* 1945, *58*(5, Whole No. 270).

Fischhoff, B. Attribution theory and judgment under uncertainty. In J. H. Harvey, W. J. Ickes, & R. F. Kidd (Eds.), *New directions in attribution research.* (Vol. 1). Hillsdale, N.J.: Lawrence Erlbaum Associates, 1976.

Harvey, J. H., Ickes, W. J., & Kidd, R. F. (Eds.). *New directions in attribution research* (Vol. 1). Hillsdale, N.J.: Lawrence Erlbaum Associates, 1976.

Jervis, R. *Perception and misperception in international politics.* Princeton: Princeton University Press, 1976.

Jones, E. E., & Davis, K. E. From acts to dispositions: The attribution process in person perception. In L. Berkowtiz (Ed.), *Advances in experimental social psychology* (Vol. 2). New York: Academic Press, 1965.

Jones, E. E., Hester, S. L., Farina, A., & Davis, K. E. Reactions to unfavorable personal evaluations as a function of the evaluator's perceived adjustment. *Journal of Abnormal and Social Psychology,* 1959, *59,* 363–370.

Jones, E. E., & McGillis, D. Correspondnet inferences and the attribution cube: A comparative reappraisal. In J. H. Harvey, W. J. Ickes, & R. F. Kidd (Eds.), *New directions in attribution research* (Vol. 1). Hillsdale, N.J.: Lawrence Erlbaum Associates, 1976.

Jones, E. E., & Nisbett, R. E. *The actor and the observer: Divergent perceptions of the causes of behavior.* Morristown, N.J.: General Learning Press, 1971.

Kelley, H. H. The analysis of common sense. *Contemporary Psychology,* 1960, *5,* 1–3. (Review of *The psychology of interpersonal relations* by Fritz Heider).

Kelley, H. H. Attribution theory in social psychology. In D. Levine (Ed.), *Nebraska symposium on motivation.* Lincoln, Neb.: Unviersity of Nebraska Press, 1967.

Kelley, H. H., Thibaut, J. W., Radloff, R., & Mundy, D. The development of cooperation in the "minimal social situation." *Psychological Monographs,* 1962, *76,* (19, Whole No. 538).

Kruglanski, A. W. Attributing trustworthiness in supervisor–worker relations. *Journal of Experimental Social Psychology,* 1970, *6,* 214–232.

Orvis, B. R., Kelley, H. H., & Butler, D. Attributional conflict in young couples. In J. H. Harvey, W. J. Ickes, & R. F. Kidd (Eds.), *New directions in attribution research* (Vol. 1). Hillsdale, N.J.: Lawrence Erlbaum Associates, 1976.

Ring, K. Some determinants of interpersonal attraction in hierarchical relationships: A motivational analysis. *Journal of Personality,* 1964, *32,* 651–665.

Snyder, M. The self-monitoring of expressive behavior. *Journal of Personality and Social Psychology,* 1974, *30,* 526–537.

Thibaut, J. W., & Riecken, H. W. Some determinants and consequences of the perception of social causality. *Journal of Personality,* 1955, *24,* 113–133.

Wachtel, P. L. Psychodynamics, behavior therapy, and the implacable experimenter. An inquiry into the consistency of personality. *Journal of Abnormal Psychology,* 1973, *82,* 324–334.

Author Index

Subject Index